# Computer control
# of flexible manufacturing
# systems

# Computer control
# of flexible manufacturing
# systems

## Research and development

*Edited by*

## Sanjay B. Joshi

*Department of Industrial and Management Systems,*
*Pennsylvania State University, Pennsylvania, USA*

and

## Jeffrey S. Smith

*Department of Industrial Engineering,*
*Texas A&M University, Texas, USA*

## CHAPMAN & HALL

London · Glasgow · Weinheim · New York · Tokyo · Melbourne · Madras

**Published by Chapman & Hall, 2–6 Boundary Row, London SE1 8HN, UK**

Chapman & Hall, 2–6 Boundary Row, London SE1 8HN, UK

Blackie Academic & Professional, Wester Cleddens Road, Bishopbriggs, Glasgow G64 2NZ, UK

Chapman & Hall GmbH, Pappelallee 3, 69469 Weinheim, Germany

Chapman & Hall USA, One Penn Plaza, 41st Floor, New York NY 10119, USA

Chapman & Hall Japan, ITP Japan, Kyowa Building, 3F, 2-2-1 Hirakawacho, Chiyoda-ku, Tokyo 102, Japan

Chapman & Hall Australia, Thomas Nelson Australia, 102 Dodds Street, South Melbourne, Victoria 3205, Australia

Chapman & Hall India, R. Seshadri, 32 Second Main Road, CIT East, Madras 600 035, India

First edition 1994

© 1994 Chapman & Hall

Typeset in 10/12 Palatino by Best-set Typesetter Ltd., Hong Kong
Printed in Great Britain by T.J. Press (Padstow) Ltd., Padstow, Cornwall

ISBN 0 412 56200 6

A catalogue record for this book is available from the British Library

Library of Congress Catalog Card Number: 94-70267

∞ Printed on permanent acid-free text paper, manufactured in accordance with ANSI/NISO Z39.48–1992 and ANSI/NISO Z39.48–1984 (Permanence of Paper).

# Contents

# Contributors

**James S. Albus**, National Institute of Standards and Technology, Building 220, Gaithersburg, Maryland, USA.

**David Ben-Arieh**, Dept. of Industrial Engineering, Kansas State University, Manhattan, Kansas, USA.

**Eric D. Carley**, Dept. of Industrial Engineering, Kansas State University, Manhattan, Kansas, USA.

**Jarir K. Chaar**, IBM Research Division, Thomas J. Watson Research Center, P.O. Box 704, Yorktown Heights, New York, USA.

**Ai-Mei Chang**, Department of Management Information Systems, University of Arizona, Tucson, Arizon, USA.

**Hyuenbo Cho**, Dept. of Industrial Engineering, Texas A&M University, College Station, Texas, USA.

**C. Thomas Culbreth**, Dept. of Industrial Engineering, North Carolina State University, Raleigh, NC, USA.

**Edward S. Davidson**, EECS Dept., The University of Michigan, Ann Arbor, Michigan, USA.

**Annap Derebail**, Dept. of Industrial Engineering, Texas A&M University, College Station, Texas, USA.

**Gabriele Elia**, Dipartimento di Automatica e Informatica, Politecnico di Torino, Corso Duca degli Abruzzi, Torino TO, Italy.

**Sebastian Engell**, Dept. of Chemical Engineering, Universität Dortmund, D-44221 Dortmund, Germany.

**Frank Herrman**, Fraunhofer-Institute IITB, D-76131 Karlsruhe, Germany.

**Janette M. Hopkins**, Dept. of Industrial Engineering, North Carolina State University, Raleigh, NC, USA.

**Leslie Interrante**, Industrial and Systems Engineering, The University of Alabama in Huntsville, Huntsville, Alabama, USA.

**Sanjay B. Joshi**, Dept. of Industrial and Management Systems Engineering, 207 Hammond Building, Pennsylvania State University, University Park, Pennsylvania USA.

**Manjunath Kamath**, School of Industrial Engineering and Management, 322 Engineering North, Oklahoma State University, Stillwater, Oklahoma, USA.

**Russell E. King**, Dept. of Industrial Engineering, North Carolina State University, Raleigh, NC, USA.

**Thomas R. Kramer**, National Institute of Standards and Technology, Building 220, Gaithersburg, Maryland, USA.

**V. Jorge Leon**, Texas A&M University, College Station, Texas, USA.

**Grace Y. Lin**, IBM Research Dvn., Thomas J. Watson Research Center, P.O. Box 218, Route 134 and Kitchawan Road, Yorktown Heights, NY 10598, USA.

**Giuseppe Menga**, Dipartimento di Automatica e Informatica, Politecnico di Torino, Corso Duca degli Abruzzi, Torino TO, Italy.

**Colin L. Moodie**, Purdue University, School of Industrial Engineering, West Lafayette, Indiana, USA.

**Manfred Moser**, ABB FLT GmbH, D-68309 Mannheim, Germany.

**Richard Quintero**, National Institute of Standards and Technology, Building 220, Gaithersburg, Maryland, USA.

**Steven R. Ray**, National Institute of Standards and Technology, Building 220, Gaithersburg, Maryland, USA.

**Anthony D. Robbi**, Discrete Event Systems Laboratory, Dept. of Electrical and Computer Engineering, Center for Manufacturing Systems, New Jersey Institute of Technology, Newark, New Jersey, USA.

**David Rochowiak**, Cognitive Science, The University of Alabama in Huntsville, Huntsville, Alabama, USA.

**M. Kate Senehi**, National Institute of Standards and Technology, Building 220, Gaithersburg, Maryland, USA.

**John P. Shewchuk**, Purdue University, School of Industrial Engineering, West Lafayette, Indiana, USA.

**Jeffrey S. Smith**, Industrial Engineering Dept., 239A Zachry Engineering, Texas A&M University, College Station, Texas, USA.

**James J. Solberg**, School of Industrial Engineering, Purdue University, West Lafayette, Indiana, USA.

**Ben Sumrall**, Acustar Inc., Huntsville Electronics Division Complex, Huntsville, Alabama, USA.

**Dharmaraj Veeramani**, Dept. of Industrial Engineering, University of Wisconsin-Madison, 1513 University Avenue, Madison, Wisconsin, USA.

**Richard A. Voltz**, Dept. of Computer Science, Zachry Engineering Center, Texas A&M University, College Station, Texas, USA.

**Theodore Williams**, Purdue Laboratory for Applied Industrial Control, Purdue University, West Lafayette, Indiana, USA.

**S. David Wu**, Lehigh University, Bethlehem, Pennsylvania, USA.

**Richard A. Wysk**, Dept. of Industrial Engineering, Texas A&M University, College Station, Texas, USA.

**MengChu Zhou**, Discrete Event Systems Laboratory, Dept. of Electrical and Computer Engineering, Center for Manufacturing Systems, New Jersey Institute of Technology, Newark, New Jersey, USA.

# Preface

With the approach of the 21st century, and the current trends in manufacturing, the role of computer-controlled flexible manufacturing will take an integral part in the success of manufacturing enterprises. Manufacturing environments are changing to small batch (with batch sizes diminishing to a quantity of one), larger product variety, production on demand with low lead times, with the ability to be 'agile.' This is in stark contrast to conventional manufacturing which has relied on economies of scale, and where change is viewed as a disruption and is therefore detrimental to production. Computer integrated manufacturing (CIM) and flexible manufacturing practices are a key component in the transition from conventional manufacturing to the 'new' manufacturing environment.

While the use of computers in manufacturing, from controlling individual machines (NC, Robots, AGVs etc.) to controlling flexible manufacturing systems (FMS) has advanced the flexibility of manufacturing environments, it is still far from reaching its full potential in the environment of the future. Great strides have been made in individual technologies and control of FMS has been the subject of considerable research, but computerized shop floor control is not nearly as flexible or integrated as hyped in industrial and academic literature. In fact, the integrated systems have lagged far behind what could be achieved with existing technology.

Most shop floor control systems are focused on information and data collection and monitoring rather than on control. Many implementations of flexible systems are soft wired versions of hard automation and fail to employ the additional capabilities available through the use of computers for control. These systems lack the required flexibility to change without major effort. Further, manufacturers with flexible manufacturing systems do not use the systems in a flexible manner. Users of FMS tend to standardize products and increase batch sizes to ease the operation of the system, and to justify the economics of using the systems. The cost of such systems is often prohibitively high and hence justification of such systems has led to their use in high volume production which does not require or exploit whatever flexibility is available. Cost is the factor that has also kept such systems from being useful to small manufacturers, which will form the nucleus of the future manufacturing environments.

The following are viewed as critical needs for computerized flexible manufacturing systems if they are to become commonplace in today's manufacturing sector:

1. *Reduction in costs of systems.* To make such systems the core of all manufacturing will require that costs to develop, implement and maintain systems be of an order of magnitude that will permit widespread use and not require large volumes to justify cost. A key component of cost is software cost; it is expected that further research will focus on the key issue of making computer control of flexible manufacturing systems feasible for the small manufacturers that will be a part of the 'distributed' manufacturing environment of the future.

2. *Increased flexibility and the ability to use it.* As the product volumes and lot sizes drop, systems will have to be designed with greater flexibility, since changes will now be the norm rather than the exception. This will impose a tremendous burden on the control system which will now have to be built to handle changes as well as increased demand in flexibility. Today's control systems, although increasingly software based, are still too inflexible to respond to the new demands that will be imposed.

3. *Seamless integration.* Another factor that will impact future control systems is the capability of providing a completely integrated environment, where individual elements are designed and used in a manner such that complete integration is possible. Similar to the trend in computers, control systems will have to work with open architectures and application program interfaces to allow continued progress in the forward direction. Current systems in place are usually specific for an installation and sold/built as a monolithic turnkey system, and end users typically have no capability of modifying control systems in house, hence they are locked into proprietary systems. The control systems of the future will provide an open architecture, with well-defined modules and application interfaces which will allow users and third party vendors to provide different functional modules which would work together with other modules in an integrated manner.

4. *Reusability.* As the trend towards recycling and reuse continues, along with the need for frequent changes, manufacturing system elements – both hardware and software – will have to be designed with reuse in mind. Turnkey systems will be a thing of the past, as the need for reconfiguration will lead to modular systems that can be easily put together and taken apart and pieces reused when need to reconfigure arises.

Unless significant improvements are made in the control and operation of computerized flexible manufacturing systems, these inflexible manufacturing systems will prove to be the Achilles' heel of future manufacturing. With this view in mind, this book attempts to provide a glimpse of several aspects of the current developments in control of flexible manufacturing systems.

The chapters in the book address several important topics in the area of computer control of FMS. Individual researchers have addressed specific problems in isolation with the assumption that solution to pieces of the whole will eventually result in solution to the whole problem. This may not be as simple as once thought, and integration as a topic of research will take on a larger role in the future.

# CHAPTER 1

# The role of CIM architectures in flexible manufacturing systems

*Theodore J. Williams, John P. Shewchuk*
*and Colin L. Moodie*

## 1.1 INTRODUCTION

The needs of world-wide industry today require manufacturers to modify their operations to ensure:

1. a better and faster response to their customers' requirements;
2. ever-higher quality for their products;
3. increased flexibility and faster response in the introduction of new products and in responding to the needs of the marketplace.

At the same time, they face further requirements to increase their overall company earnings while

1. decreasing the environmental impact of their factory's operations;
2. decreasing the plant personnel;
3. improving plant personnel working conditions and job satisfaction.

Integrated manufacturing carried out with the aid of computers has been seen by many as the means by which much of the above could be accomplished. Studies almost universally show that if a manufacturing enterprise can integrate the operations of its plants so that all available information affecting them can be used, then very large economic returns over the best present-day, non-integrated methods are both possible and likely. Projects to achieve such an integration are generally collected under the pseudonym computer integrated manufacturing, or CIM. This is because extensive use of computers appears to be a universal necessity to achieve the task undertaken.

The expected gains have been so high that numerous projects in many industries have been undertaken to achieve them, but the results have been decidedly disappointing in many cases. There have been several major causes for this. Primarily, this has been due to the fact that those planning these projects have not realized the breadth and magnitude of the overall effort necessary and the resulting capital and other resources required. They have also not developed a total plan for the overall project necessary prior to commencing implementation, and thus, have neglected to outline the **total** effort needed.

What is needed, therefore, for each company contemplating a major computer-based integration effort, is for the company to develop a master plan covering **all** of the anticipated efforts required to integrate the whole of the company or factory operation.

After this, smaller projects within the monetary and personnel resources capability of the company can be initiated with the knowledge that the sum of these and all succeeding projects will result in the final total integration of the company's activities. This will be possible provided that the requirements of the initial planning effort, or master plan, be followed in each and every one of the resulting projects.

But the detail and effort required for even the master planning activity is itself large and if done improperly will only lead to difficulties later. Thus, there is a need for a methodology to assure that the master plan is complete, accurate, properly oriented to future business developments and carried out with a minimum of resources (personnel and capital) necessary. This methodology presents a detailed description of the tasks involved in developing the master plan including its continual renewal. It gives the detail necessary both as to specifics and to quantity of information and data. It specifies the interrelationship of the informational, the human organizational and the physical manufacturing aspects of the integration considered, the management considerations and concerns and the economic, cultural, and technological factors involved, as well as the details of the computer system required. This planning effort is greatly aided by the use of a reference architecture to guide the project.

An **enterprise reference architecture** models the whole life history of an enterprise integration project, from its initial concept in the eyes of the entrepreneurs who initially developed it, through its definition, functional design or specification, detailed design, physical implementation or construction, and finally operation to obsolescence. The architecture becomes a relatively simple framework upon which all the functions and activities involved in the aforementioned phases of the life of the enterprise integration project can be mapped. It also permits the tools used by the investigators or practitioners at each phase to be indicated.

At the same time, the architecture provides the framework for all

**Table 1.1**  Benefits of the use of an architecture

1. Verification of completeness and consistency for all described functions and objects (business processes, data, material, and resources, including tools and fixtures) at any detailing level.
2. Simulation of the enterprise model at any level of detail.
3. Easy and fast change of the model in case of changing business processes, methods, or tools.
4. The use of the model to initiate, monitor and control the execution of the enterprise's daily operation.
5. Repeated resource allocation during the execution of business processes to enable better and more flexible load distribution on the enterprise's resources.
6. Model generation for existing enterprises as well as for enterprises yet to be built.

master plans and CIM program proposal activities. It also explains better than any other tool the relationships of the elements of the CIM system. It is thus the 'glue' that holds all aspects of the project together.

An architecture should illustrate clearly all the following aspects of the enterprise:

- enterprise decision making;
- enterprise activities;
- enterprise business processes;
- enterprise information exchange;
- enterprise material and energy flows.

If it does not, it cannot give a complete picture of the enterprise and its activities. The overall benefits of the use of an architecture are given in Table 1.1.

## 1.2  TYPES OF ARCHITECTURES

An architecture can be defined as a description (model) of the structure of a physical or conceptual object or entity. Thus, there are two and only two types of architecture which deal with the integration of manufacturing entities or enterprises. These are:

1. The structural arrangement (design) of a physical system, such as the computer control system part of an overall enterprise intergration system.
2. The structural arrangement (organization) of the development and implementation of a project or program such as a manufacturing or enterprise integration or other enterprise development program.

Most of the previous work on CIM architectures has involved Type 1 architectures. Some examples of these works are:

- The CAM-I Advanced Factory Management System model, developed in the late seventies by CAM-I (Computer Aided Manufacturing International), a non-profit organization promoting co-operative R and D efforts in CAD/CAM.
- The AMRF (Advanced Manufacturing Research Facility) hierarchical control model, developed by the National Bureau of Standards in the early eighties.
- The RAMP (Rapid Acquisition of Manufactured Parts) architecture, developed for the US Navy (Litt, 1990).

There are only **three** major architectures known at this time which are Type 2 architectures. These are:

1. The **open system architecture for computer integrated manufacturing – CIM-OSA**, in development by the European CIM Architecture consortium (AMICE (backward acronym)) under ESPRIT projects 688, 2422, and 5288 of the European Community. This work was initiated in 1984.
2. **GRAI-GIM**. The **GRAI integrated methodology**, developed by the GRAI Laboratory of the University of Bordeaux in France. This work resulted from the production management studies initiated at the GRAI Laboratory as early as 1974, and has its current form since about 1984.
3. The **Purdue enterprise reference architecture** and related **Purdue methodology**, as developed at Purdue University, Indiana, USA, as part of the work of the Industry-Purdue University Consortium for CIM. The work started formally in 1989, but bears on the **Purdue reference model** started in 1986 and earlier work of the Purdue Laboratory for Applied Industrial Control dating back to the mid-seventies.

In this chapter, we are concerned with Type 2 architectures and their role in FMS. Each of the Type 2 architectures is briefly described below.

### 1.2.1  CIM-OSA

CIM-OSA consists of an **architecture** and an **integrated methodology** which supports all phases of a CIM system life-cycle, from specification of requirements through system design, implementation and operation. The architectural framework is the well-known 'CIM-OSA cube' (Fig. 1.1), which specifies models in terms of three dimensions (attributes): modeling level (requirements definition, design specification and implementation description), level of solution specificity (generic, partial and particular), and view (function, information, resource and organization).

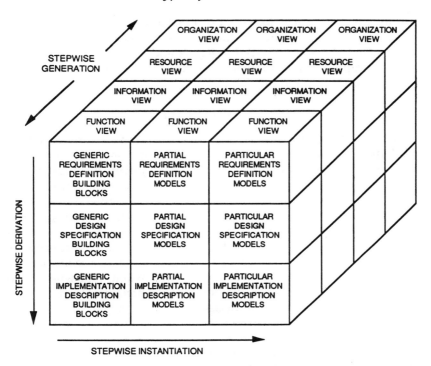

**Fig. 1.1** CIM/OSA modeling framework (after Jorysz and Vernadat, 1990a).

The integrated methodology is based upon the concept of three levels of integration in a CIM system: physical system integration, application integration and business process integration. The models which are constructed during execution of the methodology and their order of construction, are determined by three model creation processes: **instantiation**, **derivation**, and **generation**. Each of these processes corresponds to movement along a cube axis, as shown in Fig. 1.1.

The methodology is built upon a top-down approach (derivation) to achieve integration between each of the three integration levels. The business requirements of an enterprise are first captured in a requirements definition model (set of models formed by stepwise instantiation and stepwise generation at the requirements definition level). From these requirements and consideration of all specific constraints of the particular enterprise, the design specification model is then constructed. Finally, based upon this model, the implementation description model, a description of the actual CIM system to be implemented is developed. These models are constructed in the **integrated enterprise engineering environment**, a comprehensive set of modeling concepts, tools and support for model creation. Once the implementation model has been completed, it is released to the **integrated enterprise operation**

**environment**, where it is made a processable model, and can thus be used to control the operation of the CIM system at run time.

Interested readers are referred to Joyrsz and Vernadat (1990a, 1990b) for more comprehensive summaries, or to AMICE report AD 2.0 (1992) for a detailed description of CIM-OSA.

### 1.2.2 GRAI-GIM

The GRAI integrated methodology (GIM) consists of a **reference model**, a **modeling framework**, and a **structured approach**. The reference model describes the invariant parts of a CIM system: the subsystems, their relationships and their behavior. In GIM, the reference model is based upon the concepts of three activity types (decisional, informational and physical) and their corresponding executional subsystems. The GIM modeling framework (Fig. 1.2) has two dimensions: **abstraction level** and **view**. Along the abstraction level axis, conceptual models specify **what** has to be done, structural models consider the organizational point of view and answer **who**, **when**, and **where**, and realizational models consider the technical constraints to determine **how**. Along the view axis, the decisional, informational and physical views correspond to these same subsystems in the reference model; the functional view is added to provide a way to show the main functions in a manufacturing system and the flows between these functions.

The aim of GIM is to cover the entire life-cycle of the manufacturing system. The approach consists of four phases: initialization, analysis, design and implementation. Initialization consists of defining company objectives, the domain of the study, the personnel involved, etc. The analysis phase results in the definition of the characteristics of the existing system (requirements of new system) in terms of four user-oriented views. The design phase is performed in two stages: **user-oriented design** and **technical-oriented design**. User-oriented design

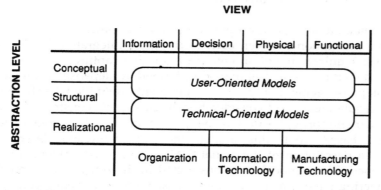

**Fig. 1.2** GRAI-GIM modeling framework (after Doumeingts *et al.*, 1992).

uses the results of the analysis phase to establish requirements for the new system (conceptual models), again in terms of the four user-oriented views. Technical-oriented design consists of transforming the user-oriented models of the new system design to technical-oriented models. Structural models are built as part of the transformation process, and the resulting technical-oriented models are at the realizational level in the modeling framework. These models express the system requirements in terms of the required organization, information technology and manufacturing technology.

For more details on the GRAI Integrated Methodology, readers are referred to Doumeingts *et al.* (1992).

### 1.2.3 The Purdue methodology

The Purdue methodology for CIM is based upon three items, the **Purdue enterprise reference architecture**, the **Purdue reference model**, and the **Purdue implementation procedures manual for developing master plans for CIM**. The Purdue enterprise reference architecture covers all the tasks involved in a CIM enterprise, starting from initial concept, through definition, design, implementation and operation to obsolescence. The architectural framework (Fig. 1.3) is based upon the concepts of two classifications of activities (informational and customer service (physical)), the classification of activities in terms of automatability and the separation of task functionality from implementation. The reference model presents a manner for specifying a CIM architecture, its tasks and their implementation. It is restricted to automatable functions only and as such, forms a subset of the Purdue enterprise reference architecture.

The Purdue implementation procedures manual for developing master plans for CIM specifies the steps which must be followed to develop a master plan for carrying a CIM system through its life-cycle stages. The master plan uses the Purdue enterprise reference architecture as a framework. The models which must be built at each stage of the process are defined in the architecture.

The Purdue enterprise reference architecture is discussed in more detail in the following section (readers are referred to Williams (1992) for full details).

### 1.3 THE PURDUE ENTERPRISE REFERENCE ARCHITECTURE

In order to understand the role CIM architectures play in flexible manufacturing systems, a basic understanding of architectures is required. Towards this end, the Purdue enterprise reference architecture is described in some detail.

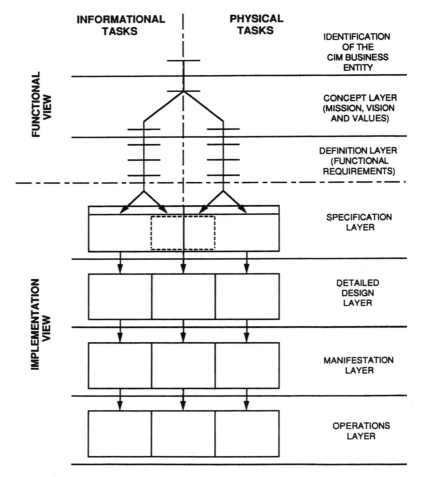

**INFORMATIONAL TASKS**

**PHYSICAL TASKS**

IDENTIFICATION
OF THE
CIM BUSINESS
ENTITY

CONCEPT LAYER
(MISSION, VISION
AND VALUES)

DEFINITION LAYER
(FUNCTIONAL
REQUIREMENTS)

SPECIFICATION
LAYER

DETAILED
DESIGN
LAYER

MANIFESTATION
LAYER

OPERATIONS
LAYER

FUNCTIONAL VIEW

IMPLEMENTATION VIEW

**Fig. 1.3**   Purdue enterprise reference architecture modeling framework.

### 1.3.1   Description of architecture

Figure 1.3 presents the modeling framework of the Purdue enterprise reference architecture to help in its explanation. Figures 1.4 and 1.5 show in more detail the components which are present in each modeling layer. The concept, definition and specification layers are presented in Fig. 1.4, while the detailed design, manifestation and operation layers are presented in Fig. 1.5.

Starting with the CIM business entity (CBE), we see that this leads first to a description of the management's mission, vision and values for the entity plus any further philosophies of operation or mandated

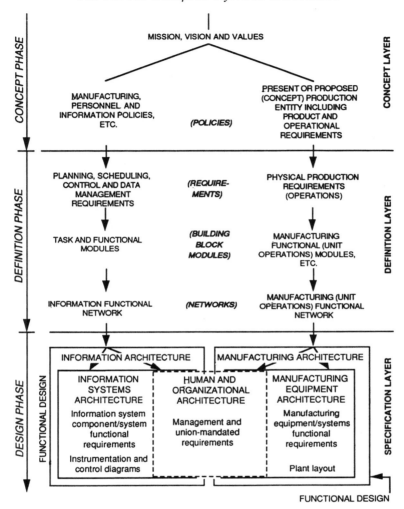

**Fig. 1.4** Components of the concept, definition and specification layers of the Purdue enterprise reference architecture.

actions concerning it such as choice of processes, vendor selection, etc. From the mission, etc. we derive the operational policies for the units for all areas of potential concern. These form the concept layer.

In the manufacturing plant, the above prescription and selection by management of possible options leads to the establishment of operational requirements for the plant. This leads to the statements of requirements for all equipment and for the methods of operation, etc. for these units. These are developed in the definition layer, as illustrated in Fig. 1.4. Note that there are two, and only two kinds

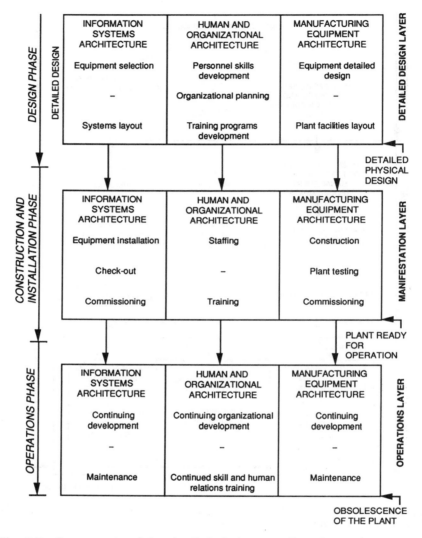

**Fig. 1.5** Components of the detailed design, manifestation and operations layers of the Purdue enterprise reference architecture.

of requirements developed from the management directives: those defining information-type tasks and those defining physical manufacturing tasks. Tasks become collected into modules or functions, and these in turn can be connected into networks of information or of material and energy flow. These latter elements then form the **information functional network** or the **manufacturing functional network** respectively.

Once implementation is considered, the first need is to define which tasks on either side of the overall architecture will be fulfilled by people. By doing so, we define the place of the human in the **information architecture** and also in the **manufacturing architecture**. These together form a **human and organizational architecture**. The remainder of the information architecture becomes the **information system architecture** (all the computers, software, databases, etc.) The remainder of the manufacturing architecture becomes the **manufacturing equipment architecture** (all the physical equipment). We have therefore converted two functional architectures into three implementation architectures. All these architectures are sub-architectures of the Purdue enterprise reference architecture itself. They are called architectures because they themselves form frameworks for extensive sets of tools, models, etc. for the development of their own contributions to the CIM or enterprise program under study.

We can now follow the life history of the implementation through its four phases – functional design or specification, detailed design, construction and commissioning or manifestation, and finally operation to obsolescence (Figs 1.4 and 1.5).

It should be noted that although the architecture has been defined above in terms of manufacturing enterprises, it can in fact be used to model any type of enterprise. Every enterprise must have some **mission** to justify its existence; this involves either the production of physical things (i.e. manufactured products), supply of purely physical services (transportation of goods, rental, lease, etc.), or supply of informational services (supply of data: books, reports, etc.). The tasks required to satisfy the mission, and the equipment and facilities used, can be referred to as the customer service architecture. In order to operate the facility to successfully attain the mission objectives, it is necessary to **control** the system via information concerning its current and future status, results attained, etc. The tasks required to execute control, and the equipment and facilities used, are defined by the information architecture. These two architectures and their life-cycle phases are defined by the right and left halves, respectively, of the Purdue enterprise reference architecture. Consideration of the customer service function to be manufacturing results in the manufacturing-specific architecture discussed in this chapter.

### Choice of human tasks

The lines separating the three implementation architectures are the **extent of automation** lines. The location of these lines depends upon two other sets of lines, the **automatability** lines and the **humanizability** lines. The origin of and relationships between these three sets of lines are described below (Fig. 1.6).

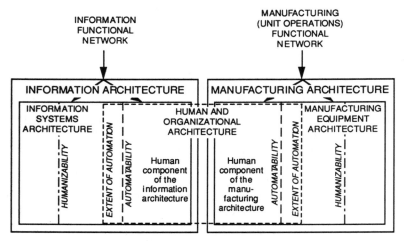

**Fig. 1.6** Relations between automatability, humanizability and extent of automation lines in the Purdue enterprise reference architecture.

The automatability lines define the absolute extent of technology in its capability of actually automating the tasks and functions of the CIM system of the CIM business entity. This is limited by the fact that many tasks and functions require human innovation, etc. and cannot be automated with presently available technology.

The humanizability lines show the extent to which humans can be used to implement the tasks and functions of the CIM system of the CIM business entity. Their placement is limited by human abilities in speed of response, breadth of comprehension, range of vision, physical strength, etc.

The extent of automation lines define the boundary between the human and organizational architecture and the information systems architecture on the one hand, and between the human and organizational architecture and the manufacturing equipment architecture on the other. The extent of automation lines show the degree of automation carried out or planned in the CIM system of the CIM business entity.

The automatability lines will always be outside the extent of automation lines. That is, not all the technological capability for automation is ever utilized in any installation, for various reasons. Thus, the human and organizational architecture is larger (i.e. more tasks or functions) and the information systems and manufacturing equipment architectures are smaller (fewer functions) than technological capability alone would allow or require. Note that for a completely automated ('lights out') plant, both the automatability line and the extent of automation line will coalesce and move to the right edge of the information architecture block and correspondingly to the left edge of the manufacturing architecture block.

A major finding of the architecture diagram is that, as long as the specifications for accomplishment of each of the tasks are honored, the three implementation architectures can be developed relatively independently.

### 1.3.2 The modular task and function representation method

The following discussion shows a method for the development of a set of generic tasks and functions which should be applicable to help prepare the concept of a generic functional architecture for any CIM system.

#### *Transformations, tasks and functions*

A **transformation process** results in the generation of outputs from inputs, as shown in Fig. 1.7. Transformations in a manufacturing enterprise are of two types:

1. An informational (data) process.
2. A physical (material and/or energy-based) process.

A **task** is what carries out (results in) a transformation process. Tasks represent the lowest level of functional decomposition of an enterprise, usually corresponding to the work of a single person or machine at a point in time.

A **function** is a group of tasks which can be classified as having a common objective within the company.

Any task and/or function within a CIM system can be classified as being either informational (i.e. moving, storing, or transforming information), or, conversely, as manufacturing or physical (i.e. moving, storing, or transforming material and/or energy). Furthermore, every task should be generic in its basic description. All task modules of either or both types can be interconnected by the flow of data (information) and commands, and/or material or energy.

**Fig. 1.7** Transformation process, Purdue modular task and function representation method.

*Implementation of tasks*

Upon implementation, all tasks will fall into one of the following three categories:

1. *Scheduling and control tasks*: i.e. information handling tasks. The transformation process for a scheduling and control task involves operations on information (data) enabled by an algorithm. The transformation may be considered a computational block.
2. *Manufacturing-based tasks.* For manufacturing processes, the basic tasks will comprise generic, irreducible operations, such as the 'unit operations' of chemical engineering, related types of operations or similar classes of operations in each of the other fields of engineering. That is, all manufacturing modules, where applicable, follow a set of basic rules like the unit operations concepts of chemical engineering, i.e. each unit operation and its associated equipment would be a module in the context used here.
3. *Human tasks.* Human tasks may result in information or material and/or energy transformation processes. They result when either or both of the following conditions occur:
   - No definable algorithm or computer program exists, as in human, thought-based, innovative processes. The transformation process may then be described by a written scope or other text which characterizes the transformation involved.
   - An algorithm exists, but for whatever reason, it has been decided to use a human-implemented method to carry out a particular function. These then become the non-automated or non-mechanized tasks or functions of the system.

*General task representation module*

To model informational tasks, manufacturing tasks and human tasks, the general task representation module (Fig. 1.8) was developed. Storage is considered to be a part of the basic module, though it could also be considered separately. Parameters (coefficients in algorithms for dynamic models) may be:

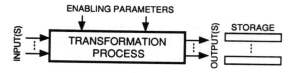

**Fig. 1.8** Task module, Purdue modular task and function representation method.

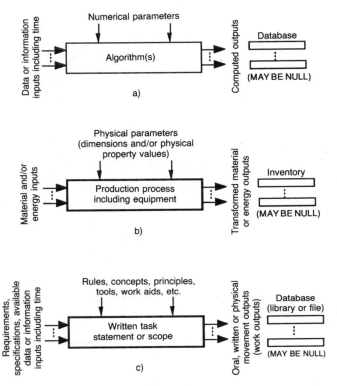

**Fig. 1.9** Example task module types, Purdue modular task and function representation method: (a) informational task module, (b) manufacturing task module, (c) human and organizational task module.

- tables of constants;
- operational variable value-dependent selections from tables of constants (precomputed adaptive control);
- results of active on-line or periodic recalculation by adaptive tuning, etc. or algorithms;
- expert system or neural net outputs;
- etc.

Examples of how the general task representation method can be used for each type of task are shown in Fig. 1.9. It should be noted that the storage element may also be considered a separate module.

### 1.3.3 Tools, etc. available for exploiting the architecture

As has been previously discussed, the Purdue enterprise reference architecture is a framework showing the various life-cycle phases of an

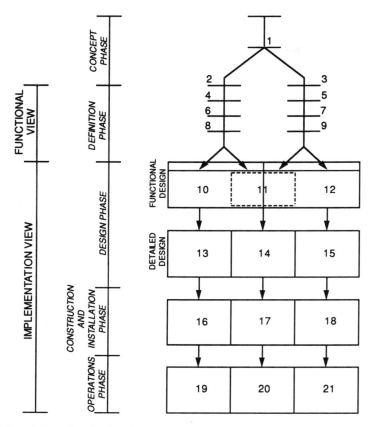

**Fig. 1.10** Abbreviated sketch showing structure of the Purdue enterprise reference architecture.

enterprise and what activities must be performed during each phase. Various tools, design aids, etc. are available for each of these activities. By mapping these tools, design aids, etc. against the activities in the framework, the Purdue enterprise reference architecture can also be used to classify and order these items. This can aid in collecting and assigning the very wide variety of manual and computer-aided tools and design aids available.

Figure 1.10 provides an abbreviated sketch of the structure of the architecture, numbering each of the important areas. Table 1.2 then lists some of the tools and aids available for the development or implementation endeavor normally associated with each of the numbered areas in Fig. 1.10. Note that these lists are examples only and are not intended in any way to be comprehensive or complete.

**Table 1.2** Examples of aids available for helping development and implementation of CIM programs and enterprise studies

| area | types of aids available |
|---|---|
| 1. | Example sets of mission, vision and values expressions from company annual reports or the basic documents themselves. |
| 2. | Generic lists of: policies and requirements related to such topics as control capabilities, degrees of performance of processes and equipment; adherence to regulations and laws (environmental, safety, etc.); quality, productivity and economic returns goals, etc. |
| 3. | Example scopes of the tasks for development and operation of specific processes and plants. |
| 4. | Generic lists of requirements necessary to carry out the policies listed in Area 2. |
| 5. | Example sets of operational requirements for specific processes and manufacturing plants, including general safety requirements, fire codes, etc., which will influence future plant design and process and equipment selection. |
| 6. | Generic lists of control and information tasks, function modules and macrofunctions. |
| 7. | Lists of generic unit process operations of the manufacturing features from group technology (discrete products industries). |
| 8. | Data flow diagram techniques. The generic data flow diagram of the Purdue reference model. |
| 9. | Example flow diagrams for commonly available processes, showing material and energy balances and example process operating procedures. |
| 10. | Generic representations of typical control and information systems. Functional design aids. Lists of sensors, actuators and control functions for various equipment. Database design techniques, entity-relationship diagrams. The Purdue scheduling and control hierarchy. Example hardware from various vendors. |
| 11. | Example lists of generally required personnel tasks. Auditing methods for skill level determination. Methods for cultural status assessement and correction. |
| 12. | Example specifications of process equipment to be required. Computer-based plant layout and design programs. |
| 13. | Computer control systems component selection aids and configuration software packages from control system vendors. |
| 14. | Example lesson plans and syllabi for necessary training courses. Example organizational charts for equivalent groups in terms of numbers of people, skill levels and tasks required. |
| 15. | Detailed design techniques for physical processes and equipment from the major handbooks (or computerized versions) of the various engineering fields. |
| 16. | Project management techniques such as CPM. |
| 17. | Continuation of the work under Area 14 in terms of training and staffing of the members of the human and organizational architecture. |
| 18. | Project management tools as noted under Area 16. |
| 19. | Continued improvement of the operation of the plant and its associated control system, involving such techniques as statistical quality control, statistical process control, total quality management and other related techniques. |
| 20. | Continued improvement of workers' skills and training of replacement workers. Same aids as for Areas 11, 14 and 17. |
| 21. | Requirements and aids available are as noted under Area 19. |

## 1.4 CIM ARCHITECTURES AND FLEXIBLE MANUFACTURING SYSTEMS

Flexible manufacturing systems (FMSs) are collections of machines and related processing equipment linked together by an automated material handling system, and all under (hierarchical) computer control. These systems are capable of producing a wide variety of items in small batches (even single items) and in random order. FMSs are meant to fill the gap between high-volume, dedicated production equipment (transfer lines) and low-volume but flexible job shops. Some of the advantages to be gained by using FMSs include increased productivity, reduced work-in-process, reduced production costs, and increased machine utilization.

Flexible manufacturing systems achieve the ability to remain flexible, yet still accommodate high production rates, by utilizing flexible processing equipment and control strategies and by utilizing a high degree of automation. As a consequence of these practices, FMSs are very complex systems and difficult to design, implement, operate and maintain. It is because of this complexity that CIM architectures are valuable. They can be used for the following:

1. *FMS specification.* A CIM architecture provides a framework within which all of the activities performed and components used in an FMS find their place. Furthermore, the framework indicates the relationships between activities, between the various components and between activities and components.
2. *FMS life cycle.* A CIM architecture which explicitly shows the life-cycle stages of a system can show what activities are required and when, in order to lead an FMS from its initial conception through design, implementation (construction and commissioning) and operation.
3. *FMS analysis.* As system functionality is represented in the CIM architecture model, this model can be used for performing simulation analysis, at varying levels of detail.
4. *FMS operation.* Because all of the control logic is contained in the CIM architecture functional models, the potential exists to employ these models for controlling the FMS.

We shall briefly examine how each of these tasks can be accomplished using the Purdue enterprise reference architecture. References to CIM-OSA and GRAI-GIM shall be made where appropriate.

### 1.4.1 FMS specification

A typical FMS consists of the following components:

- NC or CNC machine tools, usually equipped with tool magazines and tool changers;
- a part load/unload station, and other stations such as inspection or part wash;
- a set of pallets and fixtures where parts are loaded into fixtures, which in turn are mounted on pallets;
- an automated material handling system (AGVs, conveyors, tow carts, etc.);
- a tool storage area and tool transport system;
- part buffers at machines and a storage area for pallets.

The control problems for an FMS can be divided into long-term planning, medium-range planning and short-term (real-time operation) as follows:

- Long-range (strategic, company) planning: what part types to produce in the FMS, FMS capacity.
- Medium-range planning (off-line control tasks): weekly or daily decisions of what to load into the system, including both part types and tools. Tasks include batching, balancing, loading, and sequencing.
- Short-term operation (real-time control tasks): scheduling and control of machines, tooling, and material handling equipment.

A hierarchical structure is usually used for the control system architecture. At the lowest level, real-time control of machines, material handling equipment, etc. takes place; as one moves up the hierarchy, the planning period increases and the frequency of control actions decreases. To execute the decision-making processes, computers, PLCs (programmable logic controllers), machine controllers, etc. are required. Also required for control are data storage facilities, control system inputs and outputs and communications equipment. A vast quantity of data must be maintained for FMS operation, including both static data (part process plans, NC programs, etc.) and dynamic data (production schedules, system status variables, etc.). Control system input devices (sensors, transducers, encoders, limit switches, operator input consoles, terminals, etc.) and output devices (actuators, solenoid valves, relays, manual control valves, etc.) serve to interface the control system to the FMS equipment and operators. Communications are required between the various computers, etc. in the control system, and between those devices at the lowest control level and the various control system input and output devices.

The location of and relationships between the above activities and

| Information Functional Network [1] | | Manufacturing Functional Network [1] | |
|---|---|---|---|
| *Planning and Control Tasks* | Decisional tasks: long and medium range FMS planning, short-term operation. Non-decisional tasks: control signal output for operator action. | *Physical Processing Tasks* | All part processing tasks. |
| *Data Storage Tasks* | Storage of all data required for FMS operation: static (process plans, NC programs, etc.) and dynamic (status, schedules, etc.) | *Queue Tasks* | Part storage, storage of pallets, tools, etc. |
| *Data Acquisition Tasks* | Generation of information representing physical state of FMS or action results. | | |
| *Communications Tasks* | Communications between computers, PLCs, etc. Control signals to actuators, indicators, etc. Data signals from sensors, terminals, etc. | *Move Tasks* | Part load/unload, part move; tool handling, pallet and fixture movement; chip removal, etc. |

| Information Systems Architecture | Human and Organizational Architecture | Manufacturing Equipment Architecture |
|---|---|---|
| Computers, PLCs, machine controllers, etc. (decisional planning and control tasks). Indicators, operator displays, etc. (non-decisional planning and control tasks). | Humans performing informational tasks, e.g. tooling and equipment monitors, supervisors. | Part processing equipment: NC and CNC machines, wash stations, etc. (physical processing tasks). |
| Computer disc drives, magnetic tape, paper (punch) tape, ROM memory, etc. (data storage tasks). | Humans performing physical tasks in FMS, e.g. part loader, material handler. | Storage facilities and equipment: part buffers, pallet storage, tool magazines, etc. (queue tasks). |
| Wiring, connectors and related hardware, signal processing equipment, etc. (communications tasks). | | Move equipment: part transport, tool transport, chip removal, etc. (move tasks). |
| Sensors, transducers, encoders, limit switches, operator input consoles, terminals, etc. (data acquisition tasks). | Humans performing informational and physical tasks, e.g. machine tool monitor, maintenance worker. | Actuators, solenoid valves, manual control valves, etc. (all manufacturing tasks). |

[1]All tasks in the network are interconnected, and are shown separated for explanatory purposes only.

**Fig. 1.11** Mapping of FMS tasks and components onto the Purdue enterprise reference architecture framework.

components can be shown using the Purdue enterprise reference architecture framework (Fig. 1.11). The following observations are made:

1. All tasks performed in an FMS are either informational or physical (manufacturing). Informational tasks can be further divided into planning and control tasks, data acquisition tasks, data storage tasks and communications tasks. Manufacturing tasks can be further divided into physical processing tasks, queue tasks and move tasks.
2. All tasks and functions performed in an FMS can be specified independent of how they are to be implemented. That is, as long as all task requirements are met, it makes no difference whether they are performed by machines or by humans.
3. Once all tasks have been defined, we can determine, based upon a task-by-task analysis, the location of the automatability and humanizability lines. The former indicates the extent to which the proposed FMS can be automated. The objective in FMSs is usually to have the automatability lines as close together as possible (indicating that the proposed system is highly automatable). The latter indicates the extent to which humans can perform the tasks, which depends upon many factors and may vary from one installation to another. There will always be some informational tasks (e.g. execution of complex algorithms) which cannot be done by humans. Physical processing tasks must usually be done by machines (due to task time restrictions, amongst other reasons), whereas humans may or may not be capable of executing move tasks, depending upon the item size, shape, and weight, move distance, etc.
4. The extent of automation lines, and thus the relative sizes of the information systems, human and organizational, and manufacturing equipment architectures, are determined from the actual allocation of tasks to human and non-human processors. In FMSs, the human and organizational architecture will be relatively small, indicating that the majority of tasks are performed by non-human processors (i.e. automated).
5. For tasks which are to be implemented by humans, we can determine the percentage of tasks which are informational, physical, or both, as well as the nature of the task requirements (cognitive demands, physical demands). These classifications will assist in the performance of human factors activities such as task set assignment and job design. (It should be noted, however, that the benefit of this approach is not as pronounced in FMS as in other manufacturing systems, where a larger proportion of tasks are allocated to humans).

## 1.4.2   FMS life cycle

A great quantity and variety of activities are required to design, implement, operate and maintain an FMS. These activities taken together can be considered the master plan for the FMS. A CIM architecture provides a framework for:

- indicating what activities must be completed and when;
- ensuring that consistency and coherence exist between the activity inputs and outputs, and that all activities lead to accomplishment of the master plan.

The activities which must be performed (Figs 1.4 and 1.5), and what they translate into for FMS, are described as follows.

### *Concept layer*

*Establish mission, vision, and values.* Establishment of reasons for desiring an FMS: improved responsiveness to market demands, to build a high-technology profile, etc.

*Establish manufacturing, personnel, and information policies.* The following policies must be established in each of these areas:

- Manufacturing: determination of preferred processes and technologies, preferred vendors, etc. Location of the FMS in the plant.
- Personnel: determination of worker skill requirements and qualifications, relevant union mandates, and preferred levels of automation for FMS operation. Whether or not to perform FMS design and installation solely with in-house personnel, to bring in external consultants, or to sub-contract entirely.
- Information: establish preferred planning and control approaches, operating systems, communications protocols, hardware/software vendors, etc.

In addition, budgets must be set for each of these areas.

*Establish proposed production entity, including product and operational requirements.* Long-range FMS planning: what part types to make, operational requirements for parts. Establishment of FMS capacity and capabilities.

### *Definition layer*

*Establish planning, scheduling, control, and data management requirements.* Selection of strategies for planning and control (medium-range planning, short-term operation), information processing and storage, communications and for integrating FMS control with system-wide control.

*Establish physical production requirements.* Determination of process capability requirements for machines, tools, fixtures, pallets, material handling equipment and storage facilities.

*Establish informational tasks and functional modules.* Specification of planning and control, information processing and storage, communications and control system integration strategies, in terms of the basic task and functional modules.

*Establish manufacturing functional modules.* Specification of all part load/unload, processing, and movement; tool load/unload and movement, and fixture and pallet movement requirements, in terms of unit operations modules.

*Establish information and manufacturing functional networks.* Establish connectivity, precedence, span-of-control, etc. requirements for functional modules and connect modules into networks.

### Distribute functions

Following functional design of the FMS, we must determine which tasks **must** be automated and which **cannot** be automated (location of automatability and humanizability lines). The former will be executed by non-human mechanisms in the information systems architecture (informational tasks) or in the manufacturing systems architecture (manufacturing tasks), whereas the latter will be executed by humans in the human and organizational architecture. The remaining tasks are then allocated to non-human mechanisms or humans according to the personnel policy established in the concept layer.

### Specification layer (functional design)

*Determination of functional requirements of mechanisms.* Functional requirements for all mechanisms in the FMS, non-human and human, must be established. These requirements are specified as follows:

- Non-human mechanisms executing informational tasks whose requirements depend upon the nature of the informational task:
  - Planning and control tasks. For decisional tasks – computational load, processing speed, accuracy, etc. For non-decisional tasks – amount of information to present, nature of information, etc.
  - Data storage tasks such as memory, storage and retrieval speed, data integrity, etc.
  - Data acquisition tasks such as type of physical variable to measure (continuous variables such as temperature; discrete variables such as count data), required sensitivity, resolution, speed of response, etc.
  - Communications tasks such as transmission rate, bandwidth requirements, etc.

- Non-human mechanisms executing manufacturing tasks such as processing (queue, move) capabilities and capacities, production rates, accuracy, repeatability, etc.
- Human mechanisms such as cognitive capability requirements, memory requirements, etc. (informational tasks); strength, endurance, etc. (manufacturing tasks).

### Detailed design layer

*Selection and grouping of processors.* This must be performed for each of the implementation architectures as follows:

- Information systems architecture: selection of all physical mechanisms required for executing informational tasks, and grouping these mechanisms. These mechanisms include
  - computers, PLCs, machine controllers, etc. (decisional planning and control tasks);
  - indicators, operator displays, etc. (non-decisional planning and control tasks);
  - computer disc drives, magnetic tape, paper (punch) tape, computer memory, human memory, etc. (data storage tasks);
  - Sensors, transducers, encoders, limit switches, operator input consoles, terminals, etc. (data acquisition tasks).

  In FMS, the mechanisms are usually grouped into various control levels, based upon the concept of hierarchical control. Communications equipment (wiring, connectors and associated hardware, signal processing equipment, etc.) are used to link these elements together.
- Human and organizational architecture: selection of the skill requirements, job descriptions, training programs, etc. for the various jobs in the FMS which must be (have chosen to be) performed by humans.
- Manufacturing systems architecture: selection of all machines, the material handling system, the tool transport and storage system, part buffer storage at machines, and pallet storage, as well as equipment control devices (actuators, solenoid valves, manual control valves, etc.). In addition, the detailed layout of these items in the FMS is specified.

### Manifestation layer

*Construction and installation of equipment, commissioning of system.* Construction and installation must be performed for each of the implementation architectures as follows:

- Information systems architecture such as equipment programming, software development, etc., equipment installation and check-out.

- Human and organizational architecture such as staffing, training of operators, etc.
- Manufacturing systems architecture such as construction or purchase of machines, material handling equipment, etc., installation in the plant, check-out.

Once the entire system has been installed it must be commissioned. This process can take from nine months to two years, depending upon the system size and complexity (Talavage and Hannam, 1988). The architectural specification of the system can be an invaluable tool during this process, serving as the reference model for the system.

### Operating layer

*Operation of the system.* During operation of the FMS, the following should occur:

- Information systems architecture – continual development and improvement of control strategies, software and so on, and maintenance of hardware/software.
- Human and organizational architecture – continued training of workers, improved job descriptions, organizational development, etc.
- Manufacturing systems architecture – continued process improvement, etc. and maintenance of equipment.

As the FMS evolves during operation, the architectural specification of the system should be continually updated (manually or automatically), so that it is always a valid representation of the system. This is essential if the model is to be used for FMS analysis or control (discussed next).

### 1.4.3 System analysis

Two types of analysis activities exist for FMS: static analysis and dynamic analysis. Static analysis consists of ensuring that the actual system is a valid realization of the system design (new system), or that the system model is a valid representation of the actual system (existing system). Dynamic analysis consists of modeling the FMS to determine its performance and/or cost(s) for given inputs and operating scenarios. The role of CIM architectures in the execution of each of these activities shall be discussed separately.

### Static analysis

For an FMS, static analysis consists of ensuring that control software performs as intended, that the required data (machine schedules, NC programs, status data, etc.) is available at the proper time and in the proper format, and that the proper tooling, fixtures and so on are available when required so that processing activities can be executed as planned. By providing different models for activity modeling, data modeling etc., clearly showing the relationships between the elements of different models, and employing a plan stating what models to build and when, a CIM architecture can greatly facilitate static analysis. The use of CIM architectures for this purpose is an important aspect of the GRAI Integrated methodology.

In the GRAI approach, static analysis is a type of 'consistency analysis'. During the design phase, detailed analysis is performed to find and analyze the inconsistencies between the existing system and the ideal system (system design). This is accomplished by checking if a set of rules is fulfilled at each abstraction level during each stage of the design phase. The rules check for 'coherence' between the decisional and functional views, the functional and informational views, and the physical and functional views. If the rules are satisfied, the four views are said to be 'coherent', and the system is deemed to be a valid representation of the ideal system.

*Dynamic analysis*

Typical modeling approaches for dynamic analysis of FMS include rough-cut calculations, queuing models and simulation. Simulation is usually required at some stage, as the complexity of FMS limits the validity of analytical approaches. Provided that suitable modeling formalisms are used for modeling tasks (activities) and mechanisms, a CIM architecture framework and corresponding models are well-suited to the task of discrete-event simulation.

In the Purdue enterprise reference architecture, the information functional network contains all of the planning and control tasks for the FMS. These tasks are defined in terms of various attributes (precedence, connectivity, span-of-control, etc.), and execution logic (conditions for execution, task duration, frequency, etc.), and are specified independent of how they are to be implemented. The mechanisms themselves are specified in the information systems architecture, the human and organizational architecture, and the manufacturing systems architecture, all at the detailed design level. This separation of task functionality from implementation leads to models which may be used for discrete-event simulation. Possible modeling formalisms amenable to this task are:

- Purdue modular task and function representation method: modeling of task and function precedence and connectivity.
- Petri net models: modeling of task (function) execution logic.
- Object-oriented models: modeling of physical mechanisms.

Petri net transitions are linked to the corresponding task (function) in the activity models, which in turn are linked to the object-oriented models of the physical mechanisms which execute the task (function). Part arrivals, machine downtimes and other external and/or random events must be modeled externally, and their results used as inputs to the Petri net models (conditions). Execution of the Petri nets results in the execution of the various tasks (functions), and corresponding manipulation of the object-oriented models of the physical mechanisms. Statistics collection can be performed automatically by these objects (Shewchuk and Chang, 1991).

As the Purdue modular task and function representation method and Petri nets are both hierarchical modeling methods, simulation can be performed at various levels of detail.

### 1.4.4  System control

The final role which CIM architectures can play in FMS is that of controlling the FMS. Considering the Purdue enterprise reference architecture, we see that the same models of the information functional architecture can be used directly for system control. To accomplish this, the following modifications (to the manner in which these models are used for dynamic analysis) are made:

- The actual system status replaces external and randomly-generated events, such as part arrival. This status is obtained from the various sensors, transducers, operator input consoles, terminals, etc., in the information systems architecture.
- The task and function linkages to the object-oriented models of the physical mechanisms are replaced with outputs to the actual physical mechanisms, i.e. to the various actuators, solenoids, etc. in the manufacturing equipment architecture, or to the indicators, operator displays, etc. in the information systems architecture.

In an FMS, a decentralized control scheme is employed, with various control tasks and responsibilities delegated to each of the computers, PLCs, machine controllers and so on in the control hierarchy. Thus, to use the CIM architecture models for system control, it is necessary to distribute the contents of the various models of the information functional architecture amongst the various control devices. This approach can be difficult to implement with some presently-available technology due to i) the lack of a suitable design environment for developing the control models to the degree of detail required, and distributing them to the various control devices in a controlled, reliable manner, ii) differences in the types and forms of services required by the various control devices, and iii) the wide variety of formalisms used by the control devices for expressing control requirements (e.g.

ladder logic vs. NC programming languages vs. high-level programming languages).

The idea of using CIM architectural models to control system operation is one of the cornerstones of the CIM-OSA approach. An attempt is being made in CIM-OSA to create the system model in a design environment, and to then 'release' the implementable portion of the model to an operation environment, where the model is made processable and can thus be used to control the system. To cope with the aforementioned problems, CIM-OSA is planning to provide the following:

1. Separate design and operation environments. The integrated enterprise engineering environment consists of a set of CAE tools to support the design process. The integrated enterprise operation environment provides all the support services required to execute and control the business processes and enterprise activities defined in the released portion of the implementation description model.
2. A standard set of services (integrating infrastructure) to support execution of the processable model at run time. The integrating infrastructure can be considered as an 'enterprise-wide operating system', controlling all physical entities in the system (men, machines, etc.) in addition to the usual data processing requirements.
3. A reference architecture and supporting guidelines for vendors developing components which make use of and are compatible with the services provided by the integrating infrastructure.

## 1.5  CONCLUSION

CIM architectures are useful tools for the design, analysis, implementation and operation of complex manufacturing systems. They provide a framework within which all the system activities and components can be modeled, and assist in the development and implementation of a master plan for the particular system of interest. Flexible manufacturing systems are well suited for the use of a CIM architecture, due to the high level of system complexity and the high degree of automation involved. The architecture provides models which can be used not only for system specification, but for system analysis and control as well.

The Purdue enterprise reference architecture is particularly attractive for FMSs as it allows one to determine the maximum level of automation possible, as well as the extent to which the FMS actually is automated. The GRAI-GIM approach presents an example of how CIM architectures can be used for static system analysis, i.e. to ensure that the FMS is a valid representation of the designed system. Finally, the CIM-OSA approach illustrates a manner in which the models used to design the FMS can be used to control the system as well.

## REFERENCES

AMICE Consortium (1988) Open system architecture for CIM research, *Reports, ESPRIT Project 688*, Springer Verlag, Berlin.

AMICE Consortium (1989) Open system architecture for CIM research, *Reports, ESPRIT Project 688*, **1**, Springer Verlag, Berlin.

AMICE Consortium (Aug. 24, 1992) *Report, ESPRIT Project 5288*, Milestone M-2, AD2.0, **2**, Architecture description, document RO443/1, Consortium AMICE, Brussels, Belgium.

Anonymous (1986) *Information flow model of generic production facility*, The Foxboro Company, Foxboro, MA.

Anonymous (Sep. 1987) *Manufacturing Automation Protocol, Version 3.0*, General Motors Corporation, Society of Manufacturing Engineers, Detroit, MI.

Anonymous (Apr. 21, 1988) *Map in the process industries*, White Paper Developed by the Process Industries Focus Group of the MAP/TOP users Group, Draft 4.0.

Anonymous (Jan. 1988) Data management software summary, *DataPro Reports on Microcomputers*, **2**, CM45-000-501 (ed. J.R. Peck), DataPro Research Corporation, Delran, NJ.

Anonymous, Process flow and description diagrams, Annual Process Issue, *Hydrocarbon Processing*, Houston, TX.

Askin, Ronald G. and Standridge, Charles R. (1993) *Modeling and Analysis of Manufacturing systems*, John Wiley and Sons, pp. 125–35.

Bell, R.R. and Burnham, J. (1991) *Managing Productivitiy and Change*, South-Western Publishers, Cincinnati, OH.

Charles Stark Draper Laboratory (1984) *Flexible Manufacturing Systems Handbook*, Park Ridge, NJ, pp. 29–39.

Doumeingts, G. (Nov. 13, 1984) *Methode GRAI: methodes de conception des systemes en productique*, Thesis d'etat: Automatique, Université de Bordeaux I, p. 519.

Doumeingts, G., Vallespir, B., Darracar, D. *et al.* (Dec. 1987) Design methodology for advanced manufacturing systems. *Computers in Industry*, **9**(4), pp. 271–96.

Doumeingts, G. Vallespir, B., Zanettin, M. *et al.* (May 1992) GIM, GRAI integrated methodology, *A methodology for designing CIM systems*, Version 1.0, unnumbered report, LAP/GRAI, University de Bordeaux I, Bordeaux, France.

Elsayad, Elsayed A. and Boucher, Thomas O. (1985) *Analysis and Control of Production Systems*, Prentice-Hall, Englewood Cliffs, NJ, pp. 299–305.

ESPRIT Consortium AMICE (1991) Open system architecture, CIMOSA, AD 1.0, *Architecture Description*, ESPRIT Consortium AMICE, Brussels, Belgium.

IFAC/IFIP Task Force on Architectures for Integrating Manufacturing Activities and Enterprises; Williams, T.J. (ed.) (July 1993) Architectures for integrating manufacturing activities and enterprises, *Tech Rep of the Task Force, International Federation of Automatic Control*, Sydney, Australia.

Industry-University Consortium, Purdue Laboratory for Applied Industrial Control (June 1992) An implementation procedures manual for developing master plans for computer integrated manufacturing, *Report No. 155*, Purdue Laboratory for Applied Industrial Control, Purdue University, West Lafayette, IN, USA.

Jorysz, H.R. and Vernadat, F.B. (1990) CIM-OSA Part 1: Total enterprise modelling and function view. *Int. J. Computer Integrated Manufacturing*, 3(3,4), pp. 144–156.

Jorysz, H.R. and Vernadat, F.B. (1990) CIM-OSA. Part 2: Information view. *Int. J. Computer Integrated Manufacturing*, 3(3,4), pp. 157–167.

Klittich, M. (1990) CIM-OSA Part 3: CIM-OSA Integrating InfraStructure – The operational basis for integrated manufacturing systems. *Int. J. Computer Integrated Manufacturing*, 3(3,4), pp. 168–180.

Litt, Eric E. (May 1990) *The development of a CIM architecture for the RAMP program.* Proceedings of CIMCON '90, Gaithersburg, MD, USA.

Pimentel, Juan R. (1990) *Communications Networks for Manufacturing*, Prentice-Hall, Englewood Cliffs, NJ, USA, pp. 431–44.

Rembold, Ulrich (1985) *Computer-integrated Manufacturing Technology and Systems*, Marcel Dekker, New York, pp. 742–770.

Shewchuk, John P. and Chang, Tien-Chien (1991) *An approach to object-oriented discrete-event simulation of manufacturing systems*, Proceedings of the 1991 Winter Simulation Conference (eds Barry L. Nelson, W. David Kelton, Gordon M. Clark), pp. 302–11.

Talavage, Joseph and Hannam, Roger, G. (1988) *Flexible Manufacturing Systems in Practice: Applications, Design and Simulation*, Marcel Dekker, New York.

Williams, T.J. (ed.) (1989) *Reference Model for Computer Integrated Manufacturing, A Description From the Viewpoint of Industrial Automation*, Purdue University, West Lafayette, IN, Instrument Society of America, Research Triangle Park, North Carolina, USA.

Williams, T.J., and the members of the Industry-Purdue University consortium for CIM (Dec. 1991) The Purdue enterprise reference architecture, *Report No. 154*, Purdue Laboratory for Applied Industrial Control, Purdue University, West Lafayette, IN, USA.

Williams, T.J. (Oct. 1992) *The Purdue Enterprise Reference Architecture*, Instrument Society of America, Research Triangle Park, North Carolina, USA.

# CHAPTER 2

# Hierarchical control architectures from shop level to end effectors

*M. Kate Senehi, Thomas R. Kramer, Steven R. Ray,*
*Richard Quintero and James S. Albus*

## 2.1 INTRODUCTION

As automation of manufacturing systems becomes commonplace, the design, construction, and use of computerized control systems has become an increasingly vital problem. Computerized control systems are typically large, complex systems that are composed of components of lesser complexity which are developed and validated separately. Frequently, different components of the control system are made by different vendors and are designed to work with humans but not with other automated systems. Systems which are found in manufacturing environments include production machinery, communication software and hardware, computer hardware, databases, file systems, and production management software.

The purpose of a control architecture is to enable these components to work together in an integrated way to give satisfactory product quality at a reasonable price. At present, for each distinct set of components, a systems integrator defines a unique control architecture. This approach to developing control architectures is expensive, time-consuming and makes diagnosis of system problems difficult. The creation of a control architecture which can be applied to classes of machine control systems – a *reference architecture* – is therefore highly desirable.

Reference architectures often specify integration rules and standard interfaces among components. By adhering to the standard interfaces and integration rules required by the architecture, different vendors can construct components which are interoperable. Using the interoperable components, and system integration rules and methods, components may be integrated to build a machine, groups of machines and people can be integrated to form a workstation, workstations may be integrated to form cells and so on, to any degree of complexity desired. The availability of a reference architecture which defines interoperable components can improve the timeliness, reliability, safety, and extensibility of control systems.

This chapter discusses two reference architectures applicable to manufacturing developed by the Manufacturing Engineering Laboratory (MEL) at the National Institute of Standards and Technology (NIST). One reference architecture focuses on providing real-time control of equipment; the other focuses on providing information integration with factory production systems. Obviously, a general reference architecture for manufacturing must provide both of these functions. Recent work at NIST centers on determining the feasibility of combining the best features of these reference architectures into a reference architecture which spans control levels from the shop level to end effectors.

## 2.2  HIERARCHICAL CONTROL ARCHITECTURES

The following sections give a general discussion of control architectures and hierarchical control architectures. The discussion given here is meant to provide a basic understanding of the function and structure of a control architecture, and to provide a vocabulary with clear meanings so that the two NIST architectures can be discussed. (The international standards body the International Organization for Standardization, Subcommittee 5, Working Group 1 has the action item to develop a standards framework for computer integrated manufacturing architectures and a standard vocabulary. As yet this work (ISO, 1993) is too immature for use.)

In this chapter the pieces of an architecture (software, languages, execution models, controller models, communications models, computer hardware, machinery, etc.) will be called **architectural units**. The realization of an architecture in hardware and software for a particular use will be called an **implementation** of the architecture. We shall refer to the realization of an architectural unit in an implementation as a **component** of the implementation.

In section 2.2.1 we discuss a number of architectural units normally present in a control architecture. Section 2.2.2 discusses hierarchical control concepts – what a hierarchical control architecture is and its advantages.

### 2.2.1  Control architectures

The purpose of a control system is to achieve goals. A **goal** is a desired state of affairs. Goals include such items as manufacturing a part, moving a robot arm to a specific place, or navigating a vehicle from one point to another. A scheme developed to accomplish a specific goal is termed a **plan**. Typically, a plan consists of a number of discrete steps. Often the steps are sequential, but this is not necessarily so. A piece of work which achieves a specific goal – actual work, not a representation of work – is termed a **task**. A generic representation of a type of work, such as moving in a straight line from one point to another, opening a gripper, or drilling a hole, is a **work element**. The process of determining which tasks must be carried out by the control system is termed **task generation** and performing these tasks is termed **task execution**. The specification of the mechanism for task generation and execution forms a major portion of a control architecture.

In a control system, a **planner** is an agent which generates or selects plans to accomplish one or more goals, and a **controller** is the agent which directs performance of or performs specific tasks. **Scheduling** is the assignment of specific resources and times to the steps in a plan. A **scheduler** is an agent which performs scheduling. Often, the operations of scheduling and planning are combined in one architectural unit.

The specification of a control architecture normally includes:

• the functionality and interfaces of each of the architectural units;
• the organization of architectural units in the system;
• the mechanisms of communication among the architectural units;
• the mode(s) of interactions among the architectural units;
• the specification of structure of plans for the control system;
• the specification of tasks performed by the control system;
• the method of task generation;
• the method of task execution;
• provisions for error recovery in task execution and other phases of system operation.

In addition to these, manufacturing architectures must also specify resource descriptions, resource allocations and provisions for material handling. A discussion of these items is beyond the scope of this chapter. The reader is referred to Kramer (1993) for more information.

### 2.2.2 Hierarchical control

The hallmark of a hierarchical control architecture is that controllers are arranged in a hierarchy. Often, the command structure is a simple type of hierarchy, a tree, in which each controller has one superior and zero to many subordinates. A hierarchical control system explicitly disallows the exchange of commands among peer controllers, but may permit peer controllers to share information by any number of mechanisms (e.g. through information shared in a database).

In hierarchical control architectures, task execution is usually the result of a superior instructing a subordinate controller to carry out a task. Such an instruction is called a **command**. The most obvious examples of tasks initiated by commands are the physical processing (e.g. milling, cutting) and motion tasks. Other types of work, such as fetching data, navigating through the plan, or synchronizing with other plans, may also be initiated by commands.

Commands often specify the work to be performed by naming a specific work element. If a controller can carry out only one work element, the command does not need to name it. The execution of the task specified by the work element requires that parameters to the work element be provided by the superior controller.

Advantages of a hierarchical control architecture include:

- Natural modularity.
  Each controller can be treated as a module, facilitating incremental development, making the system software easier to understand and maintain, and allowing the use of templates for controller code.
- Fairly easy extensibility.
  The system may be extended by adding controllers and computers and changing the hierarchy.
- Somewhat graceful degradation.
  If something goes wrong during system operation, in a well-designed hierarchical system, only one branch of the hierarchy needs to stop.
- Allowance for different frequencies of operation of controllers on different levels of the hierarchy.
  Typically, controllers at lower levels of the control hierarchy have higher frequencies.

Disadvantages include:

- need for communications among controllers (as compared to a centralized architecture, where there is only one controller);
- need for managing interactions among controllers;
- difficulty integrating system-wide service functions, such as material handling.

Both reference architectures discussed in this chapter are strictly hierarchical.


## 2.3  REFERENCE ARCHITECTURES AT NIST

The Manufacturing Engineering Laboratory (MEL) at the National Institute of Standards and Technology (NIST), is conducting research on control of mechanical systems for use in such diverse fields as discrete part manufacturing, coal mining, under-ice submarining, and space exploration.

As a result of differing requirements in each application, the characteristics of control systems vary greatly. Nevertheless, more than 16

years of experience within the Robot Systems Division (RSD) and the
Factory Automation Systems Division (FASD) of MEL indicate that
there are aspects of control which are common to all control systems in
a broad range of applications. These aspects have been captured in a
number of control system reference architectures that provide both
specifications for the parts of the architecture and their behaviors,
and methodologies for constructing control systems according to the
prescribed specifications. One class of reference architecture is the real-
time control system (RCS) architecture (Albus, 1981, 1987, 1991 and
1992; Barbera, 1977; Herman, 1987; Quintero, 1992) and specializations
of it, such as the NASA/NBS Standard Reference Model for Telerobot
Control System Architecture (NASREM) (Albus, 1989). Other reference
architectures include the Automated Manufacturing Research Facility
(AMRF) control architecture (Jones, 1985 and 1986; McLean, 1987;
Simpson, 1982) and the manufacturing systems integration (MSI) architecture (Senehi, 1991b, 1992).

Two of the architectures under active investigation at NIST, RCS and
MSI share many common features. For example, both mandate that the
controllers in the system be arranged in a command tree. In both the
architectures each type of controller has its own specialized set of commands it can carry out. Both the architectures implement command
execution by message passing between controllers, and so forth. But
there are also some differences. Timing issues and sensory processing
receive more attention in RCS, scheduling and resource definition
issues more in MSI.

In section 2.4 and section 2.5, we discuss RCS and MSI in detail.

## 2.4  RCS DESCRIPTION

This section provides a brief description of RCS. The definition of RCS
has evolved over the years, and different people developing or using it
have different views on what it should be. The description given here
is intended to follow the mainstream, as defined primarily by the
papers by James Albus. Where there are significant variations in other
papers, they are cited, but we have not tried to describe all variants of
RCS.

RCS is intended as an architecture for complex, integrated machine
control systems which work in a changing world and keep pace with
the changes in real time. (Most of the RCS papers describe RCS as an
architecture for 'intelligent' systems. In this chapter, however, we do
not deal with the notion of intelligence.) The spectrum of intended
RCS uses includes:

• high-speed control of machines with multiple joints or axes of
  motion;

- coordinated control of several machines or large machines with several subsystems;
- computer integrated manufacturing;
- mining;
- submarine navigation;
- space station robotics;
- land vehicle driving.

A methodology for building RCS systems is given in section 2.6 of Quintero (1992), which describes the activities a systems developer should perform and the parts of the architecture that should be produced as a result.

### 2.4.1   RCS control systems and their environments

The RCS architecture provides for control of systems which react to events in the environment. Control systems are expected to have mechanisms for sensory input so that changes in the environment can be detected. The control system is constantly monitoring its sensory input to determine when events have occurred in the environment that it must react to. The processing of raw sensor data into abstract information about the condition of the environment is termed **situation assessment**. Once situation assessment has been performed, the control system makes decisions about what actions should be taken and plans reactively for the events it perceives. The execution of plans produces the external actions needed to cope with the environmental changes. An RCS controller continuously performs a sense-decide-act cycle.

### 2.4.2   Architectural units of RCS

An RCS system interacts with the environment by sensing conditions in the environment with its sensors and performing actions in the environment with its actuators. The internal representation of selected features of the environment and the state of the RCS system is termed the **world model** of the system. The world modeling architectural unit governs interactions with the world model. In addition to world modeling and the associated world model, an RCS system includes three other architectural units which are internal to the control system. The four internal architectural units of RCS are:

- sensory processing (SP);
- world modeling (WM);
- behavior generation (BG);
- value judgment (VJ).

The sensory processing, world modeling, and value judgment architectural units are involved in situation assessment, while the value judgment and behavior generation architectural units are involved in deciding what to do. Figure 2.1, *RCS view of an intelligent machine system* redrawn from Albus (1989), illustrates these conceptual architectural units. The system includes everything above the lower horizontal dotted line. The world model is central, since other architectural units rely upon it to provide and accept current information about the environment. The remainder of the architectural units are arranged in a clockwise loop, depicting the notion that the system continually repeats a sense-decide-act cycle.

In Fig. 2.1, the sensory processing function system (described in more detail below) takes sensory data from sensors, interprets the data, and passes the interpreted data to world modeling.

The world modeling function keeps a description of the environment and the internal state of the system (the world). It receives information from sensory processing for updating the world model. It also predicts events and sensory data and answers questions about the world model. The world modeling function interacts with the RCS system's database. The database is usually described as a distributed, global database – in the sense that all data is available throughout the system.

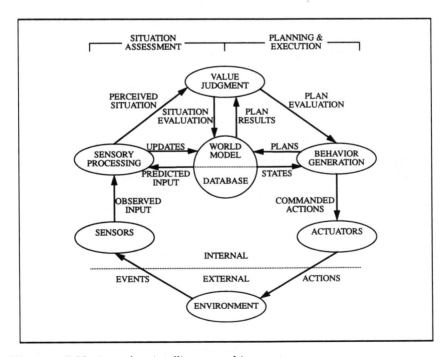

**Fig. 2.1** RCS view of an intelligent machine system.

The behavior generation function (described in more detail below) makes plans and carries them out by controlling the system's actuators.

The value judgment function evaluates both the observed state of the world and the predicted results of hypothesized plans. It computes costs, risks, and benefits both of observed situations and of planned activities. The value judgment function thus provides the basis for choosing one action as opposed to another, or for acting on one object as opposed to another.

In terms introduced earlier in this chapter, task generation in RCS is performed by the value judgment and behavior generation function, with input from the sensory processing module and sensors. Task execution is performed within the behavior generation module of RCS. The following section gives additional details on the task generation and execution in RCS.

### 2.4.3   Hierarchical levels in RCS

The behavior generation system in RCS is strictly hierarchical. That is, each controller responsible for behavior generation has at most one superior and zero to many subordinates, for the purposes of performing actions. Control levels are typically referred to by number (numbering from 1 at the bottom, on up) or by a label. Different applications of RCS have used different labels for these levels, but typically the lowest level is termed the servo level, next is the primitive level and above that is the elementary move (or e-move) level.

Superiors interact with subordinates by sending commands to them and receiving status messages from them. Each controller has a number of tasks that it can carry out, and these tasks are understood by the superior of the controller.

The RCS architecture decomposes system activities into hierarchical levels. The levels are characterized by the relative amount of time taken to perform activities and by the relative spatial extent of the activities. Roughly an order of magnitude change in spatial and temporal extent is expected between any two adjacent levels, with activities getting smaller and faster at lower levels of the hierarchy. Between levels, a corresponding change is also expected in the interval of time over which the system detects and remembers events. Approximate times corresponding to the RCS control levels are shown in Fig. 2.2 (Albus, 1989).

At each control level, the sensory processing, world modeling, behavior generation and value judgment architectural units may exist. Figure 2.3 (Albus, 1989), illustrates a six level RCS architecture appropriate for telerobotic applications. The label TD on Fig. 2.3 and elsewhere in this section stands for 'task decomposition', which is a synonym for behavior generation.

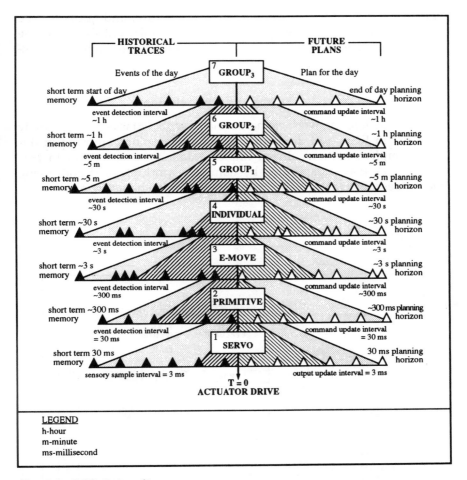

**Fig. 2.2** RCS timing diagram.

In considering Fig. 2.3, it must be understood that each rectangle labeled SP, WM or TD (except those in the top level) normally represents several separate instances of the given function. That is, there are several TD5 behavior generators that are subordinates of the TD6 behavior generator, for each TD5 behavior generator there are several TD4 subordinates, and so on down the hierarchy. The situation can be visualized by imagining that the figure is the front view of a 3-dimensional arrangement, which, when looked at from the side, is a hierarchy.

An example of a hierarchy in which there are several SP, WM, and TD boxes at each hierarchical level below the top is shown in Fig. 2.4 (a portion of Fig. 1 from Albus (1992)). This figure shows a control system for a robot with a camera, an active fixture, and grippers. The robot is subordinate to a workstation level controller not shown on the figure. Four hierarchical control levels are shown: equipment task, e-move, primitive, and servo.

### 2.4.4  Tasks and work elements

Methods of defining work elements and describing tasks are not strictly specified in RCS, and different sorts of specifications are used in different implementations. In all implementations, a command to perform a task may be specified by naming a work element and giving the values of zero to many parameters which characterize the work element.

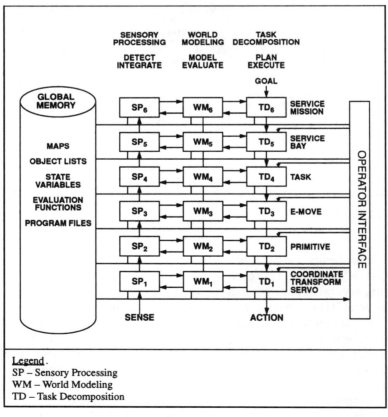

**Fig. 2.3**  RCS control system architecture.

### 2.4.5 Sensory processing

The sensory processing function of an RCS system takes sensory data at the lowest hierarchical level, interprets the data, and passes the interpreted data to world modeling. Sensory data may need to be filtered as it arrives. Sensory data may also need to be integrated over space (for constructing a map, for example) or time (for speech recognition, for example). As indicated on Fig. 2.3, the integration of data proceeds upwards from level to level. In the case of shape recognition in a vision system, for example, points might be detected at the lowest level and fed upwards where some of them may be integrated into lines; lines are fed upwards, and some of them may be integrated into boundaries of (geometric) faces; faces may be fed upwards and integrated into closed shells of solid objects.

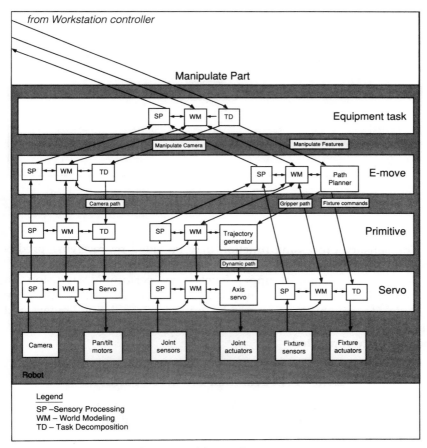

**Fig. 2.4** Example of a robot control hierarchy.

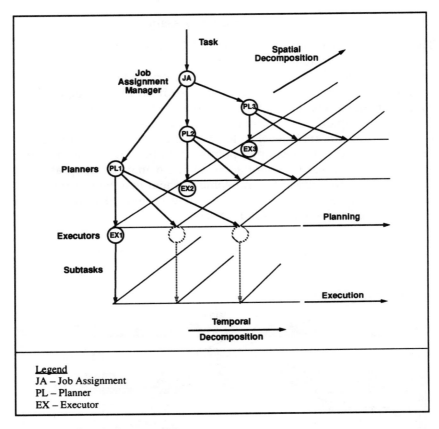

**Fig. 2.5**   RCS task decomposition.

The sensory processing function may be aided by receiving predictions of sensory data from the world modeling function.

Sensory processing at upper levels may perform data fusion, in which different sets of data which should be consistent (such as the distance to an object measured by optical triangulation, radar, and sonar) are reconciled, or different types of data (outline and color, perhaps) are correlated.

### 2.4.6   Task definition and decomposition

The behavior generation (or task decomposition) process is shown in Fig. 2.5 (Fig. 2 in Albus, 1992). Behavior generation is decomposed into three parts:

- job assignment (JA);

- planning (PL);
- execution (EX).

A task is decomposed by a job assignment manager (JA) into sub-tasks for several subordinates. The planner for each subordinate (PL1, PL2, and PL3 on the 'spatial decomposition' axis in the figure) orders the subtasks in a temporal sequence (the 'temporal decomposition' axis). Each subtask is executed by an executor (EX1, EX2, and EX3 on the figure). The same executor will execute different subtasks at different times, as indicated by the dotted circles and dotted lines on the figure. The figure shows what happens at one hierarchical level. The subtasks coming from an executor at one level become the tasks for the next level down.

### 2.4.7 Communications

RCS does not specify a standard for communications but anticipates that at lower hierarchical levels, fast communications will be required. Some RCS papers, (e.g. Quintero, 1992) state that communications must permit input and output at any time, regardless of current system activities, in order to ensure that sufficiently fast performance can be achieved. In most implementations, shared memory or some form of NIST's common memory (Libes, 1990; Rybczynski, 1988) has been used.

Standard communication protocols such as Ethernet/TCP/IP (Tanenbaum, 1988) or RS-232 (EIA, July 1991) have been used in RCS implementations for interfacing processes which are not on a common bus. For mobile applications, radio frequency communications hardware is also used.

### 2.4.8 Error recovery

Automatic error recovery for handling 'abnormal' error conditions is discussed in Albus (1989) and Herman (1987). In Albus (1989) it is anticipated that if there is a subtask failure, the executor should branch immediately to a pre-planned emergency subtask while the planner selects or generates an error recovery sequence. Herman (1987) reports an implementation of a subtask failure replanning software module.

## 2.5 THE MSI ARCHITECTURE

The MSI (manufacturing systems integration) architecture is a product of the manufacturing systems integration project which was conducted from 1990–1993 within the Factory Automated System Division. This architecture is the work of the MSI architecture committee members: Ed Barkmeyer, Steven Ray, M. Kate Senehi, Evan Wallace and Sarah Wallace.

The architecture is directly applicable to the production of discrete metal parts. Many of the concepts are more broadly applicable, but a discussion of this is not included in this summary of the architecture. The MSI architecture focuses upon the operation of a shop which receives orders and raw materials for the production of parts. (Other types of manufacturing organizations, such as rework organizations, which receive damaged pieces to be repaired and instructions on how to repair them, or shops which contain autonomous 'subshops' such as tool cribs have been considered by the MSI architecture committee. Because of differences in scheduling and error-recovery processes, these shops operate under different constraints which are not currently supported by the MSI architecture.) The architecture is able to control a shop with any combination of physical and emulated equipment. Additionally, the architecture permits the integration of systems not initially designed to work within the architecture, such as commercial products, or prototype research systems.

The architecture draws upon early work of the AMRF on hierarchical control (Albus, 1981; McLean, 1987; Simpson, 1982) and work of the manufacturing data project (MDP) (Hopp, 1988) which focused on information required for manufacturing, particularly process plans and resources.

This section provides a brief description of the second version of the MSI architecture. More details are available in Senehi (1992) and Wallace (1992). Although the second version of the architecture differs somewhat from the initial version, many concepts from the initial architecture still apply. Documentation for the initial architecture may be found in Senehi (a) and (b).

### 2.5.1  Architecture overview

The goal of the architecture is to integrate the operation of a shop which manufactures discrete metal parts. Particular emphasis is placed by the architecture on integration of shop planning, scheduling and control functions in both nominal and error situations. The architecture does not attempt to provide enterprise integration; in particular, it does not describe information needed for business decisions, such as whether to buy or manufacture a part. (It does however, permit the user to include technical information such as cost functions for use in determining the way in which an order is filled.)

The architecture approaches integration by identifying the systems in the shop which need to be integrated, examining interactions among the systems, and proposing mechanisms to ensure that these systems function in a cohesive manner.

### Architectural units in MSI

The MSI architecture identifies a number of systems which are normally part of the shop production environment. The architecture defines architectural units corresponding to each of the shop systems identified, characterizing each system by the functions which it performs. The MSI architecture avoids specifying the internal structure of any of the architectural units. This approach facilitates building implementations of the architecture which use systems which were not designed specifically to work within the architecture.

The architectural units which correspond to shop systems and their functions are summarized below. (Additional systems may, of course, be part of a manufacturing system, but these have not been considered in the formulation of the architecture.)

- Part design – which creates designs for parts, associated fixtures and jigs.
- Process planning – which creates step-by-step plans or numerical code for manufacturing the part, associated fixtures and jigs according to the part design.
- Production planning (schedulers are one type of production planning system) – which selects batch sizes, specific machines, and scheduled times to perform the tasks specified by the process plan.
- Controllers – which perform manufacturing tasks.
- Order entry – which permit entry of orders which direct the shop as to what to make and when to make it.
- Configuration management – which identifies and controls shop resources and capabilities.
- Material handling – which routes and delivers material throughout the shop.

### Interactions of architectural units

Most architectural units are loosely coupled, that is, they share information, but their activities do not need to be coordinated except to ensure integrity of the information they share. In this chapter, this type of interaction is called indirect interaction.

In indirect interaction, shared information is stored in a known location (e.g. a memory location, database, file, variable), and components (of an implementation) may be given access (e.g. read, write, no access) to the information as required. Components which have access to the same information need not be known to each other, and need not acknowledge any access or change of the information by any other component.

In order to integrate architectural units which interact indirectly, the MSI architecture specifies that it is sufficient to describe shared information at a conceptual level, and provide guidelines for access to the information. The description of the shared information is given through a number of information models. The information models and the guidelines for information access form the information architecture of MSI. This is discussed in detail in section 2.5.2.

The production planning and control architectural units are tightly coupled, that is, they must interact more closely than through the passive sharing of information. To understand the interactions of these systems and the MSI solution to integrating them, it is necessary to understand the MSI perspective on task generation and execution in a shop.

A shop's function is to manufacture products to fill orders which the shop has received. The orders are for some number of a specific product, which is described by a design. For each design, a process plan is formulated or selected. The process plan gives detailed instructions on how to manufacture the product design, usually using classes of resources. For example, a process plan would say 'This step requires a three-axis milling machine,' rather than 'This step requires machine XYZ001.' When an order is received for making a number of a specific product, an appropriate process plan is retrieved or generated, and the order is broken into batches and/or combined with similar orders to form batches for manufacturing. For each batch, specific resources for product manufacture are selected and the plan and the resources are scheduled. The end result of performing these operations is a production plan which contains all necessary information for making the product. When the scheduled time for starting the batch has arrived, the controllers in the shop interpret the production plan and perform the work of manufacturing the product.

In performing the work of manufacturing the product, the activities of controllers must be coordinated. This is accomplished by using two mechanisms. First, as described above, production plans are generated which schedule the activities of each of the controllers in the shop. Second, controllers are connected in a control structure that provides support for integrated start-up, shutdown, emergency stopping and disposition of tasks generated from the production plan. The MSI architecture requires that controllers in the shop be arranged in a hierarchical control structure. Commands are transmitted from superior controllers to their subordinate controllers, and subordinate controllers send status information to their superiors. Interaction through a command-and-status mechanism is referred to as direct interaction.

Thus, the integration of tightly coupled planning and control architectural units requires both indirect interaction through process and production plan information and direct interaction through the control hierarchy. The representation of the information for process and production plans is discussed in section 2.5.2. The control structure is discussed in section 2.5.3.

### 2.5.2 Information architecture

As previously mentioned, the MSI architecture states that for indirect system interactions, it is necessary only to describe shared information and information access characteristics (i.e. which components can access which information and what type of operations the components can perform). The following sections discuss the shared information. Shared information is described using a number of information models.

*Information models*

The information needed to integrate a manufacturing shop is highly interconnected. The Integrated Production Planning Information model describes the manufacturing environment at a high level of abstraction. This model shows the relationships among product design, shop resources, plans, shop configuration, and shop status. Detailed models were made for process and production plans, resource types, orders, tools, shop status and shop configuration. The following is a brief summary of the information models in MSI. More details of the process plan model are available in Catron (1991). Details for other models are available in Barkmeyer *et al.* (1993) and Ray (1992). The specification of product design is imported from the information models generated by the International Organization for Standardization Technical Committee 184, Subcommittee 4 (ISO TC184/SC4) (ISO).

**Plan models**
Process, production managed and production plans are key vehicles by which information is shared between planning and control systems in the MSI architecture.

A process plan designates the steps necessary to make a part, specifying the sequence(s) of operations by which a part is made and the relative timing of these operations. As received from the engineering systems, process plans may contain a number of cost effective alternatives which take into account the resources of the local production environment, but not the availability of such resources. Process plans are directed graph structures which may express both alternative and parallel paths of part production, and sets of potential resources for part production. Process plans provide for synchronization of operations by several mechanisms, and support hierarchical decomposition of operations as well. This definition differs from the traditional use of the term 'process plan' in that alternatives are expressed within a single plan and the plan may specify resources by their class instead of specific instances.

A production managed plan gives the plan for producing a specific batch of parts and is derived from the process plan for making that type of part. One or more alternatives from the process plan is selected and material handling steps are placed where needed.

A production plan is constructed from the production managed plan by selecting, scheduling and planning for the allocation of the specific resources, and refining the material handling planning necessary to move the batch of parts from one resource to another. Since production managed and production plans are constructed with reference to a process plan, their representations are logically, if not physically, linked.

Production plans are parsed by controllers. In parsing a production plan, a controller may request information from databases, make judgments on which alternative to take based on this information, produce commands to direct subordinates to perform manufacturing tasks, or perform manufacturing tasks.

**Resource model**
The resource model contains a physical and functional description of resources available in the shop. It contains templates for all such resources (e.g. machine tools, robots), information on shop floor configuration and status information on shop systems. The resource model includes consumable resources (such as coolant and solder) and logical resources which are pieces of information which have been created to assist the production management and control functions (such as batch numbers and order numbers). Items from the resource model are used in the process, production managed and production plans.

The models specified as necessary for shop integration are given in the table below.

*Data storage and access*

The MSI architecture specifies that information which must be shared among components be placed in a data storage location which is accessible by all components which need this information. The architecture does not specify the data storage mechanism. Options include files, variables, memory locations, databases, etc. MSI permits both physically distributed and centralized storage. The access method typically depends on the storage mechanism and may be different for different data, depending upon which architectural units need to share the data. Different access privileges to each item may be accorded to

**Table 2.1** MSI information models

| model | description |
|---|---|
| Product model | Specifies information needed to describe the parts being manufactured; includes information needed to create a solid model of the part, to describe manufacturing features and to specify detailed tolerancing information. |
| Process plan model | Describes plans which give the steps necessary to make a single part, specifying the sequence(s) of operations by which a part is made and the relative timing of these operations. |
| Production managed plan model | Describes plans for producing batches of parts along with routing information and is derived from the process plans for making those types of parts. |
| Production plan model | Describes fully developed plans for making batches of parts. The plans include specific resource selections, allocations and schedules for part production. |
| Resource model | Contains a physical and functional description of resources available in the shop. Resources may be physical or logical. |
| Order model | Describes information about orders; includes the type of the part to be manufactured, the quantity to be made and identifies information needed to record the engineering status and production status of the order. |
| Inventory model | Specifies information needed about stock (e.g. part blanks), consumable machining supplies, free carriers and completed parts no longer in-process. Such information includes type, quantity, location, etc. |
| Configuration model | Describes the relationships between controllers, schedulers and network entities. |
| Materials model | Describes the characteristics of raw materials, stock and consumable machining supplies. |

different systems. In some cases, multiple systems may be able to write the same data.

The architecture states that it is desirable that the physical location of the data be invisible to the components as far as practically possible within performance constraints. The architecture also permits components to make local copies of shared information, but states that in this case, the implementation is responsible for maintaining consistency between the local copy and the public copy of shared information.

At present, the architecture does not specify which systems should access each specific item of data. This omission was intentional, to give implementors of systems more freedom. Such a specification is a possible enhancement for the architecture and would aid vendors in constructing systems that could be made interoperable.

### 2.5.3   Control architecture

In the MSI architecture, the control architecture provides for integrated start-up, shutdown and maintenance of the controllers in the shop and provides a mechanism for performing operations on tasks such as starting, aborting, temporarily halting and resuming them. It is through the control architecture that unanticipated events and errors in planning and task execution are discovered and repaired. The basic tenets are discussed in the following sections.

*Hierarchical levels in MSI*

In the MSI architecture, control levels are arranged in a hierarchical tree structure. The hierarchical control tree structure has a single highest-level controller. Every other controller has exactly one supervisory controller from which it receives commands, and zero or more subordinate controllers to which it may issue commands. (The reason for this constraint on the hierarchy structure is related to the mechanisms for recovering from errors. Should multiple supervisors be allowed, there would be no guarantees that any single controller in the

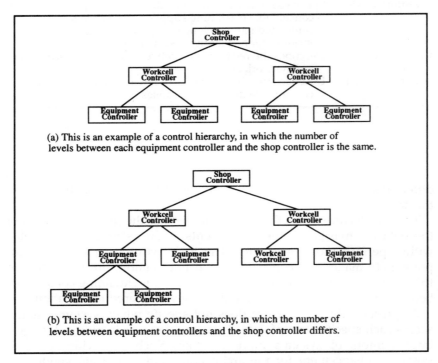

(a) This is an example of a control hierarchy, in which the number of levels between each equipment controller and the shop controller is the same.

(b) This is an example of a control hierarchy, in which the number of levels between equipment controllers and the shop controller differs.

**Fig. 2.6**   Examples of valid MSI control hierarchies.

hierarchy has a complete picture of the subordinate controller's status, making error recovery extremely difficult.) See Fig. 2.6 (Senehi, 1992) for an illustration of sample permitted control trees.

In the MSI architecture, the highest-level controller is the **shop controller**. The shop controller has general responsibility for all production processes involved in filling orders. Coordination of all orders for parts, determination of global scheduling constraints, and creation of routings for part delivery are done by the shop controller. However, many details required to fill these orders are the responsibility of subordinate controllers and are not visible to the shop controller.

Equipment is controlled by an **equipment controller**. By definition, an MSI equipment controller can execute only one task from its supervisor at a time (internal task decomposition is possible, but is not addressed by the MSI project). Equipment controller tasks are items such as loading a part, opening a vise, or manipulating the spindle of a machine tool, depending on the particular equipment.

Between the shop and equipment controllers there may be any number of controllers, called **workcell controllers**, which coordinate the activities of two or more subordinate controllers, each of which is either an equipment or a workcell controller. The number of controllers between a given equipment controller and the shop controller remains unchanged regardless of the tasks being executed. This number specifies the level of control for that controller, and may be different for different equipment controllers depending on the complexity of coordination necessary (Fig. 2.6).

In most cases it is expected that, once established, the control hierarchy will remain fixed as long as the shop is in operation. In some cases, it is possible to dynamically reconfigure the control hierarchy. It is intended that dynamic reconfiguration will only be used to remove dysfunctional equipment or to bring new equipment on line. In any case, at any fixed time the MSI architecture specifies that there be a *single* control hierarchy originating at the shop controller.

### *Task generation and execution process*

In the MSI architecture, execution of manufacturing tasks is a result of parsing production plans. For each controller involved in making a product, a production plan must be available to tell the controller what task to perform and when to perform it. Production plans are generated from corresponding process plans. Therefore, the hierarchical organization of process and production plans mirrors the hierarchy of control entities in the MSI architecture.

For each part design, process plans must be constructed for each level of the manufacturing hierarchy. Although process plans contain similar structure at all levels, distinct types of operations are performed at each level which are unique to that level. Listed below are descriptions of the operational characteristics of process plans at levels pertinent to the MSI architecture.

- Shop level.
  At the shop level, process plans primarily address movement of workpieces and sequencing of different types of machining operations, such as turning, milling, etching, etc.
- Workcell levels.
  Workcell level process plans prescribe the coordination of controllers subordinate to a given workcell, such as the use of a robot to load a machine tool table. Such plans can require extensive use of synchronization between process plans for subordinate controllers.
- Equipment level.
  Equipment level plans describe the most detailed level of operation that a process planner would generate. In this case, the activity called process planning in MSI terminology overlaps with what is usually termed off-line programming. The steps within such a plan provide instructions which are carried out by individual pieces of equipment. Examples in metal cutting might include steps such as **drill hole** or **chamfer edge**. The degree of detail required in such a step depends on the capability of the controller. If the controller possesses sophisticated capabilities, higher level instructions such as those above, or even as abstract as **load part** and **fixture part**, might be sufficient. If the controller is less capable, instructions to the level of numerical code may be required.

While specification of a task is gleaned from process and production plans, a controller will not perform the specified task unless it is instructed to do so by its supervisor. (The shop level controller is a special case. The architecture specifies that there be an entity which checks to see if a new order has arrived and which then selects and passes the appropriate production plan to the shop controller. In an implementation, the previously described entity may be a separate component, or may be within the shop controller.) This permits the supervising controller to remain in control of execution of a task by its subordinate. The ability of the supervising controller to manipulate execution of the task by its subordinate on a gross level (such as stopping or aborting the task) is necessary for the handling of scheduling and execution errors.

*Error recovery*

In an error-free environment, the relationship between planning and control is straightforward. When errors occur, this relationship is greatly complicated and the ability of the control system to recover from an error is intimately related to the capabilities of the planning systems and controller. The MSI architecture explores error recovery from specific types of errors in the shop in detail, extracts requirements for planning and control systems and devises interfaces which support error detection and recovery. These aspects of the architecture are detailed in the following sections.

**Error scenarios**

Errors can be grouped into three different classes based upon their cause: resource error, task error, and tooling error. A resource error occurs when a piece of equipment, whose controller is part of the control hierarchy, becomes impaired (e.g. a machine tool changer jams). A task error is an error which affects a specific task only: the resource on which it is being performed is unaffected (e.g. if a robot drops a workpiece, the robot is unaffected). While a resource error usually causes a task error, task errors may occur without a resource error. A tooling error occurs when a tool is damaged (e.g. a cutter breaks) or unavailable (e.g. a tool was not delivered at the proper time). Tools differ from other resources in that they are not permanently associated with any member of the control hierarchy, but are moved from resource to resource as needed.

The MSI architecture committee examined a number of error scenarios from the task and resource error categories. It was observed that the use of a hierarchical control system facilitates the localization of task error handling. When an error occurs in executing a task, if it is possible to resolve it by affecting only subtasks of the task, controllers at all levels of control above the supervisory controller are unaffected by the error. If localized error recovery is not achieved at this level of control, the recovery for the error is handled by the next higher control level in the hierarchy. At each level, there is potential for error resolution. Only in the event that the error cannot be resolved at any lower level is a global solution to the error required.

For the error scenarios considered, methods of recovery from scheduling errors and equipment failures were examined and incorporated into the specifications for the functionality of architectural units and the interfaces of the architectural units in the control architecture. This is discussed briefly in the following sections. More details on error handling in MSI are available in Wallace (1992) and Senehi (1992).

**Planners, controllers and control entities**

Analysis of error scenarios reveals two capabilities which a production planner must have in order to be effective. The first and more general requirement is that the production planner must be able to do re-planning. Re-planning is the ability to localize an error to a subset of the tasks and only re-plan those which are affected. If re-planning capability does not exist, automated error recovery will be extremely limited. The production planner must schedule for the entire shop again. The availability of resources can be fed back into the production planner, but the production planner cannot plan for the completion of partially executed production plans. Human intervention is required to avoid scrapping everything in execution when an error occurs.

The second requirement is the need for the production planner to work with the hierarchical control system. In order to localize an error at a given level of the hierarchy, the production planner must be able to plan for that level. Additionally, it must be possible for the production planner to be informed that a resource or task error has occurred at a given level, and that re-planning may be necessary. Error information which is needed to re-plan must be made available to the production planner (e.g. how many minutes late a machining task is expected to be). The architecture does not specify whether the production planner has an interface to each controller or whether error-related information transfer is accomplished through a database interface. As a minimum, however, the production planner must be able to be notified by the shop controller that a resource or task error has occurred and re-planning may be necessary. It is the controller's responsibility to notice the error and inform the production planner; the production planner does not monitor either the health of the controller, or the execution of tasks. Beyond these interface requirements, the internal architecture of the production planner is not specified.

When an error occurs, a controller may apply any strategies it has available to repair the problem. If these local efforts at correcting the problem fail, the controller must report the problem to its superior for correction. In order for a controller to participate in error recovery involving its superior, it must be able to:

- detect when a subordinate has failed;
- detect when a subordinate's task is late;
- abort task execution;
- halt task execution and retain information to restart later;
- restart task execution from previous point;
- halt task execution and discard all information related to the task;
- halt task execution and regard the task as complete;
- estimate task completion time;
- alter task execution based on new parameters (e.g. new start, completion times).

The inability of either the production planner or the controller to perform any of the indicated functions does not prevent a production planner or controller from being integrated into a control system for a shop using the architecture, but it does weaken the recovery ability of the system.

*Control entities*

Since effective participation in the error recovery mechanism requires both a production planner (that is, the production planner architectural unit, referred to as planner in the remainder of the chapter), and a controller, the MSI architecture defines an architectural unit called a control entity, which consists of a planner and its associated controller. (The architecture allows for hierarchies of planners only, without associated controllers. These hierarchies would only be needed for 'what if' scenarios and would not need error recovery capabilities.) It should be emphasized that the control entity is a logical, rather than a physical architectural unit. The planner in the control entity is required to support scheduling of plans, and may support process planning and batching. The controller specified must support task execution and may have any level of intelligence desired.

Since process planning systems, production systems, and controllers are not likely to be capable of fully supporting error recovery in the near future, a mechanism for external intelligent intervention is included in the MSI architecture. Throughout this chapter, the intelligent agent will be referred to as the **guardian**.

The MSI architecture requires interfaces for any control entity in the architecture and contains detailed specifications for each interface. In the following sections, the communications mechanisms of the control entities, the interfaces of the control entity, and the physical distribution of the control entity will be discussed in turn.

**Communication of control entities**

All communication between control entities which is direct (section 2.5.1, *Interactions of architectural units*), is required by the MSI architecture to be via a command and status interface. Such interfaces require communication channels between architectural entities. The MSI architecture requires that the communication channels for command and status messages use a point to point, guaranteed message communication paradigm. One communication mechanism that provides such a communications service is the manufacturing automation protocol (MAP) (MAP (a) and (b)), with the manufacturing messaging specification (MMS) application layer (ISO).

Since message delivery is guaranteed, messages can rely on information conveyed in previous messages. This means that messages need not contain all the information required for a complete picture of the situation, reducing the amount of data which must be transferred with each message. As a consequence of point to point communications, the communication pairs must be set up when the connections are established, and it is not possible to hide the way in which communicating control entities are physically distributed.

**Control entity interfaces**

A control entity may have as many as five types of direct interfaces, in addition to the indirect interfaces. These direct interfaces are:

1. A planning interface – which governs interactions of supervisors and subordinate planners concerning the selection, generation and scheduling of process, production managed and production plans.
2. A controller interface – which governs interactions of supervisors and subordinate controllers concerning task execution.
3. A guardian to planning interface – which governs how an intelligent agent may interact with the planner.
4. A guardian to controller interface – which governs how an intelligent agent may interact with the controller.
5. A planner to controller interface – which governs how the planner and the controller may interact in both ordinary and error situations.

A detailed specification of each of these interfaces is found in Wallace (1992). A conceptual view of the potential direct interfaces is shown in Fig. 2.7 (Wallace, 1992).

In an implementation of the architecture, the determination of which interfaces must actually be supported is based on the physical and logical distribution of the control entity. The general rule is that, if the two interacting components are physically or logically distributed,

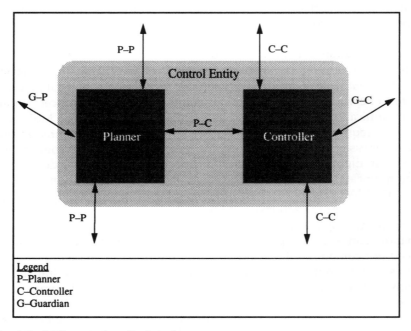

**Fig. 2.7**   MSI control entity interfaces.

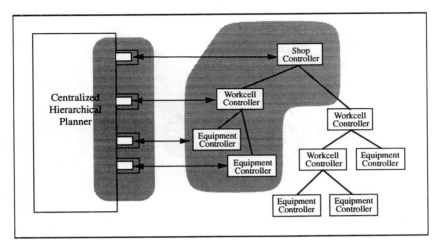

**Fig. 2.8** An MSI system with a centralized planner.

the exposed interface must conform to the corresponding interface specification.

**Physical and logical distribution of control entities**
Permitting flexibility in the physical and logical distribution of the control entity allows the MSI architecture to accommodate a number of common configurations for planners and controllers. Examples are:

- A centralized planning system may be used, provided that each controller has a logically distinct interface to the planning system and that the planning system can plan for each member of the hierarchy. In this case, the internal functioning of the planner is not made public, but the interfaces between controllers and between a controller and its planner are exposed. This configuration is shown in Fig. 2.8 (Wallace, 1992). (Note that, in Figs 2.8, 2.9 and 2.10, the interfaces between the planner and the controller are shown only for the highlighted figures.)
- A distributed planning hierarchy which mirrors the control hierarchy may be used. In this case, the interfaces between planners, between controllers, and between each controller and its planner are public and must conform to the MSI interface specification. Figure 2.9 (Wallace, 1992) shows this configuration.
- The planner and controller functions may be embedded in a control entity, resulting in a single hierarchy of control entities. In this case, the interfaces between the planner portion of a control entity and the planner portion of both its supervisory and subordinate control entities are public, and the corresponding controller interfaces are

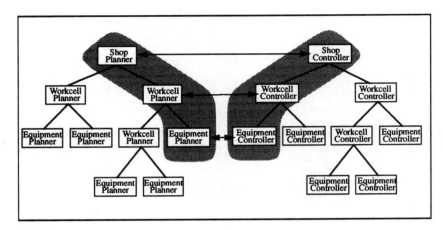

**Fig. 2.9**  An MSI system with a planning hierarchy.

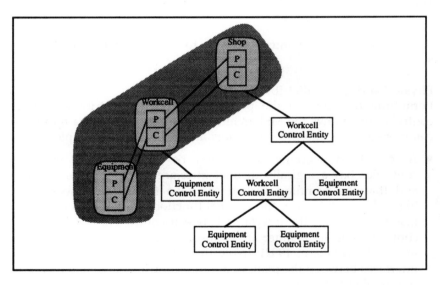

**Fig. 2.10**  An MSI system with embedded planners.

public, but the interface between the planner and the controller of any one control entity remains private. Figure 2.10 (Wallace, 1992) shows this configuration.

In Figs 2.8, 2.9 and 2.10, the control entities are homogeneous (i.e. they all split or combine their component planners and controllers the same way). The architecture also allows the use of heterogeneous control entities.

### 2.5.4   Summary of MSI

The MSI architecture provides an architecture for a shop which manufactures discrete parts, that supports information integration for the major shop systems and provides specifications for an integrated production planning and control environment. The MSI architecture can be used with a centralized or a distributed planner, and other combinations of control and planning systems.

The operations of a shop are guided by schedules generated for its current orders. This mode of operation permits global optimization of the shop. The architecture provides for schedule maintenance via detailed sets of command and status messages for controller and planner interactions. The use of hierarchical control aids in localizing and recovering from scheduling errors. It is anticipated that the MSI architecture will be useful both for the integration of current shops and in future research. The information models are immediately applicable to aid in shop integration, while the interface specifications provide direction for implementations of MSI and further research in automated replanning and control.

## 2.6   COMBINING ARCHITECTURES AT NIST

While MSI and RCS are both hierarchical control architectures, they are not interoperable. That is, a controller developed to work in one architecture cannot be plugged into a hierarchy of controllers from the other architecture. Much of the difference between the two is the result of requirements placed upon the architectures by their intended uses.

Recent comparative work at NIST has shown several characteristics to affect control system architecture greatly. The first important characteristic is the degree to which the environment is known in advance. The more the environment is known in advance, the less sensory processing and adaptive capabilities the system need have. Another important feature is the degree to which the environment is structured. A highly structured environment permits the control system to make certain assumptions, but may force the system to be able to handle systems of constraints. Finally, the degree of variability greatly affects the control system architecture. An environment with rapidly varying features requires more rapid response. The MSI architecture is formulated to function in an environment very different in many of these factors from the RCS architecture.

We are undertaking to build a joint architecture that combines the strong features of MSI and RCS. A feasibility report (Kramer, 1993) has been prepared which includes both a general and a detailed comparison of the two architectures. MSI and RCS strengths and weaknesses complement each other. We have concluded that a joint architecture is feasible. The report gives a broad description of its outlines. The second phase of the project is in progress and is charged with filling in more details.


## 2.7 CONCLUSION

In this chapter we briefly introduced control architectures. The bulk of the chapter was then given to describing the real-time control system (RCS) architecture and the manufacturing systems integration (MSI) architecture, two hierarchical control architectures developed at NIST. After a study was performed it was determined that it was feasible to develop an architecture combining the best features of those architectures. A project to construct such an architecture is in progress.

The full development of a broadly applicable, readily usable, reference architecture for machine systems control remains an unattained goal. Further research and development is required.


## REFERENCES

Albus, J., Barbera, A. and Nagel, R. (Sept. 1981) *Theory and practice of hierarchical control*. Proceedings of the 23rd IEEE Computer Society International Conference.

Albus, James S. and Blidberg, D. Richard (June 1987) *A control system architecture for multiple autonomous vehicles*. Proceedings of the Fifth International Symposium on Unmanned, Untethered Submersible Technology. Merrimack, NH, USA.

Albus, James S., McCain, Harry G. and Lumia, Ronald (1989) NASA/NBS Standard Reference Model for Telerobot Control System Architecture (NASREM). *NIST Technical Note 1235*, National Institute of Standards and Technology.

Albus, James S. (1991) A Theory of Intelligent Systems. *Control and Dynamic Systems*, **45**, pp. 197–248.

Albus, James S. (May 1992) RCS: A Reference Model Architecture for Intelligent Control. *IEEE Journal on Computer Architectures for Intelligent Machines*, pp. 56–9.

ANSI/EIA/TIA/232-E (July 1991) *The Interface between Data Terminal Equipment and Data Circuit Terminating Equipment Employing Serial Data Binary Interchange*. (Available from Global Engineering Documents, 15 Inverness Way, E. Englewood, Colorado 80112-5704.)

Barbera, Anthony J. (Dec. 1977) *An Architecture for a Robot Hierarchical Control System*. NBS Special Publication 500-23; National Bureau of Standards.

Barkmeyer, Edward J., Ray, Steven, Senehi, M. Kate, *et al.* (1993) *Manufacturing Systems Integration Information Models for Production Management*, NISTIR (to be published); National Technical Information Service, Springfield, VA 22161.

Catron, Bryan and Ray, Steven R. (1991) ALPS – A Language for Process Specification. *International Journal of Computer Integrated Manufacturing*, **4**(2), pp. 105–13.

Herman, Martin and Albus, James S. (Dec. 1987) *Real-time Hierarchical Planning for Multiple Mobile Robots*. Proceedings of DARPA Knowledge-Based Planning Workshop. Austin, Texas, pp. 22-1 to 22-10.

Hopp, T. (June 1988) Proceedings of the Manufacturing Data Preparation (MDP) Workshop, unpublished report. Factory Automation Systems Division, National Institute of Standards and Technology, Building 220, Room A127, Gaithersburg, MD 20899.

ISO 9506, *Industrial Automation Systems Manufacturing Message Specification, Part 1: Service Definition*. International Organization for Standardization, Geneva, Switzerland.

ISO 10303, *Product Data Representation and Exchange, Part 1: Overview and Fundamental Principles*. ISO TC184/SC4/Editing: Document N11 (working draft) IGES/PDES/STEP Administration Office, National Institute of Standards and Technology, Building 220, Room A127, Gaithersburg, MD 20899.

ISO (1993) *Framework for Enterprise Modelling*, ISO TC184/SC5/WG1: Document N-282 Version 3.0 (working draft). Available from National Electrical Manufacturers Association, 2101 L Street, NW Washington, DC 20037.

Jones, Albert T. and McLean, Charles R. (Aug. 1985) *A Production Control Module for the AMRF*. Proceedings of the 1985 ASME Computers in Engineering Conference.

Jones, Albert T. and McLean, Charles R. (1986) A Proposed Hierarchical Control Model for Automated Manufacturing Systems. *Journal of Manufacturing Systems*, **5**(1), pp. 15–25.

Kramer, Thomas R. and Senehi, M.K. (1993) Feasibility Study: Reference Architecture For Machine Control Systems Integration. NISTIR 5297. *National Institute of Standards and Technology Interagency Report*. Available from the National Technical Information Service, Springfield, VA 22161.

Libes, Don (1990) *NIST Network Common Memory User Manual*; NISTIR 90-4233; National Institute of Standards and Technology.

*Manufacturing Automated Protocol Version 3.0*, August 1, 1988. North American MAP/TOP Users Group, ITRC, P.O. Box 1157, Ann Arbor, MI 48106.

*Technical and Office Protocols Version 3.0*, August 31, 1988. North American MAP/TOP Users Group, ITRC, P.O. Box 1157, Ann Arbor, MI 48106.

McLean, C.R. (1987) *Interface Concepts for Plug-Compatible Production Management Systems*. Proceedings of the IFIP WG5.7 Working Conference on Information Flow in Automated Manufacturing Systems; Gaithersburg, MD; August 1987. Reprinted in *Computers in Industry*, **9**, pp. 307–18.

Quintero, Richard and Barbera, Anthony J. (Oct. 1992) A Real-Time Control System Methodology for Developing Intelligent Control Systems. *NISTIR 4936*. National Institute of Standards and Technology.

Ray, Steven R. and Wallace, Sarah (Sept. 1992) A Production Management Information Model for Discrete Manufacturing, submitted for publication to *Production Planning and Control*.

Rybczynski, S. *et al.* (June 1988) AMRF Network Communications. *NISTIR 88-3816*. National Institute of Standards and Technology.

Senehi, M.K., Wallace, Sarah and Barkmeyer, Edward J. *et al.* (June 1991) Control Entity Interface Document. *NISTIR 4626*. National Institute of Standards and Technology.

Senehi, M.K., Barkmeyer, Edward J. and Luce, Mark E. *et al.* (Sept. 1991) Manufacturing Systems Integration Initial Architecture Document, *NISTIR 4682*; National Institute of Standards and Technology.

Senehi, M.K., Wallace, Sarah and Luce, Mark E. (Apr. 1992) *An Architecture for Manufacturing Systems Integration*. Proceedings of ASME Manufacturing International Conference, Dallas, Texas.

Simpson, J., Hocken R. and Albus, J. (1982) The Automated Manufacturing Research Facility, *Journal of Manufacturing Systems*, **1**(1).

Tanenbaum, Andrew S. (1988) *Computer Networks*, 2nd edn. Prentice Hall, Englewood Cliffs, New Jersey, USA.

Wallace, Sarah, Senehi, M.K. and Barkmeyer, Edward J. *et al.* (Oct. 1992) Control Entity Interface Document. *NISTIR 5272*. National Institute of Standards and Technology.

# CHAPTER 3

# Characteristics of computerized scheduling and control of manufacturing systems

*V. Jorge Leon and S. David Wu*

## 3.1 INTRODUCTION

This chapter discusses the unique characteristics of computer controlled scheduling (CCS). The uniqueness arises from the assumption that only limited human intervention is allowed during the operation of the system. In particular, emphasis is given to the complexity resulting from the explicit consideration of general resource models, the use of flexible and programmable technologies, and the short-term nature of the decisions.

Given a set of production requirements and a physical system configuration, scheduling deals with the timing and coordination of activities which are competing for common resources. Scheduling is important for the efficient operation of manufacturing systems. It typically corresponds to a difficult combinatorial optimization problem where the objective function comprises multiple and conflicting performance criteria. Due to its complexity, manufacturing scheduling and control is commonly decomposed into a hierarchy of decision levels. An example is the hierarchical control system proposed by the National Institute of Standards and Technology (NIST) (Jones and McLean, 1986). The discussion in this chapter focuses on the scheduling and control decisions at the operational level of the hierarchy. At this level, it is assumed that a supervisory computer is responsible for the operation of a number of programmable machines, automated material handling systems, automated storage systems, and auxiliary workstations. The equipment under the control of the supervisory computer may include CNC machining centers, conveyors, automatic storage and retrieval systems, and degreasing stations. In the time domain, the control decisions deal with the immediate future (minute-to-minute) and are directly used in the coordination and execution of part flow and processing.

The scheduling, coordination and control of the above systems is clearly a complex task. The lack of a generic scheduling model applicable to this automated environment is in fact an inhibitor to its widespread implementation. Fortunately, the advances in computing and manufacturing technology now make it possible to deal with this problem sufficiently.

This chapter is organized as follows. Section 3.2 provides a brief background on traditional job shop scheduling. Section 3.3 describes the characteristics that make the computer scheduling problem unique. Section 3.4 discusses the scheduling approaches used in practice and research aimed at the efficient utilization of automated manufacturing systems.

## 3.2  THE CLASSICAL JOB SHOP SCHEDULING PROBLEM (JSP)

Simply stated, a job shop scheduling problem involves the determination of start time for each operation in a finite and given set, $N$. Associated with each operation $i \in N$ there is a processing time $p_i$. The operations in $N$ are partitioned into $n$ mutually exclusive and exhaustive subsets $J_k$, where $J_k$ is called job $k$. Also, the operations in $N$ are partitioned into $m$ mutually exclusive and exhaustive subsets $M_r$, where $M_r$ is the set of operations to be processed on machine $r$. Also given are precedence relations, or technological constraints, between the operations in a job. A pair $(i, j)$, $i, j \in J_k$, indicate that operation $i$ precedes operation $j$ in job $k$. Let $A_k = \{(i, j) \mid i, j \in J_k$ and $i$ must precede $j\}$ denote the set of pairs that represent the technological constraints associated with job $k$. Associated with each operation $i \in N$ there is a processing time $p_i$. The traditional job shop scheduling problem is to determine the processing order of the operations in $M_r$ for $r = 1$, $2, \ldots, m$, such that some objective function is optimized.

This problem consists of finding the vector of operation start times, $t = (t_1, t_2, \ldots, t_{|N|})$, that minimizes a given objective function $Z(t)$. The problem can be formulated mathematically as follows,

(JSP): Minimize $Z(t)$

  s.t.

$$t_i - t_j \geq p_j, \qquad\qquad (i, j) \in A_k, \, k = 1, 2, \ldots, n \quad (3.1)$$
$$t_i - t_j \geq p_j \vee t_j - t_i \geq p_i, \qquad i, j, \in M_r, \, r = 1, 2, \ldots, m \quad (3.2)$$
$$t_i \geq 0, \qquad\qquad i \in N.$$

Equation (3.1) ensures that the technological constraints are satisfied. The disjunctive relations in (3.2) ensure that the capacity constraints on the machines are not violated; i.e. a machine can process only one operation at a time. JSP has been extensively studied and it is well known that it belongs to a class of the most difficult combinatorial optimization problems.

Typical assumptions made in job shop scheduling include the following:

**A.1** All jobs require only one machine at a time; i.e. $M_r \cap M_s = \emptyset$, $r \neq s$.

**A.2** One machine processes one job at a time; i.e. $|J_k \cap M_r| = 1$, $k = 1, \ldots, n$ and $r = 1, \ldots, m$. Also, this implies that a job visits each machine exactly once; i.e. two operations in the same job cannot use the same machine.

**A.3** The order in which a job visits different machines is predetermined by technological constraints; i.e. the set $A = A_1 \cup A_2 \cup \ldots \cup A_n$, is given and fixed. $A_k$ can be viewed as the process plan or machine routing for job $k$.

**A.4** No explicit consideration is given to auxiliary resources such as material handling, tooling and buffer space.

These assumptions are appropriate for manufacturing environments in which human intervention is significant and the equipment used is manual or hard automation. It is also appropriate in environments characterized by batch production, in which every part type has a predetermined and fixed process plan. A given machine or workstation is pre-assigned to perform each step in the process plan.

It must be noted that in many existing models some of the above assumptions are partially relaxed. For instance, the literature on parallel machine scheduling studies workcenters consisting of $m$ processors that may be able to process $m$ jobs simultaneously. For detailed treatment of classical machine sequence the interested reader is referred to Baker (1974), Rinnooy Kan (1976) and French (1982).

## 3.3 CHARACTERISTICS OF COMPUTER CONTROLLED SCHEDULING (CCS)

CCS models must explicitly consider all the equipment under the supervisory computer control. This includes machines, material handlers, auxiliary stations, and finite capacity storage space. The uniqueness of the CCS problem originates mainly from the following:

- general resource models;
- part contact states;
- deadlocking;
- dynamic machine routing.

The following sections will discuss each of the above characteristics.

### 3.3.1   General resource models

Machines, material handling equipment, tooling and buffer space must be included in CCS under a unified model. The explicit consideration of these resources results in additional decisions to be made in the schedule and consequently increased problem complexity. For instance, common in computer controlled systems is the simultaneous requirement of two or more resources for the execution of a single activity. An example is the task of loading a part onto a machine using a robot. At some point of time in the loading operation, the robot must hold the part in position for the machine to secure the part in its machining position (i.e. in a chuck or vise). Clearly, the robot and the machine are used simultaneously in this operation. In the case of tooling, the problem is further complicated by the fact that tools may be shared between different machines. Hence, in making sequencing decisions one must schedule both part movement and tool movement; i.e. tools have 'part' attributes since they will also consume material handling resources.

In reference to JSP model, assumptions A.1, A.2 and A.4, must be relaxed to capture the nature of CCS. Specifically, A.1 is relaxed because if some operation (i.e., loading) requires more than one resource at the time, then $|M_r \cap M_s| \geq 1$. A.2 is relaxed because when loading and unloading are disaggregated from the processing time, the result is that more than one operation in a job requires the same machine, or $|j_k \cap M_r| \geq 1$. Clearly, A.4 is relaxed since other resources such as material handlers must be considered.

Once the assumptions mentioned above are relaxed, the machine scheduling problem becomes a more general Resource-Constrained Scheduling Problem (RCSP). Readers interested in RCSP are referred to Davis (1973), Boctor (1990) and Slowinski and Weglarz (1989). It must be noted that, although models in RCSP allow for quite general resource/task relationships, their application to the computer controlled manufacturing environment is not trivial. As discussed in the following sections, part contact states and deadlocking must be explicitly considered.

### 3.3.2   Part contact states

The loading, unloading and movement of parts through the manufacturing system must be scheduled by the supervisory computer. In contrast with classical scheduling, one can no longer assume that the parts will move to the next machine immediately after it has completed processing. Rather, the part will wait on the machine until the material handler (i.e. a robot) unloads it. Clearly, no part can be scheduled on the machine under consideration until the previous part has been unloaded.

In order to model this situation, it is necessary to explicitly consider loading and unloading activities, in addition to the processing activity. Let $l_i$ and $u_i$ denote the load and unload operations associated with the processing operation $i$. The technological relationships between these operations can be represented by the pairs $(l_i, i)$ and $(i, u_i)$. Hence, the set of operations contains $2|N|$ additional operations, and each arc set $(i, j) \in A_k$ is transformed into $(l_i, i)$, $(i, u_i)$, $(u_i, l_j)$, $(l_j, j)$ and $(j, u_j)$. The material handling resource $v$ can be modelled as the set, $H_v$, consisting of operations that require resource v. Consequently, in CCS $|M_r \cap H_v| > 0$, indicating that there exist operations in $N$ that require both machine $r$ and the handling resource $v$ for execution; i.e. loading and unloading operations.

Given that operation $i$ has been loaded on machine $r$, part contact constraints must ensure that no other operation is scheduled on this machine until the unloading operation, $u_i$, has been completed. Mathematically, the part contact constraints can be represented by the following set of constraints:

$$t_i - t_{li} \geq p_{li} \wedge t_{ui} - t_i \geq p_i \qquad \forall\ i \in M_k,\ k = 1, \ldots, m \qquad (3.3)$$
$$t_{lj} - t_{ui} \geq p_{ui} \vee t_{li} - t_{uj} \geq p_{uj} \qquad \forall\ i, j \in M_k,\ i \neq j,\ k = 1, \ldots, m \quad (3.4)$$
$$t_w - t_s \geq p_s \vee t_s - t_w \geq p_w \qquad \forall\ w, s \in H_v,\ v = 1, \ldots, h \qquad (3.5)$$

where, the constraints in (3.3) represent the technological relations and that no operation can be scheduled between $l_i$, $i$ and $u_i$, the constraints in (3.4) represent the capacity constraint for the machine and the constraints in (3.5) represent the capacity constraints for the material handling resources. It is assumed that each resource processes one operation at a time.

Notice that part contact state consideration has resulted from the relaxation of the JSP assumptions A.2 and A.4. Also, the problem has increased in size, the number of operations to schedule is up from $|N|$ to $3|N|$ and the number of resources is up from $m$ to $m+h$. Further, additional complexity results from the fact that two resources may be required per operation.

### 3.3.3 Deadlock states

Deadlock is a system state in which any part flow is inhibited due to inappropriate scheduling decisions made by the computer controller. Once a system is in a deadlock state, there is no possible movement of parts and external intervention is required to reestablish the product flow. Deadlock arises from the explicit recognition of material handling and buffer space resources. In other words, a part in a system must be 'in contact' with some resource considered by the model, rather than assuming that it is in some implicit queue.

A characteristic of a deadlock state is that more than two resources may be involved in a deadlock. Consider the case of $M$ machines and two parts. The order in which one of the parts visits the $M$ machines is exactly the opposite order in which the second part visits the same machines. In this case deadlock will occur if both parts are allowed, at the same time, in the system. It can be shown that the use of buffers can attenuate but not eliminate deadlock situations (Wysk *et al.* 1991).

If only deadlock between two resources is considered, it is easy to describe deadlock avoidance constraints that extend the basic JSP formulation. The constraints must prevent the scheduling of two operations (i.e. $i$ and $j$) of different jobs on their required machines if their corresponding next operations (i.e. $i+1$ and $j+1$) require the resource occupied by the other job. Specifically, consider two jobs, $J_k$ and $J_l$, and two consecutive operations in each job, $i$ and $i+1 \in J_k$, and $j$ and $j+1 \in J_l$. Further, operations $i$ and $j+1$ require machine $r$, and operations $i+1$ and $j$ require machine $s$. In addition to the part contact constraints (3), (4) and (5), the following disjunctive constraints will inhibit deadlock situations between any two machines:

$$t_{li} - t_{uj+1} \geqslant p_{u,j+1} \vee t_{lj} - \vee t_{lj} - t_{u,i+1} \geqslant p_{u,i+1}$$

for,

$$i, j+1 \in M_r; \ i+1, j \in M_s; \ M_r, M_s \in M; \ (i, i+1) \in A_k; \ (j, j+1) \in A_l \quad (3.6)$$

where, $M = M_1 \cup M_2 \cup \ldots \cup M_m$. It must be noted that the equations in (3.6) only consider pairwise relations between machines and as a result they do not model the case for three or more machines. The derivation of deadlock avoidance constrains for the latter case is still a matter of further research and will not be considered here.

### 3.3.4  Dynamic machine routing

Machine routings specify the machines that are required for each operation of a given job. In the JSP model these precedence relations, $A$, were referred to as technological constraints and have been assumed predetermined and fixed. However, the flexible and programmable nature of the typical equipment used in computer controlled systems make machine routing a dynamic decision process. The latter characteristic make assumption A.3 inappropriate for CCS.

Consider a feature on a part that may be machined on either of two different machines. In making a routing decision, one may prefer to route the part based on the state of the machines at that time. For instance, given two machines, $r$ and $s$, assume a part can be processed faster on $r$ than on $s$; however at a given time, it may be desirable to process it on $s$ because $r$ is busy processing other jobs.

Alternative machine routings can be represented using OR-graphs. In the OR-graph $G(N, A)$, each node in $N$ is associated with a processing operation, and the arcs in $A$ are associated with the technological precedence relations between them. An important characteristic of OR-graphs is that only one outgoing arc must be contained in any solution. Hence, at each node in the graph, the associated decision problem is determining which alternative (outgoing arc) to select. Note that, in the traditional JSP, $G(N, A)$ will consist of a simple path of operations and no decisions concerning alternative plans need consideration.

Previous work in this area suggests that benefits result from using alternate plans given machine breakdowns or other system state information (Wilhelm and Shin, 1985). By incorporating alternative process plans during the schedule generation, one should expect better results than attempting to solve process planning and scheduling independently. Clearly, the computational burden will increase. This is because of the additional disjunctions associated with the technological constraints, in addition to the disjunctive constraints associated with machine capacity, part contact and deadlock avoidance.

It is worthwhile noting that it is quite possible to have alternative process plans (i.e. processing steps) to be executed on the same machine. This additional complexity may also be modeled by extending the OR-graph representation mentioned above.

## 3.4 PROBLEM STATEMENT AND SOLUTION APPROACHES

In the previous sections we have discussed characteristics that make computer controlled scheduling unique when compared with traditional scheduling. In summary, the computer controlled scheduling problem CCSP can be stated as (CCSP):

Minimize $Z$

s.t.

  (i) technological constraints;
 (ii) general resource constraints;
(iii) deadlock avoidance constraints;

where, (i) may consider alternative process plans, (ii) assume that each resource can process only one job at a time and explicitly consider machines and auxiliary equipment, and (iii) must inhibit deadlocking between two or more resources. The objective function is typically multi-criteria and can include measures related to completion time of jobs, due-dates, and schedule robustness. It should be clear that CCSP is at least as complex as the traditional JSP. This is true even when the formulation above does not explicitly consider uncertainties, buffer space restrictions, and tooling.

Not surprisingly, current implementations decompose the problem into more tractable subproblems and use simplistic dispatching rules as decision policies. CCSP is commonly decomposed into three sub-problems (Askin and Standridge, 1993). First, decisions are made concerning when to release parts into the system. Second, the parts are routed within the system according to some internal priorities. Third, control policies are devised to recover the system from unforeseen events.

### 3.4.1  Part release

Given the limited degrees of freedom for part movement and staging, it is of critical importance to carefully decide when to introduce a part into the system. If too many parts are allowed into the system at the same time, the system may rapidly get congested resulting in longer lead times and increased likelihood of deadlock situations. On the other hand, if too few parts are allowed, low performance may result due to the under-utilization of equipment. From the scheduling stand-point this poses additional complexities since part release decisions require the use of non-regular measures of performance. In other words, inserting idle time (or delaying job completion times) may be beneficial for system performance.

A common approach is to use simple dispatching rules. For instance, one may prioritize jobs based on their due-dates, processing times, or some other criterion. The advantages of dispatching are their simplicity and minimal computational requirements. Their disadvantages are that the performance of each rule under different conditions must be tested experimentally and is difficult to generalize. However extensive experimental evidence exists in both the FMS literature (see for example, Stecke and Solberg, 1981) and classical job shop scheduling (Panwalker, 1977) which may be readily used as guidelines in some implementations.

Analytical approaches based on the hierarchical decomposition of the problem have been proposed for the establishment of optimal loading policies. In these approaches, uncertainties originate mostly from processing time variations and machine breakdowns. Decisions to load the system take into consideration in-process inventory or machine availability. For an example of this type of approach and a recent literature review, the interested reader is referred to Maimon and Gershwin (1988).

### 3.4.2  Internal part flow

Once the part enters the system, detailed scheduling decisions must be made for the use of machines and material handling. The synchronization of part movements and processing is critical for high performance and continuous part flow through the system.

Again, the most common approach used is dispatching. At this level, dispatching is used both for machines and material handling. Deadlock conditions are avoided by fixing the logic at the programmable logic controller (PLC) level and allowing for human intervention when required. This results in a loss of flexibility, limited system performance, and a lesser degree of automation. The explicit consideration of deadlocking conditions during the scheduling decision-making may result in more efficient and automated manufacturing.

The alternative process routings can be exploited by careful scheduling for higher system performance. Wilhelm and Shin (1985) suggested that part flow times (i.e. the time the part is in the system) can be significantly reduced by choosing an alternative machine equipped with the required tools. Sherali *et al.* (1990) proposed a problem formulation for job selection, routing and scheduling. They developed two heuristics to solve this problem.

In order to achieve full computer control of the system, deadlocks must be explicitly considered. Wysk *et al.* (1991) formulate the deadlocking problem and propose a graphic representation. In the graph, each node is associated with a resource and directed arcs are associated with possible movements of parts. Sufficient and necessary conditions are given for deadlocking. These conditions are then used for system deadlock detection.

### 3.4.3 Short-term scheduling and control

The main objective at this level is to ensure the smooth and efficient operation of the system in the event of unforeseen disturbances. These disturbances are generated by inherent variability associated with the events under consideration and disruptive events such as equipment failure or tool breakage. Eventually, any planned sequence of activities will require updating during its execution. The solution approaches can be classified into three main groups: dispatching, simulation and rescheduling.

Dispatching in its simplest form consists of using rules-of-thumb to recover the system from disruptions. A more sophisticated form of dispatching makes explicit use of stochastic information, such as processing time or machine breakdown probability distributions. Examples of this kind of dispatching can be found in the stochastic scheduling literature (c.f. Pinedo, 1983, and Glazebrook, 1987). Another line of research combines mathematical programming and dispatching. Roundy *et al.* (1989) solved a higher level planning problem using mathematical programming and used the cost information from the solution to establish dispatching priorities at a given time for each machine.

Simulation is often proposed as a real-time evaluation technique of the proposed control actions. Shanthikumar and Sargent (1983) suggested using hybrid simulation and analytical models. Wu and Wysk (1989) used an expert system to select a list of potential dispatching rules which will in turn be evaluated using simulation. A problem in using simulation is the computer time requirements to get significant differences between alternatives. In 'perturbation analysis' the system is modeled as a discrete event dynamic system (DEDS) and estimates of the desired statistics are obtained using a small number of replications (Ho and Cao, 1983).

Another line of research proposes 'rescheduling' as a means to recovering the system from a disruption. It implies the existence of a precomputed schedule and set of control actions. For instance, a control action may cause re-routing of a part initially destined for processing on a machine that has just failed. Rescheduling decisions must be aimed at maintaining high system performance. A desirable characteristic of rescheduling is that of minimizing the deviations from the original plan. The latter can be critical in computer controlled systems, since a change in sequence may require the reordering of parts already in queue, or unnecessary machine setup changes. Total or partial rescheduling is performed to recover the system from disruptive events. One approach is total rescheduling (Yamamoto and Nof, 1985). The disadvantage of the latter is that no consideration is given to the impact that rescheduling may have on the system. Match-up scheduling attempts to mitigate the effect of rescheduling by determining a time period within which the system will 'match-up' with the original plan (Bean *et al.*, 1991). Leon *et al.* (b) modeled the control problem as a DEDS model that minimizes a linear combination of part completion times and deviations from the original schedule. This representation allowed the incorporation of uncertain information into this dynamic decision problem.

An important aspect of rescheduling is that of deciding the time and frequency of rescheduling. Leon *et al.* (b) suggest making rescheduling decisions when the system's reliability reaches a given threshold level or when a machine breakdown occurs. Church and Uzsoy (1991) studied the worst-case behavior of periodic rescheduling in dynamic parallel machine systems. Their results suggest advantages associated with the use of periodic, rather than continuous, rescheduling.

Leon *et al.* (a) proposed measures of robustness and a methodology for robust scheduling. Given that the scheduling decisions under consideration are directly implemented for execution, random disturbances will eventually disrupt the operation of the system. These disturbances are originated from a wide range of sources including demand variability, equipment reliability and the uncertainty associated with model parameters. A desirable characteristic of a schedule in a computer-controlled environment is that it is insensitive, or robust, to minor random disturbances. By robust it is meant that the system performance will be maintained with minor adjustments to the schedule, if any. Examples of minor disturbances include variations in processing times, short machine breakdowns, short part jamming, and others. The use of robust schedules should minimize the efforts required by rescheduling and execution control. Notice that, in systems where human intervention is allowed, most 'minor' disruptions can be easily fixed by the operator and no special consideration may be required during scheduling.

Readers interested in a more detailed review of scheduling and control approaches to scheduling and control of automated systems are referred to Buzacott and Yao (1986), Maimon and Gershwin (1988) and Wu and Wysk (1991). Buzacott and Yao provide a comprehensive review of analytical models for flexible manufacturing systems. Maimon and Gershwin review recent efforts in hierarchical control of manufacturing systems using control theory principles. Wu and Wysk describe real-time scheduling and control issues and research in automated systems.

## 3.5 CONCLUSION

In this chapter we have summarized the characteristics of computerized scheduling in an automated manufacturing environment. Various extensions to the classical job shop scheduling model were discussed which allow the consideration of material handling, deadlock avoidance and alternative machine routings. Solution methods proposed for different aspects of the problem were briefly discussed.

## REFERENCES

Askin, R.G. and Standridge, C.R. (1993) *Modelling and Analysis of Manufacturing Systems*, John Wiley and Sons Inc.

Baker, K. (1974) *Introduction to Sequencing and Scheduling*, Wiley.

Bean, J.C., Birge, J.R., Mittenthal, J. and Noon, C. (1991) Match-up Scheduling with Multiple Resources. *Operations Research*, **39**, pp. 470–83.

Boctor, F.F. (1990) Some Efficient Multi-heuristic Procedures for Resource-Constrained Project Scheduling. *European Journal of Operational Research*, **49**, pp. 3–13.

Buzacott, J.A. and Yao, D.D. (1986) Flexible Manufacturing Systems: A Review of Analytical Models. *Management Science*, **32**(7), pp. 890–905.

Church, L.K. and Uzsoy, R. (1991) Analysis of Periodic and Event-Driven Rescheduling Policies in Dynamic Shops. *Research Memorandum No. 91–13*, August 1991, School of Industrial Engineering, Purdue University, West Lafayette, IN, USA.

Davis, E.W. (1973) Project Scheduling under Resource Constraints – Historical

Review and Categorization of Procedures. *AIIE Transactions*, December 1973, 297–313.

French, S. (1982) *Sequencing and Scheduling*, Ellis Horwood.

Glazebrook, K.D. (1987) Evaluating the Effects of Machine Breakdowns in Stochastic Scheduling Problems. *Naval Research Logistics*, **34**.

Ho, Y.C. and Cao, X.R. (1983) Perturbation Analysis of Discrete Event Dynamic Systems. *J. Optim. Theory and Appl.*, **40**, pp. 559–82.

Jones, A.T. and McLean, C.R. (1986) A Proposed Model for Automated Manufacturing Systems. *Journal of Manufacturing Systems*, **5**(1), pp. 15–25.

Leon, V.J., Wu, S.D. and Storer, R.H. (a) Robustness Measures and Robust Scheduling for Job Shops. *IIE Transactions*, to appear.

Leon, V.J., Wu, S.D. and Storer, R.H. (b) A Game-theoretic Control Approach for Job Shops in the Presence of Disruptions. *IJPR*, to appear.

Maimon, O.Z. and Gershwin, S.B. (1988) Dynamic Scheduling and Routing for Flexible Manufacturing Systems that have Unreliable Machines. *Operations Research*, **36**(2), pp. 279–92.

Panwalker, S.S. and Iskander, W. (1977) A survey of scheduling rules. *Operations Research*, **25**, 45–61.

Pinedo, M. (1983) Stochastic Scheduling with Release Dates and Due Dates. *Operations Research*, **31**, 559–72.

Rinnooy Kan, A.H.G. (1976) *Machine Scheduling Problems*, Martinus Nijhoff, The Hague.

Roudy, R., Herer, Y. and Tayur, S. (1989) Price-directed Scheduling of Production Operations in Real-time. *Proc. 15th Conference of NSF Production Research and Technology Program*, Berkeley, CA.

Shantikumar, J.G. and Sargent, R.G. (1983) A Unifying View of Hybrid Simulation/Analytic Models and Modeling. *Operations Research*, **31**, pp. 1030–53.

Sherali, H.D., Sarin, S.C. and Desai, R. (1990) Models and Algorithms for Job Selection, Routing and Scheduling in a Flexible Manufacturing System. *Annals of Operations Research*, **26**, pp. 433–53.

Slowinski, R. and Weglarz, J. (1989) *Advances In Project Scheduling*, Elsevier.

Stecke, K.E. and Solberg, J.J. (1981) Loading and Control Policies for a Flexible Manufacturing System. *IJPR*, **19**(5), pp. 481–90.

Wilhelm, W.E. and Shin, H. (1985) Effectiveness of Alternate Operations in an FMS. *International Journal of Production Research*, **23**(1), pp. 65–79.

Wu, S.D. and Wysk, R.A. (1989) An Application of Discrete-Event Simulation to On-line Control and Scheduling in Flexible Manufacturing. *IJPR*, **27**, pp. 1603–23.

Wu, S.D. and Wysk, R.A. (1991) Scheduling, Optimization and Control in Automated Systems. *Control and Dynamic Systems*, **47**, Academic Press, Inc.

Wysk, R.A., Joshi, S. and Yang, N.S. (1991) Scheduling and Control of Flexible Manufacturing Systems – some Experiences and Observations. *Proceedings of the Joint US/German Conference in Operations Research in Production Planning and Control*, July 30–31, 1991, Gaithersburg, MD.

Yamamoto, M. and Nof, S.Y. (1985) Scheduling/Rescheduling in the Manufacturing Operating System Environment. *IJPR*, **23**(4), pp. 705–22.

# CHAPTER 4

# Priority rules and predictive control algorithms for on-line scheduling of FMS

*Sebastian Engell, Frank Herrmann and*
*Manfred Moser*

## 4.1  INTRODUCTION

Flexible manufacturing systems (FMS) are one of the key ingredients of modern flexible production systems where small batches of specialized products are produced to satisfy the specific demands of specific customers. To deliver high quality products with short response times has become a key factor for competitiveness. Short response times to fluctuating demands can always be achieved by large spare capacities in the production process. The high investment which is necessary to install flexible highly automated manufacturing systems however renders this solution unacceptable. Thus scheduling policies are necessary which make sure that under the constraint of a high average load of the system, the due dates of the production jobs are met.

The optimal allocation of the resources of complex manufacturing systems to a large number of competing jobs exceeds the capacities of humans using simple decision aids as pen and paper, wall charts etc. The implementation of similar tools on a computer which is only used for the graphical representation of the actual state of the decision process, as it is frequently offered by the vendors of computer-based factory control systems does not remove this bottleneck. Scheduling should be done on-line, i.e. depending on the actual situation in the production process, by suitable algorithms and only be controlled and eventually modified by the dispatchers.

The resource allocation problem in manufacturing systems of the job-shop type is known to be NP-hard. This means that the computational effort to find the optimal solution grows exponentially with the number of operations and the number of machines considered, and a true optimization becomes unfeasible on-line even for very small systems. The standard solution to the scheduling problem in practice is to generate suboptimal schedules using priority rules for the individual queues. Most priority rules are computationally extremely simple and can be implemented easily. For a test problem described in detail in section 2, we present the results of a detailed study on the performance of all usual priority rules in section 4.

Observations of the scheduling errors caused by the known priority rules motivated the introduction of a new rule, the WLS (weighted loss of slack) rule. This new rule is described and compared with the conventional rules in section 5. To overcome the deficiencies of priority rules in general, we then investigated predictive strategies for multi-machine problems.

Our approach is based on the idea of predictive control as it has emerged in the context of standard continuous control problems (Richalet *et al.*, 1978; Clarke *et al.*, 1987; Garcia *et al.*, 1989). The basic idea of predictive control is: assume the present state of a system and a model of its dynamics are known, and a desired trajectory of some variables (outputs) is prescribed. Then, at a given instant of time, the effect of all possible control inputs on the future evolution of the system can be evaluated, and the input sequence which yields the best fit to the desired trajectory can be determined. As both disturbances and changes of the desired trajectory may occur, this process is iterated, and only the first or the first few control inputs are used.

In the control of standard continuous-time or discrete-time dynamical systems with continuous variables, the computation of the optimal input which minimizes a given cost function over a finite or infinite horizon is relatively simple in many cases, e.g. for quadratic cost functions. In our case, due to the discrete nature of the problem, an analytical solution of the optimization problem is not possible and the computational effort increases exponentially with the length of the horizon which is considered. The key factor for the applicability and the performance of predictive scheduling algorithms is an adequate restriction of the search problem to critical decisions and/or promising candidate control sequences.

We have investigated two different strategies for predictive control algorithms for FMS scheduling. One strategy aims at the avoidance of a specific, frequently occurring scheduling error produced by simple priority rules by a partial simulation of the future evolution of the system. The other uses a restricted branch-and bound search technique to investigate a promising part of the full decision tree for the next decisions. We present the results of the application of both techniques to the test problem for FMS scheduling under different load levels, queue-lengths, and time-pressure levels, and compare the performance with those priority rules which were found to be Pareto-optimal for the same problem.

The investigation of scheduling policies for FMS has to be seen in the context of decentralized-hierarchical production scheduling as discussed in Engell and Moser (1990) and Engell (1992). Local scheduling must be complemented by a global assignment of tasks to the subsystems with earliest possible starting times and local due dates. After the local decisions are made, the overall system must be coordinated because the local completion times in many cases determine the earliest possible starting times in other subsystems. This coordination process is disturbed much more severely if a significant fraction of the jobs is finished with large delays than if almost all jobs are finished with an approximately equal delay (Engell, 1992). So the width of the tardiness distribution is very important and not only the average tardiness. We therefore simultaneously consider three measures of tardiness: mean, rms and maximal tardiness.

## 4.2 A TEST PROBLEM IN ON-LINE SCHEDULING

### 4.2.1 Origin of the problem

The problem is a modification of a flexible manufacturing cell at ABB's plant for the manufacturing of turbochargers in Baden, Switzerland. The original production data for one week were given in Solot and Bastos (1987). We modified the problem slightly to make it more demanding by removing one loading station (otherwise the average utilization of the loading stations is below 40% for full utilization of all machines) and by changing the part mix to get a more uniform distribution of the load in the system (Engell and Moser, 1992). We consider two different cases. In the first case, the original loading times and continuous order release during the shifts are assumed. This results in moderate average queue lengths. In the second case, the loading times are increased, different load levels for the machines are generated, and the orders are released only at the beginning of each shift. This results in a considerably more difficult decision problem at the loading station where now long queues exist and the so-called shading effect occurs.

### 4.2.2 Description of the system: machines, part types, operation times, part mix

The structure of the flexible manufacturing system is shown in Fig. 4.1.

In the industrial system, there are five machines, two of them identical, two loading stations, a deburring station, and a washing station. We assume here that only one loading station is available because otherwise utilization of the loading stations is only 40%. All parts have to be loaded first, then one or two machining operations follow, then

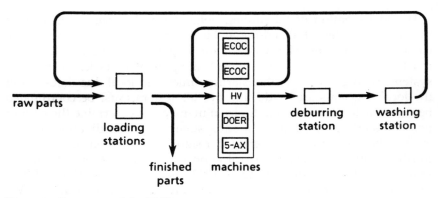

**Fig. 4.1**  Structure of the FMS.

deburring, washing, reloading, another one or two machining operations, deburring, washing, and unloading.

There are 11 part types. The maximum number of operations per
part type is 11, the minimum is nine. Flow diagrams for all 11 part
types are shown in Fig. 4.2, the operation times for the machining
operations are given in Table 4.1. Operation times at the stations
LOAD, WASH, DEBUR are identical for all parts. In **case 1**, the operation times are:

- initial loading        8 min.
- unload/reload        13 min.
- final unloading        5 min.
- deburring (each time)  8 min.
- washing (each time)    6 min.

In **case 2**, all operation times on the loading stations are increased by
five minutes, the other operation times remain unchanged. This produces a bottleneck at the loading station through which all parts must
pass three times, and hence a more demanding decision problem at the
loading station results.

### 4.2.3  Order sequences and order release

For this system, sequences of orders of single parts (batch size 1) are
generated in the following manner: for each machine, there is, for each
shift, a prescribed maximal load level. The maximal load level L is
computed as

L(machine type, shift number)
$$= L_0(\text{machine type}) + \Delta L(\text{shift number}).$$

$\Delta L$ is a random parameter, the load variation.

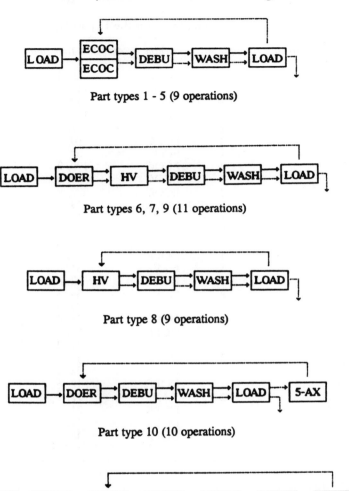

Part types 1 - 5 (9 operations)

Part types 6, 7, 9 (11 operations)

Part type 8 (9 operations)

Part type 10 (10 operations)

Part type 11 (10 operations)

**Fig. 4.2** Machining sequences for the 11 part types.

Due to the structure of the processing sequences, the average load $L_0$ can only be chosen freely on three of the four types of machines. The chosen values are:

**Case 1**
98% on HV, DOER
98% on ECOC

**Case 2**
98% on HV, DOER,
70% on ECOC

**Table 4.1**  Machining times for the 11 part types

| part type | cycle # | machine | operation time [min] |
|-----------|---------|---------|----------------------|
| 1 | 1 | ECOC | 65, 4 |
|   | 2 | ECOC | 149, 4 |
| 2 | 1 | ECOC | 42, 6 |
|   | 2 | ECOC | 153, 6 |
| 3 | 1 | ECOC | 51, 0 |
|   | 2 | ECOC | 270, 0 |
| 4 | 1 | ECOC | 69, 0 |
|   | 2 | ECOC | 190, 2 |
| 5 | 1 | ECOC | 63, 0 |
|   | 2 | ECOC | 277, 8 |
| 6 | 1 | DOER | 59, 4 |
|   | 1 | H–V | 18, 6 |
|   | 2 | DOER | 76, 8 |
|   | 2 | H–V | 103, 8 |
| 7 | 1 | DOER | 21, 0 |
|   | 1 | H–V | 31, 8 |
|   | 2 | DOER | 72, 6 |
|   | 2 | H–V | 34, 8 |
| 8 | 1 | H–V | 39, 0 |
|   | 2 | H–V | 35, 4 |
| 9 | 1 | DOER | 20, 4 |
|   | 1 | H–V | 31, 2 |
|   | 2 | DOER | 43, 2 |
|   | 2 | H–V | 32, 4 |
| 10 | 1 | DOER | 12, 6 |
|    | 2 | 5-AX | 77, 4 |
|    | 2 | DOER | 31, 8 |
| 11 | 1 | DOER | 24, 0 |
|    | 2 | DOER | 45, 6 |
|    | 2 | 5-AX | 51, 6 |

The load amplitude $\Delta L$ is assumed to be either 0, $-30\%$, or $+30\%$.

The actual part mix for a simulation run is generated from a basic random part type sequence with average part type frequencies of 10% for part 1 and 9% for the other parts. For each shift, there is a load account for each machine. An order for a certain part type is accepted as long as the load accounts of all required machines for this shift do not exceed the maximum load level L for these machines and this shift. Else, it is skipped. If no more part types can be accepted for this shift, the order sequence for this shift is complete. The next order in the basic sequence becomes the first new order for the next shift. The actual average part frequencies are thus different from the values in the basic sequence.

As the load level can be above 100% machine capacity, there are backlogs. They are carried over to the next shift. Over the first 5, 10, 15 etc. consecutive shifts, the load level variations average to zero.

The actual order sequences are generated from five different basic part-type sequences and five basic load variation sequences for each of the situations described below.

In case 1, orders are released with equal spacing over the duration of a shift (every 19 minutes). In case 2, all orders are released at the beginning of a shift. This gives rise to considerably longer queues for the first half shift, in particular at the loading station.

Case 1 is thus characterized by:

- equally distributed high loads on the machines with temporary overload;
- short to moderate queues at the stations;

whereas in case 2:

- part types 1–5 only require machines which are not bottlenecks;
- only part types 6–11 use machines which may be overloaded;
- the queues are long;
- the loading station is a critical resource.

### 4.2.4  Due dates

The due date distribution is also a critical parameter which affects the performance of the scheduling policies. The due dates were generated as multiples of four hour (½ shift) intervals from order release. The percentage of orders for each value of this demanded throughput time was determined such that under scheduling according to FIFO a prescribed value of the percentage of late jobs of 30, 50, 70 and 85% was achieved. The resulting distributions are indicated in Tables 4.2(a) and (b). There also, the resulting ratios of demanded to actual throughput times for the different situations under the FIFO rule are given.

The situations A–D correspond to case 1, the percentage of late jobs increasing from 30 to 85, and situations E–H correspond to case 2 with the same levels of late jobs under the FIFO rule.

## 4.3  GENERAL PROBLEM TYPE AND PERFORMANCE CRITERIA

### 4.3.1  Problem type

The example described above belongs to the class of non-cyclic scheduling problems characterized by the following features:

- the flexible manufacturing system consists of $M$ stations (machines, workplaces) each of which can process at most one operation at a time;

**Table 4.2a**   Due data distributions for situations A–D

**Situation A: Case 1, 30% late jobs under the FIFO scheduling rule**

| demanded throughput time | 240 | 480 | 720 | 960 | 1200 |
|---|---|---|---|---|---|
| % of parts | 10 | 10 | 10 | 35 | 35 |

average throughput time: 900 min.
demanded/achieved flow factor: 1.29

**Situation B: Case 1, 50% late jobs under the FIFO scheduling rule**

| demanded throughput time | 240 | 480 | 720 | 960 | 1200 |
|---|---|---|---|---|---|
| % of parts | 10 | 25 | 30 | 25 | 10 |

average throughput time: 720 min.
demanded/achieved flow factor: 1.03

**Situation C: Case 1, 70% late jobs under the FIFO scheduling rule**

| demanded throughput time | 240 | 480 | 720 | 960 | 1200 |
|---|---|---|---|---|---|
| % of parts | 24 | 34 | 32 | 5 | 5 |

average throughput time: 559 min.
demanded/achieved flow factor: 0.80

**Situation D: Case 1, 85% late jobs under the FIFO scheduling rule**

| demanded throughput time | 240 | 480 | 720 | 960 | 1200 |
|---|---|---|---|---|---|
| % of parts | 60 | 25 | 5 | 5 | 5 |

average throughput time: 408 min.
demanded/achieved flow factor: 0.57

- the inventory, which may change at any time, consists of $N$ jobs with known earliest possible starting times and due dates;
- a job is a sequence of operations with precedence constraints, i.e. most operations can only be started if one or several preceding operations have been completed;
- for each operation it is known a priori on which stations (one or more) it can be performed and how long this takes (including set-up and transportation times); these durations are assumed to be independent of the scheduling policy and deterministic;
- operations that have been started cannot be interrupted (non-preemptive scheduling);
- the idling times between the operations are not restricted, neither are the buffers in front of the stations or at the output.

So the main restriction is the capacity of the stations. Other limitations are assumed to be of secondary importance. They may be represented by the earliest possible starting times of the operations or included in the operation times, e.g. by adding a constant average transportation

**Table 4.2b** Due data distributions for situations E–H

| Situation E: Case 2, 30% late jobs under the FIFO scheduling rule | | | | | |
|---|---|---|---|---|---|
| demanded throughput time | 240 | 480 | 720 | 960 | 1200 |
| % of parts | 5 | 5 | 5 | 35 | 50 |

average throughput time: 1008 min.
demanded/achieved flow factor: 1.20

| Situation F: Case 2, 50% late jobs under the FIFO scheduling rule | | | | | |
|---|---|---|---|---|---|
| demanded throughput time | 240 | 480 | 720 | 960 | 1200 |
| % of parts | 5 | 10 | 35 | 27 | 23 |

average throughput time: 847 min.
demanded/achieved flow factor: 1.00

| Situation G: Case 2, 70% late jobs under the FIFO scheduling rule | | | | | |
|---|---|---|---|---|---|
| demanded throughput time | 240 | 480 | 720 | 960 | 1200 |
| % of parts | 19 | 25 | 31 | 15 | 10 |

average throughput time: 653 min.
demanded/achieved flow factor: 0.78

| Situation H: Case 2, 85% late jobs under the FIFO scheduling rule | | | | | |
|---|---|---|---|---|---|
| demanded throughput time | 240 | 480 | 720 | 960 | 1200 |
| % of parts | 30 | 45 | 15 | 5 | 5 |

average throughput time: 504 min.
demanded/achieved flow factor: 0.60

delay. Of course, alternatively, the transportation system can be included by adding transport operations to the work plans and defining one or more resources which provide transport and have to be allocated as well. The only restriction implied by our assumptions is that the resource utilization times are independent of the processing sequences.

In the test problem, the choices on which station an operation is performed are only among identical stations. Thus for each operation there is only one value for its (nominal) net operation time. This assumption however is not necessary for the algorithms discussed later.

### 4.3.2 Performance criteria

We assume that an FMS is part of a larger production process. So the performance of the FMS is not a primary goal but a means to achieve cost-efficient timely production in the overall system. On the present state of technology, complex production systems must be controlled by hierarchical decentralized systems (Engell and Moser, 1990; Engell, 1992). The decisions in the subsystems must be based on local due dates and local earliest possible starting times which are provided by the higher-level scheduling system. For a smooth operation of the overall system, a narrow distribution of the deviations from the planned process is essential.

Our general performance criterion therefore is minimal tardiness. We neither honour nor punish earliness, because in a complex system the benefits of local earliness are doubtful. The tardiness of all jobs which are finished on time or earlier is zero. We do not use throughput time or machine utilization as criteria, because on the long run (we use very long horizons for the evaluation) the task of a production system is to produce the goods to be delivered at the right time where the demanded throughput times may vary very much between the jobs; thus the average throughput time may be a misleading indicator. Of course, tardiness must always be seen relative to a certain demanded flow factor distribution, indicated by the time pressure level in our case.

There is a number of different reasonable measures of tardiness. It is known that these different measures can in general not be minimized simultaneously. In cases of high loads of the system, algorithms that prefer shorter operations give good values of average tardiness $T_{\text{mean}}$ but at the expense of a small fraction of the jobs which suffer from very large delays. If an FMS is integrated into a larger overall production process, not only a small value of $T_{\text{mean}}$ is desired but also a narrow distribution of the delays around this value, because large discrepancies between planned and achieved completion times create problems for the other local schedules. So we use the three following criteria:

- average tardiness (over **all** jobs)

$$T_{\text{mean}} = \frac{1}{N} \sum_{i=1}^{N} T_i$$

where

$$T_i = \max(0, t_{fi} - t_{di})$$

$t_{fi}$ is the completion time of job $i$ and $t_{di}$ its due date.

- rms-tardiness

$$T_{\text{rms}} = \left( \frac{1}{N} \sum_{i=1}^{N} T_i^2 \right)^{1/2}$$

- maximal tardiness

$$T_{\text{max}} = \max_i T_i.$$

Note that $T_{mean}$ is not substracted in the definition of $T_{rms}$, so $T_{rms}$ is not a variance but a measure of relative tardiness, and the difference of $T_{mean}$ and $T_{rms}$ indicates the width of the distribution. The relative emphasis given to either of the three criteria depends on the specific situation. In general, $T_{rms}$ will be the most appropriate single criterion.

In Fig. 4.3, the values of mean, rms, and maximal tardiness are plotted versus the duration of the experiment for the test problem (case 2) if the jobs are scheduled according to the FIFO rule. It can be seen that the mean and the rms tardiness reach stable equilibrium values after about 600 8-hour-shifts, whereas the maximal tardiness increases for up to 2000 shifts. As a compromise we use a simulation horizon of 1000 shifts. To avoid errors from startup und rundown, the first and the last 10 shifts are disregarded, i.e. only those jobs which were released in shifts 10–990 are evaluated.

## 4.4 CONVENTIONAL PRIORITY RULES

### 4.4.1 Motivation

The standard solution for production scheduling (sequencing and routing) problems is the use of priority rules. A priority rule is an ordering of the queues in front of the machines (units, servers, work-places, cells, etc.) according to some parameter which is calculated from the information available about the respective operation or job. Priority rules are easy to understand and simple to implement, which explains their widespread use. A large number of priority rules have been proposed (Panwalker and Iskander, 1977; Blackstone *et al.*, 1982). But so far, no comprehensive study existed which compares the quality of these rules for a realistic FMS over the full range of operation conditions (load and time pressure levels). Consequently, contradictory results have been published (Baker, 1984). In Moser (1992), a comprehensive empirical study of the behaviour of priority rules is reported. Results on the single-machine case can be found in Moser and Engell (1992a).

One main result of our studies is that the rules can be classified into two categories:

- rules which achieve a small value of the average tardiness but under high pressure produce very large delays of a small fraction of the jobs;
- rules which avoid excessive delays at the expense of a markedly increased average tardiness.

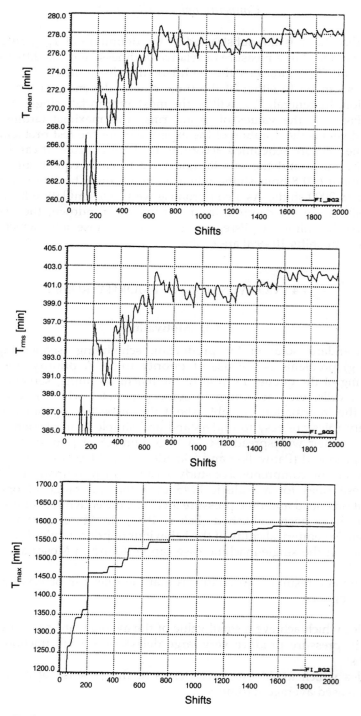

**Fig. 4.3** Performance measures versus simulation horizon. Case 2, FIFO rule.

The best-known representative of the first class is the SPT (shortest processing time first) rule (which we all tend to use to avoid the overflow of our desks), whereas the EDD rule (earliest due date first) belongs to the second class.

### 4.4.2 Priority rules

A priority rule is an algorithm which calculates a priority number $PN_j$ for each operation $j$ that enters the queue in front of a station (machine, workplace, cell, processing unit). This calculation can be based on any of the following quantities:

- arrival time in the queue $t_{aj}$;
- duration of the operation $o_j$;
- (static) due date of the operation $t_{dj}$;
- due date of the corresponding job $t_{dj}^*$;
- number of future operations in the job $n_j^*$;
- remaining work content of the job $w_j^*$.

$t_{dj}$ is computed by backwards scheduling from the due date of the job with constant flow factor.

Priority rules can be classified into

- static or dynamic rules;
- a priori or a posteriori rules.

For a static priority rule, the value of PN is fixed when the job is introduced up to a linear shift corresponding to the increment of the system's clock. Dynamic priority numbers depend on the actual evolution of the system and cannot be computed beforehand without simulation. A simple dynamic priority rule is FIFO: $PN_j = t_{aj}$; a simple static priority rule is $SPT$: $PN_j = o_j$. But also the slack rule:

$$PN_j = t_{dj}^* - t - w_j^*,$$

where $t$ is absolute time, is static according to our definition.

A priori priority rules calculate $PN_j$ from the information which is available about operation $j$ alone. In contrast, a posteriori rules consider the situation which would arise if operation $j$ would be scheduled as the next operation. They make a comparison of the damage which is done to the operations which have to wait further, whereas a priori rules only compare the benefit obtained for operation $j$. Obviously, a posteriori rules cannot be static.

The following rules have been proposed and were reported to be successful in the literature:

**1. Simple rules**

| | | |
|---|---|---|
| FIFO | (first in first out) | $PN_j = t_{aj}$ |
| SPT | (shortest processing time) | $PN_j = o_j$ |

| ODD | (earliest operation due date) | $PN_j = t_{dj}$ |
|-----|-----|-----|
| EDD | (earliest (job) due date) | $PN_j = t_{dj}{}^*$ |
| SL | (slack) | $PN_j = t_{dj}{}^* - t - w_j{}^* = t_{sj}$ |
| CR | (critical ratio) (Putnam *et al.*, 1971) | $PN_j = (t_{dj}{}^* - t)/$ $(t_{dj}{}^* - t_{dj} - o_j)$ |
| SL/OPN | (slack per remaining operation) | $PN_j = t_{sj}/n_j{}^*$ |
| SL/RPT | (slack/remaining processing time) | $PN_j = t_{sj}/w_j{}^*$ |

## 2. Combined rules

| MOD | (modified operation due date) (Baker and Bertrand, 1982) | $PN_j = \max(t + o_j, t_{dj})$ |
|-----|-----|-----|
| SPT-T | (truncated SPT) | $PN_j = \min(o_j + r, t_{sj}/n_j{}^*)$ |

*r* is a free parameter which influences the weight of the SPT component relative to the slack component

| SL+SPT | (max of slack and operation time) | $PN_j = \max(o_j, t_{sj}/w_j{}^*)$ |
|-----|-----|-----|
| CR+SPT | (dynamic version of the MOD-rule) (Anderson and Nyirenda, 1990) | $PN_j = \max(t + o_j,$ $t + o_j \cdot (t_{dj}{}^* - t)/w_j{}^*)$ |

## 3. Split queue rules

SPT*    (Fry *et al.*, 1987)
Two queues prioritized according to SPT. The operations for which $t_{sj} - u$ is negative form a high priority queue which is processed first. The order within the two queues is according to the SPT-rule. *u* is a design parameter which defines when an operation becomes urgent.

SPT/SL    (Fry *et al.*, 1987)
Same as SPT* except that the high-priority queue is ordered according to SL instead of SPT.

CEXSPT    (Schultz, 1989)
Three queues are considered in front of each station. The first queue contains the operations with negative slack $t_{sj}$ (already delayed, tardiness inevitable), the second those with negative local slack (operation due date has passed), and the third the remaining (uncritical) operations. All queues are prioritized according to SPT. The first queue has the highest priority. However, operations from this queue are only scheduled, if none of the remaining operations would have negative slack afterwards. Operations from the second queue are only scheduled if no

operation from the third queue would have a negative local slack afterwards.

## 4. Throughput-oriented rules

WINQ     (smallest work in next queue)

COVERT   (cost over time) (Russel *et al.*, 1987)

         Decision based on an estimation of the waiting time in the next queues.

### 4.4.3   Theoretical results

In the single machine case, some theoretical results are known or can be derived (French, 1982; Moser, 1992):

- if all operations can be scheduled such that each one is finished before its due date, then EDD achieves this;
- if the schedule which is generated by EDD is such that only one operation finished late, then EDD minimizes the mean tardiness $T_{mean}$;
- if all arrival times $a_j$ or all operation times $o_j$ are equal, then EDD minimizes the maximal tardiness $T_{max}$;
- if all due dates $t_{dj}$ are equal, then SPT minimizes $T_{mean}$ and FIFO minimizes $T_{max}$;
- if all operations are inevitably finished late, then SPT also minimizes $T_{mean}$ because it minimizes the mean throughput time;
- if only two operations exist which have the same arrival time, then MOD minimizes $T_{mean}$. This is no longer true for three or more operations in the queue.

These results only hold for static problems where a fixed queue has to be ordered and no new jobs arrive. Nonetheless, they give interesting indications for the general situation:

- with respect to $T_{mean}$ and $T_{max}$, ODD or similar rules can be expected to give good results for a situation where most of the work can be finished in time (low load);
- with respect to $T_{mean}$, SPT and rules that use SPT for urgent jobs can be expected to give good results for situations where most of the work is finished late.

### 4.4.4   Results for the test problem (Moser and Engell, 1992b)

The rules mentioned above and some variants of these rules which turned out to be inferior were tested for the eight different situations of the FMS for the production of turbochargers described in section 2. All three performance criteria are evaluated for 25 different simulation runs and then averaged. The 25 different simulation runs are produced from five different random basic part type sequences and five different random due date sequences. For a significance analysis of the difference of two policies, the outcome of all 25 runs was used in a Wilcoxon test.

First, the simple rules were compared. It turned out that for the three criteria $T_{mean}$, $T_{rms}$, and $T_{max}$ and the eight cases considered, the following rules outperformed the others:

- ODD            (best rule with respect to all criteria under high time pressure);
- SL/OPN         (best rule with respect to all criteria under low time pressure);
- CR             (best mean and rms tardiness in case 2 under medium pressure).

Among the combined and split queue rules, the best are

- CR+SPT         (best rule with respect to $T_{mean}$ in all situations);
- SPT-T (r = 0)
- SPT/SL (u = 0).

The last two rules are the best ones with respect to $T_{max}$ in all situations, and with respect to $T_{rms}$ under medium and under high pressure. For case 2 under low pressure, CR+SPT gives better results. The differences between SPT-T and SPT/SL are not significant, thus only the results for CR+SPT and SPT-T are shown in the figures.

The results for these rules are compared with the results obtained with the throughput-oriented rules WINQ and COVERT in Figs 4.4 and 4.5 for the two cases. It can be concluded that for the system considered here the more sophisticated rules are outperformed by the simpler ones.

Obviously, the known priority rules for job shop scheduling can be divided into two classes depending on their behaviour under high pressure:

- rules which use SPT for critical jobs;
- rules which use SL or similar criteria for critical jobs.

The first class of rules achieves the best values of $T_{mean}$ at the expense of high values of $T_{max}$. The second class prevents excessive delays of a small fraction of the jobs at the expense of a larger value of $T_{mean}$. Within these classes, several rules show a comparable performance. In the first class, CR+SPT for our problem gave the best results, in the second the best were ODD and SL/OPN. These three rules produced Pareto-optimal results with respect to our three criteria. In the single machine case (Moser and Engell, 1992a), the Pareto-optimal rules were EDD and MOD which in this case are identical with ODD respective CR+SPT.

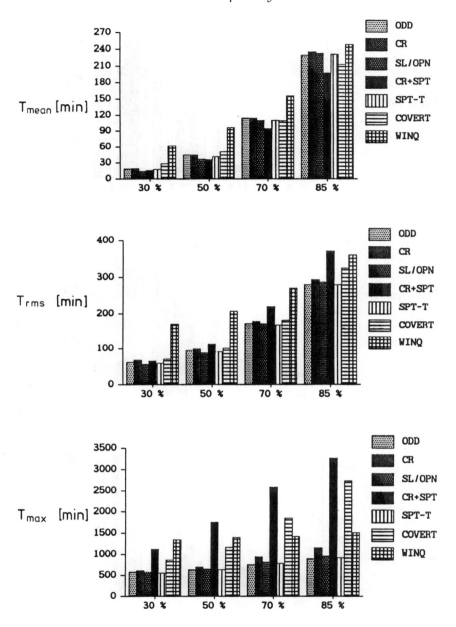

**Fig. 4.4** Performance of priority rules versus time pressure (% of late jobs under FIFO), case 1.

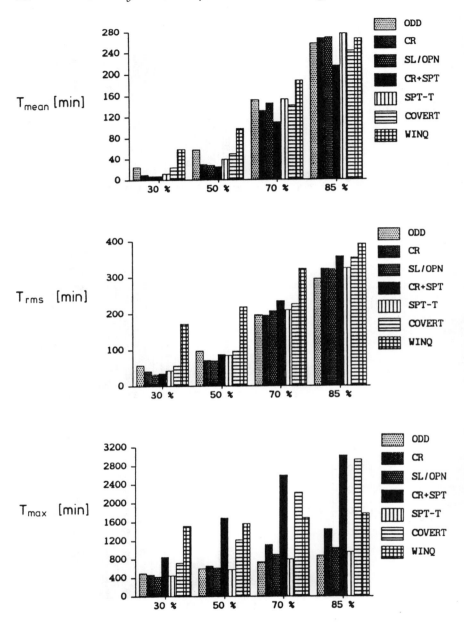

**Fig. 4.5** Performance of priority rules versus time pressure (% of late jobs under FIFO), case 2.

### 4.5  A NEW RULE WITH LOCAL ONE-STEP AHEAD PREDICTION: THE WLS-RULE

The observation of this tradeoff motivated the invention of a new dynamic a posteriori rule, the WLS (weighted loss of slack) rule (Moser 1992; Moser and Engell, 1992a,b).

The weighted loss of slack rule compares the situation of **all** other operations in the respective queue if operation $j$ would be scheduled. This decision will cause a loss of slack for all other operations waiting in the queue. This is most critical if an operation would have negative slack $t_{si}$ after scheduling of operation $j$. Thus we define as **the critical loss of slack of operation $i$ due to the scheduling of operation $j$ before $i$**:

$$\Delta t_{si} = \begin{cases} o_j & \text{if } t_{si} \leqslant 0 \\ o_j - t_{si} & \text{if } o_j > t_{si} \\ 0 & \text{if } t_{si} \geqslant o_j. \end{cases}$$

The priority number $PN_j$ is then calculated as a sum of the weighted losses of slack:

$$PN_j = \sum_i e^{-c \cdot \min(0, t_{si})} [e^{c \cdot \Delta t_{si}} - 1].$$

In this formula, the differences of the **exponentially weighted slack** before and after the scheduling of operation $j$ are summed up. The exponential weighting provides an incentive to prefer long operations (which cause larger losses of slack than short ones) when their slack decreases and thus helps to keep the maximal tardiness small while still providing small throughput times on the average.

The queue is again divided into two subqueues. The high-priority queue contains those operations for which $t_{sj} < \max_{i \neq j} o_i$, i.e. the operations which have negative slack or might get negative slack due to the scheduling of some other operation before operation $j$. These critical operations are ordered according to the above formula. The uncritical operations are scheduled according to the ODD rule.

The best rules from Figs 4.4 and 4.5 are compared with the new WLS rule in Figs 4.6–4.9. The parameter $c$ was set to 0.01 (this corresponds to roughly $1/(2$ average machining operations), the same value was found to be optimal in a single-machine example with similar operation times).

It is obvious that the WLS rule is a very good compromise between the two classes of rules. For long queues and high pressure, it dominates SL/OPN both in $T_{mean}$ and $T_{rms}$, and it performs well in all situations while avoiding extreme values of $T_{max}$. In our view, it is the prime candidate for a good simple priority rule especially in decentralized scheduling.

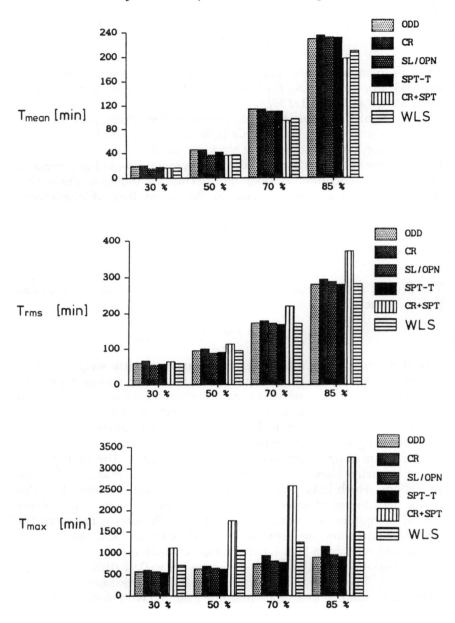

**Fig. 4.6**  Comparison of the new MLS-rule with the best conventional priority rules for the test problem, case 1.

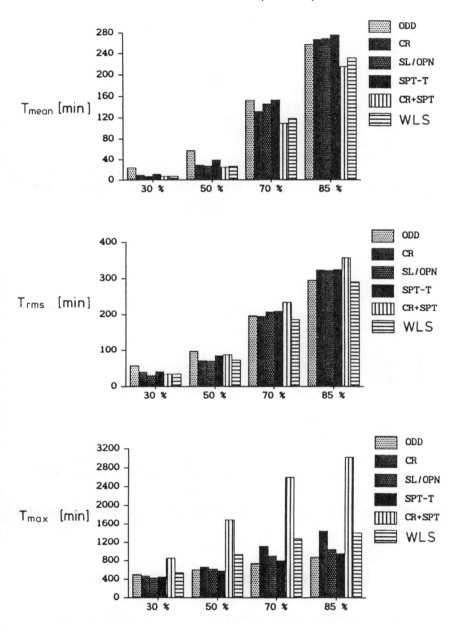

**Fig. 4.7** Comparison of the new WLS-rule with the best conventional priority rules for the test problem, case 2.

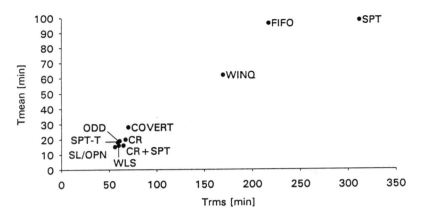

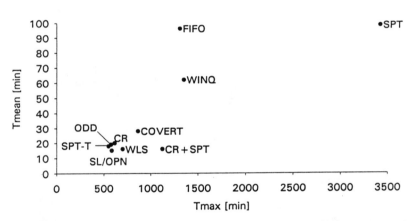

**Fig. 4.8** Overview of the results for case 1 under low time pressure (situation A).

## 4.6 PREDICTIVE CONTROL ALGORITHMS

### 4.6.1 Avoiding scheduling errors by partial look-ahead

As already mentioned above, the heuristic priority rules which are commonly used in industrial practice only consider the local status of the jobs in the queue to be sequenced. No information about the future passing of the jobs through the network of queues is used.

Consequently, priority rules lead to scheduling errors. One parti-

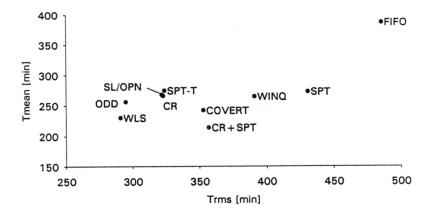

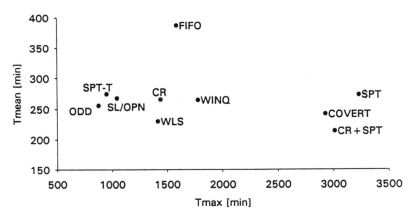

**Fig. 4.9**  Overview of the results for case 2 under high time pressure (situation H).

cular, frequently occurring type of error is called by practitioners the shading effect. It can be understood very easily from the following example.

Let us assume that in the queue in front of a highly loaded station there are two types of jobs, A and B, say. Operations belonging to jobs of type A have a high priority because they are already critical. Operations of type B have lower priority. So a priority-based local scheduling algorithm would first assign the operations of type A and then those of type B, provided that nothing else happens. Now it may occur that the jobs of type A next have to visit a station which is also highly loaded and on which they compete with jobs of type C, all or some of which are even more critical. However, the jobs of type B next visit a machine which is ready to process them and might even idle as long as no jobs of type B are finished on the first machine. From the information available to the scheduler of the first machine, to prefer operations of type A to help them meet their due dates is correct. However, due to the situation downstream, the result is to cause additional delays for jobs of type B without any gain for jobs of type A.

In practice, the situation may be more complicated. The bottlenecks in the system may be dynamic, i.e. the reverse situation on the downstream stations may also occur, so a simple bonus for jobs of type B will not help. Or at some instance there may be competition by urgent jobs of type C, but at another, not. And the error may not become visible in the next processing step but only later.

As said above, to explore all possible sequences on all stations even over a very short horizon is unfeasible. The solution which we propose is of a predictor-corrector type. The essential idea is not to start with the uncontrolled system, but with the system controlled by a basic control algorithm. The basic control algorithm is a priority rule. Then we focus on the potential errors. This means we start on the (dynamic or static) bottleneck machines and simulate the situation on the stations which the jobs which are presently waiting will encounter on the next station. This requires the stimulation of all other stations from which potentially competing jobs may originate. But still only a small part of the system is considered. This analysis yields the probable earliest possible starting times for the next processing steps and, hence, the real (in practice at least improved) due dates for the operations waiting in front of our bottleneck station. Scheduling on this station is then done based on these due dates rather than on the global slack of the jobs or on the operation due dates which were determined beforehand. The result of the partial simulation is a **dynamic operational due date** (DODD) for each operation in the queue considered.

In order to optimize the job sequence of the queue, the DODDs can be used in several ways. We tested three strategies:

A. The simplest possible scheduling rule on the station for which the simulation is done is to use the DODD as priority criterion, i.e. to schedule the operation with the earliest DODD first. This is consistent with the results reported above that the ODD-rule is a good choice if the mean, rms and max tardiness are all relevant.

B. The queue is first ordered according to a local priority rule. An operation from the original queue is only moved to the top of the queue if
   – it has an earlier DODD than the present top candidate;

- no operation which originally was scheduled before this operation misses its DODD.
C. Again the queue is ordered according to a local priority rule. If the machine for the next operation of the same job is a static bottleneck, then this operation is marked. All marked operations are shifted to the top of the queue; then the same moving procedure as under B is applied to the modified queue.

Strategy B may cause some idling of downstream machines in order to avoid additional delays of critical jobs. Strategy C puts a higher emphasis on avoiding idling of bottleneck resources because this in a high-load situation is severely detrimental to overall performance.

### 4.6.2 Restricted search over all possible schedules

An alternative to this approach is to compute a schedule for a certain number of operations and then to schedule the first operation or some operations and to iterate this process, i.e. rolling horizon optimization.

All possible sequences of operations can be represented by a decision tree. Each edge or branch of the tree corresponds to the assignment of an operation to a station, each node representing a partial schedule which is determined by the sequence of edges that lead from the root to this node. From each node, there are many edges which represent all possible next assignments of operations to stations. The idea of branch-and-bound algorithms is to construct a partial decision tree which contains the optimal solution. The tree is reduced by cutting off those branches which will only lead to solutions which are worse than the best solution found so far or can be excluded by a bound of the cost functional at the optimum which can be estimated from the solutions obtained so far. In our case, i.e. for scheduling with respect to minimal tardiness, no efficient bounds are known so that a consideration of all potentially optimal solutions is only feasible in cases with a few jobs and a few machines. In particular, it is not possible to give effective bounds in dynamic situations.

The algorithm which we have developed and tested performs the following steps:

**Step 1: Continuation**
We assume that a partial schedule (which may be the empty schedule) is given. For this partial schedule, a number of possible continuations is generated. In order to cope with the combinatorial explosion, the continuation is based on a limited set of candidates for each station. This set is determined by priority rules.

**Step 2: Evaluation**
All continuations of the partial schedule which are generated in step 1 are ranked according to a cost function. This cost function may be

different from the performance criterion because a cost has to be assigned to the partial completion of a job.

**Step 3: Selection**
Of the $p$ best continuations $q$ (first) operations are scheduled. This gives $p$ new partial schedules. Then step 1 is repeated.

After the generation of a number of new partial schedules, $r$ operations of the best continuation found so far are definitively scheduled and the process is restarted from this point.

These steps are described in more detail below.

**1. Continuation**
It is crucial to use a strategy which ensures that in one continuation step operations are scheduled on all or at least most machines. Otherwise, the evaluation of the alternative schedules becomes very difficult. The algorithm proceeds as follows:

1. From the partial schedule, there are at most $N$ operations which can be started. Each operation is assigned the station where it would have the earliest possible starting time (we assume here that if there are different stations for this operation, their efficiency is identical, otherwise the earliest possible completion time has to be computed).
2. The resulting set of potential operations on station $m$, $S(m)$ is reduced by discarding those operations which would lead to an idling period of more than $D$ time units if they were the next operation on this machine, and ordered according to a priority rule. The $s$ prime candidates in $S(m)$ are then considered further, denoted by $S'(m)$.
3. The machine $m^*$ is determined which has the earliest possible starting time of the next operation among all machines. One operation is chosen from $S'(m^*)$, and its completion time $t_e(m^*)$ is calculated.
4. For all other machines, the set $S''(m)$ is determined which contains all potential operations from $S'(m)$ which can be started before $t_e(m^*)$. From all these sets, one operation is scheduled.

This procedure allows the construction of a number of continuations of length $L$, $1 \leqslant L \leqslant M$, until all operations in $S'(m^*)$ are used as initial operations and each combination of operations in the sets $S''(m^*)$ for this initial operation has been generated. This process is iterated until all generated continuations have a certain length.

**2. Evaluation**
The schedules are evaluated according to the criterion weighted loss of slack. Let $t_{si}^o$ be the slack (time buffer) of job $i$ according to the initial schedule and $t_{si}(n)$ be the slack of job $i$ after the continuation number $n$. Then the cost of the continuation is

$$Q(n) = \sum_{i=1}^{N} \left( e^{-c \cdot t_{si}(n)} - e^{-c \cdot t_{si}^o} \right).$$

The exponential weighting was used for the same reasons as in the WLS rule.

### 3. Selection

The parameter $p$ was set to 10 and $q$ was set to the length of each continuation. Starting from the partial original schedule, 100 different continuations were produced in this manner, i.e. two continuation steps were performed. The first operation of the best schedule was assigned to give the new partial schedule.

### 4. Heuristics

The heuristics which were tested in the reduction of the sets $S(m)$ were slack per remaining operation (S/OPN) and simple slack (SL). S/OPN turned out to be more effective.

#### 4.6.3 Application of the algorithms to the test problem

*Results using partial look-ahead strategies*

The simplest variant of the partial look-ahead strategy – strategy A – did not succed at all. The simple priority rule SL/OPN for example outperformed the more refined algorithm in all three criteria (mean, rms and maximal tardiness).

Why does this happen? The reason is that in the critical queue, some operations will miss their DODD. So a critical job may suffer from additional delays on the next stations, increasing its tardiness to the benefit of uncritical jobs finishing early. This however does not lead to improvements in tardiness-related performance criteria.

In the case of a short queue at the loading station and balanced loads of the machine tools (case 1), strategies B and C performed only slightly better than the priority rules. This is not surprising because the shading effect does not occur very often under these conditions.

Significant improvements were obtained in the case of a long queue at the loading station and unbalanced loads on the machine tools (case 2). Obviously, the shading effect occurs frequently and thus the improved strategy clearly outperformed all priority rules tested (Figs 4.10, 4.11).

With strategy B, a reduction of both the mean and the rms tardiness by 10–15% was generally achieved. Under medium and high time pressure, strategy C provided an even bigger reduction by 20–25% compared with the priority rules. But, unfortunately, under low time pressure strategy C performed much worse. This result clearly indicates that the benefit for some expedited high priority jobs outweighs the global damage of idling periods of bottleneck machines only under low time pressure.

In most situations both strategies also reduced the maximum tardi-

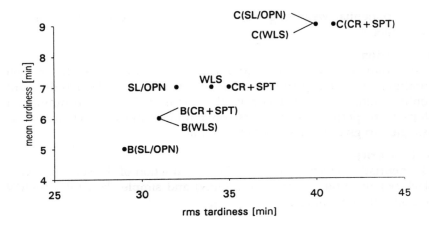

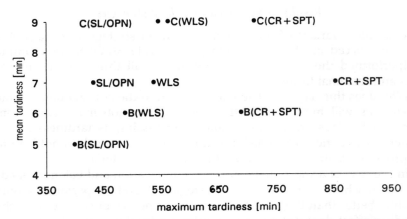

**Fig. 4.10** Results of the look-ahead algorithm (strategies B and C) compared to the priority rules SL/OPN, CR + SPT and WLS for the test problem. Case 2, low time pressure.

ness by 10–15%, only in a few cases a slightly higher value was observed.

### *Results obtained with the restricted branch-and-bound algorithm*

The parameters $c$ (i.e. weighting constant in the evaluation process) and $s$ (i.e. number of candidates in the reduced set $S'(m)$ in the continuation process) were varied to gain insight into the behavior of the algorithm.

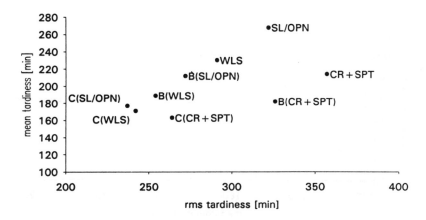

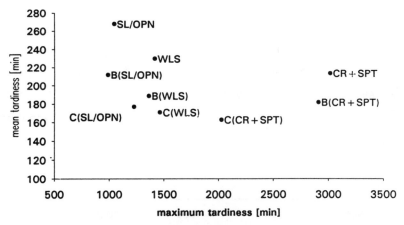

**Fig. 4.11** Results of the look-ahead algorithm (strategies B and C) compared to the priority rules SL/OPN, CR + SPT and WLS. Case 2, high time pressure.

In the cases with relatively short average queue lengths, the optimal choice turned out to be $s = 2$. If more candidates are considered, the average tardiness and the rms-tardiness increase. This shows that the WLS cost function used in the selection step does not exactly reflect our goal to minimize tardiness. The maximal tardiness is hardly affected by the choice of $c$ and $s$. Increasing values of $c$ lead to smaller values of $T_{rms}$ relative to $T_{mean}$, but larger mean values, as expected. The results are considerably better than those obtained with the best heuristic priority rules alone, including the WLS rule. Figure 4.12 compares the results of the branch-and-bound algorithm with $c = 0.01$ and $s = 2$ with

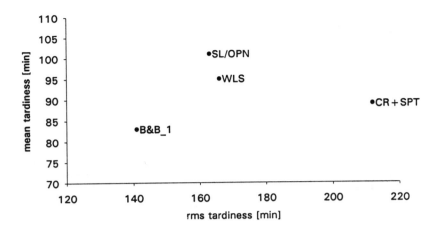

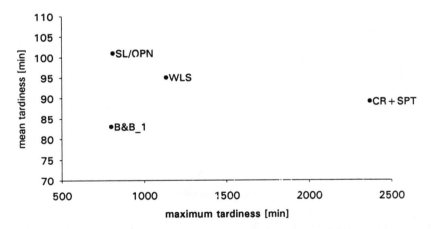

**Fig. 4.12** Comparison of the restricted branch-and-bound algorithm and the best priority rules for the test problem. Case 1, high time pressure.

those of the best three priority rules. The application of the partial look-ahead strategy does not improve the results of the priority rules significantly in this case, since the shading effect does not occur.

For the system with longer queues in front of the load station, $s = 3$ turned out to be optimal. The results are better than with priority rules but not as good as those obtained with the partial look-ahead strategies. This is due to the fact that the depth of the decision tree which is evaluated in each step is not sufficient to cope with the shading-effect.

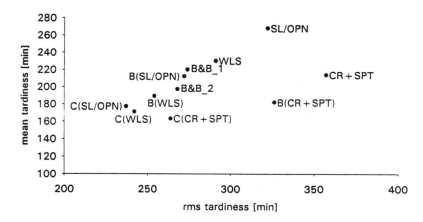

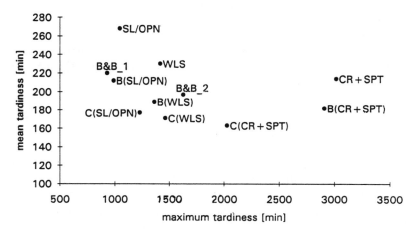

**Fig. 4.13** Comparison of both predictive control strategies and the best priority rules for the test problem. Case 2, high time pressure.

In Fig. 4.13, the results of three conventional priority rules, two priority rules with partial look-ahead strategy (anti-shading algorithm), and two branch-and-bound algorithms are presented. The parameters in the branch-and-bound strategies have these values: $c = 0.01$ and $s = 2$ in B&B-1 and $c = 0.005$ and $s = 3$ in B&B-2.

## 4.7  CONCLUSION

Starting from a realistic example, we gave a survey of the performance of known priority rules for FMS-scheduling. The analysis of the shortcomings of these rules led to the introduction of a new rule and to two new scheduling algorithms based on the concept of predictive control. The first algorithm was designed to tackle a well-known shortcoming of the conventional priority rules: the shading effect. The basic idea is to perform a partial look-ahead to the next manufacturing stage by means of simulation. The second algorithm is a rolling horizon branch-and-bound approach, which performs a restricted search over all possible continuations of the actual schedule.

Both predictive algorithms turned out to perform better than priority rules with respect to all performance criteria (up to 25% lower mean and rms tardiness than priority rules). For unbalanced machine loads and long queues, when the shading effect occurs frequently, the partial look-ahead clearly performed better than the branch-and-bound algorithm. Otherwise, i.e. for problems with little structure, the branch-and-bound algorithm gave the best results.

The new algorithms presented here are a significant improvement over conventional scheduling rules. With all algorithms described, a scheduling decision, i.e. the selection of the next job when a machine becomes idle, is obtained within a few seconds on a PC. Therefore the algorithms are able to control an FMS on-line.

## ACKNOWLEDGEMENT

The research reported here was mainly done while all authors were with the Fraunhofer-Institut für Informations-und Datenverarbeitung (FhG-IITB), Karlsruhe, Germany. It was supported by the Deutsche Forschungsgemeinschaft by a research grant to the first author (DFG En152/3) and by the Fraunhofer-Gesellschaft. This support is very gratefully acknowledged, as well as the contribution of numerous students, in particular M. Herrmann, E. Heß, K. Müller, and A. Schroff, to the development of the simulation software, the implementation of the algorithms, and the performance evaluation.

## REFERENCES

Anderson, E.J. and Nyirenda, J.C. (1990) Two new rules to minimize tardiness in a job shop. *Int. J. Prod. Res.*, **28**, pp. 2277–92.
Baker, K.R. (1984) Sequencing rules and due-date assignments in a job shop. *Mgmt Sci.* **30**, pp. 1093–1104.

Baker, K.R. and Bertrand, J.W.M. (1982) A dynamic priority rule for sequencing against due-dates. *J. Ops Mgmt*, **3**, pp. 37–42.

Blackstone, J.H., Phillips, D.T. and Hogg, G.L. (1982) A state-of-the-art survey of dispatching rules for job shop operations. *Int. J. Prod. Res.*, **20**, pp. 27–45.

Clarke, D.W., Mohtadi, C. and Tuffs, P.S. (1987) Generalized predictive control – part 1: the basic algorithm. *Automatica*, **23**, pp. 137–148.

Engell, S. (1992) *Hierarchical decentralized scheduling using min-max algebra*. IFAC Symposium Large Scale Systems 92, Beijing, Preprints 567–72.

Engell, S. and Moser, M. (1990) *Two-layer Online Scheduling of Flexible Manufacturing Systems*. Proc. Rensselaer's 2nd Int. Conference on Computer Integrated Manufacturing, IEEE Computer Society Press, pp. 435–42.

Engell, S. and Moser, M. (1992) *A Benchmark Problem in Online Scheduling*. Proc. IEEE Int. Conference on Decision and Control, Tucson, 1992, pp. 386–91.

French, S. (1982) *Sequencing and Scheduling: An Introduction to the Mathematics of the Job-Shop*. John Wiley, New York.

Fry, T.D., Philipoom, P.R. and Blackstone, J.H. (1987) A simulation study of truncated processing time dispatching rules. *J. Ops Mgmt*, **7**, pp. 77–92.

Garcia, C.E., Prett, D.M. and Morari, M. (1989) Model predictive control: theory and practice – a survey. *Automatica*, **25**, pp. 335–48.

Moser, M. (1992) Regelung der Maschinenbelegung in der flexiblen Fertigung (Control of the Machine Allocation in Flexible Manufacturing Systems). Dissertation, Universität Karlsruhe, Dept. of Electrical Engineering (in German).

Moser, M. and Engell, S. (1992a) *Comprehensive Evaluation of Priority Rules for On-line Scheduling: The Single Machine Case*. Proc. Rensselaer's 3rd International Conference on Computer-Integrated Manufacturing, IEEE Computer Society Press.

Moser, M. and Engell, S. (1992b) *A Survey of Priority Rules of FMS Scheduling and their Performance for the Benchmark Problem*. Proc. 1992 IEEE International Conference on Decision and Control, Tucson, pp. 392–97.

Panwalker, S.S. and Iskander, W. (1977) A survey of scheduling rules. *Ops Res.*, **25**, pp. 45–61.

Putnam, A.O., Everdell, R., Dormann, G.H. *et al.* (1971) Updating critical ratio and slack time priority scheduling rules. *Production and Inventory Management*, pp. 51–73.

Richalet, J., Rault, A., Testud, J.L. and Papon, J. (1978) Model predictive heuristic control: applications to industrial processes. *Automatica*, **14**, pp. 413–28.

Russel, R.S., Dar-El, E.M. and Taylor, B.W. (1987) A comparative analysis of COVERT job sequencing rule using various shop performance measures. *Int. J. Prod. Res.*, **25**, pp. 1523–40.

Schultz, C. (1989) An expediting heuristic for the shortest processing time dispatching rule. *Int. J. Prod. Res.*, **27**, pp. 31–41.

Solot, Ph. and Bastos, J.M. (1987) Choosing a Queueing Model for an FMS. Brown Boveri Forschungszentrum, *Report CRB 87-40 C*, Baden, Switzerland.

# CHAPTER 5

# Scheduling of automated guided vehicles for material handling: dynamic shop floor feedback

*Leslie Interrante, Daniel Rochowiak*
*and Ben Sumrall*

## 5.1  INTRODUCTION

### 5.1.1  Material handling and manufacturing

In a modern manufacturing facility, flexibility and dynamic recon-figuration are rapidly becoming the norm. At the same time there is a growing realization that material handling is a key component in accomplishing the goals of flexible, dynamic, agile manufacturing. Such manufacturing is critical in placing industries in an efficient and competitive position. Advanced material handling capabilities are essential, since without this ability to supply the required material to the appropriate manufacturing station at the right time and in the right quantity the entire facility will 'bog down', become less efficient and produce less profit and/or operate at higher costs.

### 5.1.2  Automated guided vehicles

One material handling technology currently addressing these concerns is automated guided vehicle (AGV). An AGV is a general material-carrying mobile device designed to receive and execute instructions, traverse a path, and accept and deliver materials. The vehicles generally follow a track which can take many forms including wire guides and chemical tracks. The chemical tracks can be easily reconfigured as the plant is reconfigured. The instructions to an AGV indicate where it is to go, how it should get there, and what it should do when it gets there. Additionally, the AGV is able to identify its location within track sections and broadcast information that allow the AGV's position to be mapped globally. AGV technology can meet the demands of the new manufacturing approach since the track system is flexible and the commands to the AGVs can reflect shop floor conditions.

Although this technology has met with success there is still room for improvement. A flexible information system is needed to make full use of the mechanical flexibility provided by an AGV system. The central AGV controlling system must be able to respond to shop floor conditions that, while generally manageable, are not amenable to specific, 'hard coded' solutions. Taking care of such unusual conditions with a system that can 'reason' its way to a solution would be an important addition to the technology and to the productivity and efficiency of the plants that use it.

In this chapter we examine the ways in which current AGV material-handling technology can be extended. Although we draw from our experiences at Chrysler's Acustar plant in Huntsville, Alabama, we believe that our approach is sufficiently general that it can be applied in many contexts.

### 5.1.3 Goal of current research

The goal of our research is to demonstrate the value of integrating different theoretical approaches in a practical context to address the needs of AGV material handling technology. Both artificial intelligence (AI) and operations research (OR) techniques are employed to control the AGVs so that the control strategy is dynamic and dependent on the current shop floor situation. The key mechanism is selective attention: the ability to direct the reasoning/planning focus intelligently in order to respond dynamically to changing events in monitoring the plant (Interrante, 1991a; Rochowiak and Interrante, 1991).

Each type of analysis (AI and OR) has much to offer. The careful identification of the appropriate places at which each should be used should prove fruitful, where the notion of 'appropriateness' is highly informed by practical concerns, including:

- experience in manufacturing which generates the particular problems that need to be addressed and the general concerns that must be addressed to devise an acceptable practical solution, and
- data which serves as input for both the basic analyses and for computer simulation studies.

It should be noted that simulations are an essential part of this research since it is not cost effective to perform the testing of hypotheses and models directly at the shop floor level.

The following issues are addressed in the course of this research:

1. Sensor-based monitoring of shop floor state. This requires an adequate computer description of the plant and its interacting subsystems, such as material handling, process lines, receiving, shipping, maintenance, etc.
2. Assessment of the current state so that the appropriate AGV control strategy can be chosen. This requires incorporating knowledge of empirical results, determining which AGV strategies for both dispatch (assigning jobs to AGVs or vice versa) and routing (AGV path planning) work well for given shop floor conditions. In addition, the changing of strategies in mid-production must be accomplished efficiently to maintain progress in meeting master production goals.
3. The blending of AI and OR technologies to achieve global production goals. These goals require the effective use of limited resources under constraints. The use of AI techniques alone, while effective for such tasks as constraint propagation and domain description, fails to incorporate the results of years of mathematical development invested in OR techniques for manufacturing. On the other hand, the use of OR techniques alone can be too limiting in assumptions and may fail to be flexible enough to reflect actual shop floor conditions. By incorporating both types of analyses, the strengths of each can overcome inherent weaknesses of the other.
4. An understanding of interacting manufacturing subsystems. It is desired to effectively control the whole system rather than optimize small portions of the system in an uncoordinated fashion. The manufacturing floor is a large system with many interacting components. The material handling subsystem is not an end unto itself. The effect of material handling control decisions on other manufacturing subsystems (e.g. inventory, shipping, etc.) must be modeled and incorporated into the global assessment of the impact on the master production schedule.

### 5.1.4 Approach

Research began at Chrysler by gaining an understanding of their existing material handling control system (developed by Litton Corporation). Data flow diagrams were developed to assure an adequate understanding of information flow for AGV control. A dynamic, stochastic simulation was developed in SIMAN for modeling the AGV subsystem which delivers parts to the process lines. Simulation input distributions were developed based on plant data for time between AGV orders, number of parts ordered, machine breakdowns, time to repair, and backorders. A distributed, agent-based design was developed for reasoning, including selective attention capability (Interrante *et al.*, 1992). Empirical simulation studies of AGV strategies for both dispatching and routing are being conducted, to gain knowledge of the conditions under which strategies perform best. This knowledge is incorporated into the distributed reasoner for use in determining when to change the AGV control strategy.

Our research focuses on the concerns of an operating manufacturing facility. Such a focus provides benefits and difficulties. The benefits are obvious, if one is concerned with the practical problems of manufacturing. The operating plant provides data and information that otherwise can only be conjured by the researcher. The researcher's conjuring can have the effect of missing significant practical and operational issues, while emphasizing issues that are of little practical importance. However, simply addressing research to a particular plant often means that the solutions and innovations derived from the research are valid only for that plant or for very similar plants. Further, a more theoretical approach can provide a source of new ideas, solutions and technologies that might not otherwise be examined within the demanding, practical world of the shop floor. Thus, the judicious integration of information from the shop floor and ideas from theory provide an environment that is both innovative and grounded. To achieve viable plant automation, manufacturing research must include theory applied to an adequate plant description which incorporates the results from empirical studies such as simulation. Computer-controlled manufacturing must be based on a model generated from such an approach to be successful.

Given a general approach that attempts to integrate theory and practice, we present several components of our research in this chapter. First is an overview of the basic ways in which AGVs operate drawn from published literature and from information provided by various AGV vendors (not specific to any particular vendor). Second we examine the concerns of the shop floor at our target facility. The concerns are expressed in a general way so as to gain wider applicability for the analysis techniques and models that are generated and to protect proprietary information. These components provide a general set of constraints for the research. In the third section we present the way in which the basic plant data was analyzed. The fourth section presents the development of the simulation. The fifth component is the investigation and development of a model that incorporates OR and AI in an effort to address the issues identified through the examination of AGV strategies and the shop floor. Our results and suggestions conclude this chapter.

## 5.2 AGV STRATEGIES

In this section we examine strategies from the published literature and those that were gleaned from AGV vendors. The first group of strategies is for AGV dispatching: the assigning of jobs to vehicles or vehicles to jobs. The second group of strategies is for AGV routing: path planning.

There are two types of AGV dispatch strategies. The first type is vehicle-initiated dispatching, where a task is selected from a set of requests for AGV service. Such strategies include: earliest pickup (EP), extracted rule erlang (ERE), modified first come first serve (MFCFS), rule, shortest travel time/distance rule, and maximum outgoing queue size rule. The first strategy, EP, minimizes pickup time by calculating when an AGV can pick up a part at a particular station (Wilson, Hodgson, and King, 1991). ERE is a heuristic rule that predicts future demands utilizing the tracked stages of arrivals at each station, and positions the AGVs accordingly (Wilson, Hodgson, and King, 1991). The third strategy, MFCFS, attempts to sequentially assign vehicles to jobs in chronological order as requests are received from the stands (Egbelu and Tanchoco, 1984). The goal of rule is to maximize job throughput (Hodgson, King, and Monteith, 1987). The shortest travel time/distance rule dispatches the vehicle to the work center that is closest to the vehicle by time or distance (Egbelu and Tanchoco, 1984). The final strategy, maximum outgoing queue size rule, dispatches a vehicle to the work center that has the maximum outgoing queue size (Egbelu and Tanchoco, 1984).

The second type of AGV dispatch strategy is workstation-initiated dispatching, where a vehicle is selected from a set of idle vehicles to be assigned the next task in the queue of tasks. Such strategies are chosen based on the AGV system type in use. For example, some strategies work better on uni-directional tracks than on bi-directional tracks. The location of pickup and delivery (P/D) stands does not matter in some dispatch strategies, whereas others incorporate the notion of distance. The workstation-initiated dispatch strategies seen most often in manufacturing are vehicle look for work (VLFW), first in first out (FIFO), nearest vehicle (NV), farthest vehicle (FV), priorities, and combinations thereof.

Newton (Newton, 1985) defines VLFW as the strategy where a vehicle consistently makes a pickup close to its last deposit point, and FIFO as the strategy where the vehicle always serves the oldest move request first. He also defines a hybrid FIFO/VLFW strategy referred to as critical request. A critical request is a request that has waited a certain amount of time for services, and then is assigned to be the next job to be serviced by the first available AGV.

Vehicle look for work is used primarily when the floor is divided into zones with uni-directional tracks. Zones can be defined as lines, sections of the floor, or loops. Some AGV systems require that only a defined number of AGVs can travel in a zone at one time, while others require that an AGV work exclusively in one zone. Bartholdi and Platzman (1989) expanded VLFW and Mahadevan and Narendran's work (Mahadevan and Narendran, 1990) with loops, and assigned a vehicle to a specific loop to look for work. The AGV, using the VLFW strategy, travels the zone looking for a job. The AGV will process the first job that it comes to, and the destination of the job becomes the starting point in looking for another job.

One common scheduling strategy used by AGV vendors is a combination of VLFW, zones, and priorities. The VLFW strategy is employed, except that the AGV accepts the job in its zone that has the highest priority. One such AGV system sends a list of all P/D stands in a zone to the AGV that has just entered the zone. The list contains the station numbers listed in decreasing order according to station priority, the critical times for the station, and the job status. The list of pending jobs is examined to determine whether a job has become critical. A job is critical if the preset critical time has expired before the job has been serviced. If no jobs are critical then, starting at the top of the list, the AGV looks for work. The AGV takes the job at the station with the highest priority in its zone. The zone of the destination of the job becomes the AGV's new zone and the process is repeated. If a job becomes critical and no AGV is in its zone, the AGVs in the surrounding zone must report their job status and location. A vehicle with appropriate status and location is immediately assigned to the job. If a vehicle cannot find a job in its zone, it proceeds to the next zone and looks for work. If no work is found, the AGV utilizes this time for a maintenance check or battery charge.

Another strategy used by vendors employs the concepts of priorities and age rates. Age rates are used to increase the stand's priority every set interval of time during which the stand is not serviced. The priority increase amount is not the same for every station; it varies depending on the importance of the stand and its product. The age rate can be zero. This strategy ensures that the most important jobs are handled first, and that all jobs are eventually serviced. This strategy is used when vehicles are not assigned to specific zones. Vehicles are free to respond to the next call according to the strategy rules, regardless of the current location of the vehicle.

Nearest vehicle and farthest vehicle strategies are defined by Egbelu and Tanchoco (1984). In the systems that utilize the NV and FV strategies the entire shop floor is usually referred to as a single zone. Jobs are handled most often in a FIFO manner by the vehicle that is nearest to the station. If this choice of vehicle will result in blockage, then the farthest vehicle, calculated by distance, is chosen for the job. The NV strategy can be used in combination with the concept of stand priorities. The nearest vehicle strategy can be modified to refer to nearest in travel time rather than distance. Since the nearest vehicle according to distance is not always the vehicle that can service the job most quickly, this strategy is sometimes preferred over NV distance.

Several strategies exist for AGV routing. Beasley (1984) devised a strategy in which two heuristic solution methods were used to address the fixed routes problem with daily repetition:

1. An adapted savings algorithm, the Clarke and Wright algorithm, is used to construct a route based upon a savings measure. If the route produces an unfeasible link, it is rejected, else it is added to the solution set.
2. An adapted r-optimal algorithm is employed which behaves in a similar manner to the adapted savings algorithm, except that it examines a set of routes, rather than single links, and checks for feasibility.

Kaspi and Tanchoco (1990) devised flexible flow lines in which vehicles follow 'virtual' flow paths. The direction of the flow paths must be able to change as the need arises. A branch-and-bound technique is employed that utilizes depth-first search and backtracking.

Sinriech and Tanchoco (1991) developed an intersection graph method for solving the AGV flow path optimization model developed by Kaspi and Tanchoco (1990). The objective of the model is to direct the arcs in an unidirected graph with intersection nodes which represents the flow path network so that the total vehicle travel distance is minimized.

Kim and Tanchoco (1991) worked on the conflict-free shortest-time AGV routing problem: to find a path for an AGV which allows the vehicle to arrive at the destination as early as possible without disrupting other active travel schedules. The proposed algorithm maintains a list of free time windows or time intervals at each node and routes the vehicle through the free time windows.

Desrochers *et al.* (1987) defined the vehicle routing problem with time windows (VRPTW), in which a number of vehicles located at a single depot must serve a number of geographically-dispersed customers. Dynamic programming and branch-and-bound techniques are used to minimize the total travel costs.

Egbelu (1987) solved the problem of selecting the home positions of vehicles in a single loop-type AGV network using heuristic algorithms and mathematical representations. The objective is to minimize the vehicle dead travel time (during which the AGV is empty).

Desrochers *et al.* (1992) viewed the VRPTW as a generalized vehicle routing problem where the service of a customer can begin within the time window or interval. The linear programming (LP) relaxation of the set partitioning formulation of the VRPTW is solved by column generation. Feasible columns are added as needed by solving a shortest path problem with time windows and capacity constraints using dynamic programming. The LP solution uses a branch-and-bound algorithm to solve the integer set partitioning formulation.

Fisher and Jaikumar (1981) developed a heuristic to determine which of the demands are satisfied by each vehicle and what route each vehicle would follow in servicing its assigned demand in order to minimize total delivery cost. The heuristic technique assigns a vehicle to a customer by solving a generalized assignment problem with an objective function that approximates delivery cost.

Goetschalckx and Jacobs-Blecha (1989) were concerned with the vehicle routing problem with backhauls, a pickup and delivery problem where all deliveries on each route must be made before any pickups can be made. A two-phased solution was proposed. In the first phase, a high-quality initial feasible solution is generated based on spacefilling curves. In the second phase, this solution is improved based on optimization of the subproblems identified in a mathematical model of the problem.

Shop goals and the shop floor itself are less than clearly understood when taken in their specificity. We can make many notions 'clear enough' to work with at a sufficient level of abstraction. That level of abstraction is often needed to perform effective research. Such research is, however, unfruitful when it is not informed by actual plant conditions.

## 5.3 THE PLANT FLOOR

### 5.3.1 Description of Acustar plant

Chrysler Acustar's Huntsville Electronics Division is located in a 131.5 hectare industrial park complex near the airport in Huntsville, Alabama. The complex is in a free trade zone and it is adjacent to a rail-truck-air intermodal transportation facility. The Acustar facility is located on 50.6 hectares. It is an integrated complex consisting of an 11 148 square meter research and development building, a 9290 square meter administration building, and a 51 373.7 square meter manufacturing building.

The factory is 202.7 meters wide and 243.8 meters long, and is designed for flexibility and expandability. The plant is a highly automated and productive manufacturing operation utilizing the latest state-of-application printed circuit board insertion equipment and state-of-the-art assembly, test, data collection, and inventory control techniques.

Major products are electronic controller systems, entertainment systems, and electronic and electro-mechanical instrument panels for the automotive industry. Raw materials are received and processed through the factory using JIT/zero defects principles and philosophies. Receiving docks and the distribution system are located at one end of the factory and the finished goods storage and shipping docks are located at the other end. Most of the products are assembled and shipped daily, reducing the need for extensive finished goods storage.

There are 15 assembly lines with a standard length of 182.9 meters and a standard width of 7.6 meters. The production lines are standardized to provide the advantage of a clean, clear and streamlined operation. This layout renders production easily visible and manageable, and provides ultimate flexibility for future line re-layout. Each production line is relatively independent of other lines such that alterations can be performed without creating a domino effect.

A state-of-the-art material handling system utilizes:

- an automatic storage and retrieval system (ASRS) for storage and picking (retrieving) of small, commonly-used electronic parts;
- radio-controlled fork trucks for storing and picking full pallets of bulk items such as housings from an east and west static rack area;
- an automated guided vehicle (AGV) system (18 vehicles) for distributing raw materials from the distribution center to points of use;
- an AGV system (nine vehicles) for supplying the production lines with packaging material and moving finished product from the lines to the shipping department;
- radio-controlled, manually operated fork trucks for pulling finished product from inventory and delivering to the designated shipping dock.

Delivery of all parts requires standard static protective returnable containers because of the nature of the products and the automation of the material handling system. All parts must meet bar code labeling requirements which allow for efficient inventory control and data collection, part lot traceability, and quality control intervention when required.

The automated material handling system is a demand pull system. Bar coded point-of-use storage locations are established on each line. Production material handlers monitor the locations to determine when additional material is required by the line. When the line material position reaches a pre-established level, a replenishment order is generated. The order is made by entering it at one of the line's fixed terminals or by using a portable bar code reading data collection device. Some of the high volume lines are also equipped with automatic ordering stands for bulk items. Sensors are used to determine when material is used and a new order is automatically generated. If the portable device is used, the material handler can monitor a wide area of the line and store several replenishment requests before uploading to the system control computer. The system control computer prioritizes orders and automatically generates picks (withdrawals from the distribution center). Material is currently picked based on first in first out (FIFO) routines.

When material is picked at the Miniload ASRS, the operator interfaces with the system by appropriate bar code scanning operations and then places the material onto a distribution conveyor. The material then flows to an appropriate AGV pickup stand where an operator scans the bar code to inform the computer system that the appropriate material has been received. The computer automatically combines all available orders going to the same stand and monitors to assure the arrival of all the ordered material. When it is confirmed that all the material has arrived at the stand, the computer determines whether there are any available dunnage (trash, empty boxes/containers, and/or unused parts) pickups for the AGV's return trip. If available, the computer combines the orders and downloads them to the next available AGV based on a priority system. Through an infrared communication system, the computer can also download orders to empty AGVs on the shop floor.

Bulk items from the rack areas are similarly picked. The orders are processed by the computer and downloaded to the radio-equipped fork trucks. The operator picks from the designated location and delivers the material to the designated pickup stand. Once the operator confirms that the material is at the appropriate stand, an AGV is dispatched to deliver the material to the proper location on the shop floor.

The finished goods system works in a similar fashion. When the finished goods pallet is complete with bar coded labels attached, the material is declared ready for inventory. An AGV is dispatched to pick up the completed pallet for delivery to shipping. At the same time that the material is declared ready, the computer generates a replacement order for packaging material for delivery to the line. When the replacement order is ready, it is delivered to a packaging pickup stand and confirmed to be there through a bar code scanning operation. The system dispatches and AGV to deliver the material to the proper location on the production line. The finished goods AGV control system combines pickup (of finished goods) and delivery (of packing materials) orders for more efficient system operation.

Once the AGV delivers the material to shipping, another bar code scan operation buys it in (processes acceptance) to finished goods inventory and designates a storage location. The material is then stored and confirmed to be in the designated location. Material is later picked using a FIFO routine.

The material handling control system is connected to the management information system (MIS) so that data can be exchanged. Through the implementation of this system and other management, process, and inventory control innovations, investment in inventory has been significantly reduced, inventory turns have significantly increased and inventory accuracy is maintained at approximately 99%.

### 5.3.2    View from the battlefield

Although systems in theory seem to be perfect, the reality of the factory floor is a process of constant exceptions and change, including:

1. The supplier cannot deliver on time for any number of reasons: manufacturing process problems, quality problems, the supplier's source has a problem, weather, labor problems, etc.
2. A quality problem occurs on the production line and material is declared unusable, creating instant demand for additional material that may or may not be available.
3. The customer increases his/her requirements, expecting (and demanding) an instant increase in shipments. Conversely, requirements may be lowered.
4. Engineering changes are required in the product and they must be implemented without creating excessive obsolete material.
5. Process lines must be relocated and modified for new product introductions which must be accomplished without affecting the current product or production.

In addition, a reality of the manufacturing plant is that quality must be maintained or everything comes to a standstill!

The bottom line in manufacturing today is flexibility. Any systems developed and installed for today's fast-paced and high-quality manufacturing environment must be flexible in their ability to deal with the exceptions that will occur. One must always anticipate the unexpected and design for flexibility. What is expected to happen is only going to be partially true. It is how well the exceptions are handled that is the true test of a good, efficient system.

## 5.4    EMPIRICAL ISSUES

There is a tradeoff in manufacturing research between the need for real-world approaches and the need to generalize to extend the applicability of results. To ensure that research results are useful, it is desirable to perform the research in a true manufacturing setting as opposed to an idealized world. On the other hand, to the extent that the research is directed to a particular plant, the results may not be useful for any other plant setting. One of the most difficult aspects of manufacturing research is to determine what can be generalized (from research performed in a specific plant environment) in terms of assumptions made, modeling components emphasized, problems identified, and solutions developed.

In our case, we have limited our research to the following:

- automated guided vehicles alone as the method of delivery of parts from the distribution center in receiving to production;

- light assembly-type operations with small parts such as printed circuit board manufacture;
- a production layout consisting of manufacturing process lines;
- a relatively low level of dependency among lines (very few line-to-line AGV transfers of parts or assemblies), interactions consisting primarily of requirements for limited AGVs and the need for the same raw material parts;
- two-way AGV travel allowed for each line, with turnaround points on each line;
- relatively stable long-term demand for finished goods (seasonal shifts in demand are ignored, since we are interested in peak-period production levels);
- orders for parts generated at the point of need from the production line;
- master production schedule requirements which can be met by adding more shifts of operation (e.g. weekend shifts) if needed.

It was necessary to determine how to model the effect of AGV strategies on the global production goals. The material handling system is viewed as one of several interacting manufacturing subsystems, each of which contributes to the global production goals. Other subsystems include receiving, shipping, maintenance, and production lines. Interactions are studied in terms of schedules and schedule interactions. Scheduling is the use of limited resources to accomplish goals under constraints in time. The manufacturing subsystems interact in that they may compete for resources and conflict in goals. To achieve efficient computer control of the manufacturing environment, the subsystem interactions must be understood and modeled such that a change in one subsystem is represented with corresponding effects on other subsystems. With this sort of model, the global production goals may be kept in view and each subsystem's contribution to the goals, in concert with the interactions among other subsystems, can be defined.

Suppose that a part is on backorder. It is known that the backordered parts will arrive within the next fifteen-minute time interval. When the backorder arrives, the parts must be expedited to the shop floor in order to keep production interruptions due to lack of the part to a minimum. The AGV schedule is altered to allow for the expediting process. This schedule change results in less AGV capability to deliver other parts which have been ordered, which in turn slows production on some lines. In our model, the AGV schedule is altered to minimize a delay in the production schedule for some lines, whereas the schedule alteration may cause a slight delay in production for other lines. This action 'balances' or mitigates the effect of the part stockout on the achievement of the master production schedule. The production lines competed for limited AGV resources. The goal of maintaining production for some lines was in conflict with the goal of expediting the backordered parts. Note that a schedule change in a subsystem may cause a ripple effect on other subsystem schedules. Subsystem interactions must be accounted for to achieve realistic computer modeling of manufacturing systems.

We can draw from much former work in both OR and AI analyses for manufacturing. Formal, mathematical models have been developed for manufacturing (e.g. linear and nonlinear programming). Heuristic models have also contributed to manufacturing research (e.g. dispatching rules, branch-and-bound techniques, etc.). From a modeling standpoint, mathematical models and heuristics represent two opposite ends of the spectrum. We seek to develop a base-level, semi-formal approach to plant description as a substrate from which both formal and heuristic analysis techniques can be employed in solving manufacturing problems.

## A Semi-formal approach to plants

### Basic descriptions and definition

1. A plant is a collection of lines.
2. Lines are composed of paths and stations.
3. A station gets materials, modifies materials, produces a product and produces waste.
4. A station may get materials from an upstream source or an external source.
5. A station may modify material either by composing from sources or actually altering a source.
6. The result of a station's modifications is its product.
7. Any material that flows into the station, but is not part of the product is waste.

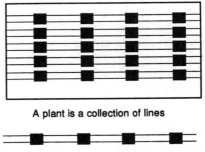

**A plant is a collection of lines**

**A line with four stations**

### Flow of materials

1. A material has a point of origin.
2. Points of origin are either supplies (stores) or stations.
3. Materials flow from the supplies to the final product through stations.
4. Stations produce products and waste.
5. If no materials or products are available to a station the line of which it is a part stops.
6. If waste becomes too great, the station stops.
7. If a station breaks (becomes inoperable) it stops.
8. If the quality of materials or products is unacceptable, then the station stops.
9. The goal of rationalization for the flow of materials is to avoid the stopping of the line.

### Evolving product description

1. A product can be described as a collection of predicates.
2. Each station adds or subtracts a predicate from the description of the material and product it gets.
3. The description of the final product of the line is the set of predicates generated by the stations.

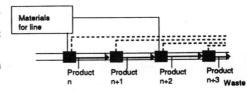

**Product description**

at n: P1(n)

at n+1: P1(n) & P2(n, n+1)

at n+2: P1(n) & P2(n, n+1) & P3(n+1, n+2)

at n+3: P1(n) & P2(n, n+1) & P3(n+1, n+2) & P4(n+2, n+3)

**Fig. 5.1**  A semi-formal approach to plants.

Figure 5.1 represents such an approach to plant description, from the standpoint of material handling analysis. The plant is viewed as a collection of lines with paths and stations. This description provides the computer with a way to express what can and cannot occur in the manufacturing arena with regard to AGV behavior. For example, stations get materials, modify them, and produce products and waste.

Flow of materials through stations is described. The goal of rationalization for the flow of materials is to avoid the stopping of the line (from lack of materials). The product is also described semi-formally as a collection of predicates. As stations modify the product, predicates are added or subtracted from the product description.

This type of description provides the reasoner with a foundation for common-sense reasoning about plant operations. Goals, resources, and constraints can be added to the model in terms of baseline descriptions. The reasoner represents cause-and-effect interactions in terms of changes expressed in the description language. The description language can generate inputs to mathematical models, and the outputs can be incorporated into the reasoner's decision-making process.

Empirical studies must be performed in order to tie research developments and results to practical manufacturing problems. Typically, it is reasonable to begin with one or several hypotheses concerning the manufacturing system. Data is collected and analyzed in order to gain more information concerning the hypotheses. In a laboratory setting it is difficult to collect enough data points to carry out such experiments. In a plant setting, the problem is just the opposite: determining what data from the vast amount of available information is worth analyzing. Such analysis is extremely time consuming, and without an appropriate plan for experimentation, much valuable research time can be wasted without producing any reasonable conclusions. The advantage that we had at Chrysler Acustar was that a well-functioning computer-controlled ASRS/AGV system was already in place. Our task was to render a current computer-controlled AGV system more effective by dynamically adjusting strategies to meet shop floor contingencies. We are able to 'tap into' the existing system to obtain a myriad of data points. Examples of the available data include: number of AGV trips per P/D stand, mean time to arrive, AGV down time for maintenance, etc.

The following hypotheses represent a sample of those which were of interest:

1. Delays in receiving schedules require offsetting changes in material handling and production schedules in order to minimize effects on global production goals.
2. The pattern of delivery points for pending part movement orders greatly affects the usefulness of particular routing strategies. Only one AGV at a time can pass through a track intersection. Intersection rules should be tied to particular routing strategies.

3. Priorities in the master production schedule for finished goods greatly affects the usefulness of particular dispatching strategies; choice of strategy should incorporate both a local component (relative to the collection of orders) and a global component (relative to the priorities for finished goods).
4. Strategies should not be changed too frequently. Chaos will result. There are some 'allowable' transitions from one strategy to another and some transitions which are not practical.
5. Characteristics of the current shop floor state can be used as indicators that a strategy change should occur.
6. A measure of prediction of needs for the near future can be obtained by an examination of the master production schedule and other shop floor indicators. Some events cannot be predicted. An appropriate balance between the predictable and the unpredictable can be incorporated into the AGV reasoner.
7. The effect of AGV strategies on global production goals can be determined through modeling subsystem schedule interactions; a purely mathematical model is not adequate for such a determination.
8. Some combinations of dispatch and routing strategies are reasonable; others are not.

To gather data concerning these and additional hypotheses, the following systems study was conducted (Romero, N.R., 1993). The listing below represents a subset of the systems study. Space does not permit a complete description.

1. Simulation goal: given a production schedule and associated plant scenario, determine the best AGV strategy. The strategy consists of both dispatch (assigning AGVs to jobs or vice versa) and routing (path planning) techniques.
2. State variables
   (a) AGV: status (idle, busy; on shop floor, busy; maintenance);
   (b) distribution center: AGV queue length, order pickup time, order dropoff time, etc.;
   (c) shop floor: request made for AGV job, P/D stand status (empty or loaded), production line status (running or down).
3. Random variables
   (a) time between order creation (delivery of parts of shop floor, return of trash/containers/spare parts, also known as 'dunnage', to distribution center);
   (b) order amount;
   (c) order pickup time (distribution center, P/D stands);
   (d) time between AGV breakdowns;
   (e) AGV repair time;
   (f) time between line breakdowns;
   (g) line repair time;

(h) time between part backorders;

(i) time to fill backorder.

4. Measures of effectiveness

   (a) AGV utilization;

   (b) throughput of finished goods;

   (c) average AGV idle time;

   (d) average AGV blockage (from traffic-hold) time;

   (e) maximum AGV waiting time in queue;

   (f) average time to process order.

The simulation was designed based on the systems study.

## 5.5  SIMULATION ISSUES

### 5.5.1  Modeling decisions

Simulation studies were employed to conduct the empirical analysis. The results were used to confirm, refute, or gain information concerning the hypotheses mentioned in the previous section. Once again, the sheer volume of data was overwhelming. Several decisions had to be made:

1. What should be simulated? Obviously, not every aspect of a large plant can be modeled in a single simulation. Typically, simulation studies focus on a particular manufacturing subsystem, such as a single production line or the material handling system. Our requirements of understanding the global effects of AGV strategy changes created a modeling problem. We needed to have the ability to determine how AGV strategy changes affected line throughput (and thus, the master production schedule goals). It was determined to simulate the ordering of parts from production lines, the movement of parts from the DC area to lines, a limited amount of machine breakdown information, and the completion of finished goods at the end of lines. The ASRS area and shipping areas were not considered in the initial simulation studies.

2. How should it be simulated? Several difficulties arose in this respect. Most of the difficulties arose because of the volume of information required for the simulation. The 15 production lines contain 296 AGV pickup and delivery (P/D) stands. Routing and dispatch strategies both depend to a certain extent on the spatial position of the stands from which part delivery orders are generated. AutoMod (Auto-Simulation, Inc.) was chosen as the simulation tool because of the ability to import CAD files. This saved much time in the placement of P/D stands for simulation development. It also caused a problem, however. The memory requirements for loading in the CAD file were high (over 16 RAM) on the SUN machine.

3. What is the appropriate level of granularity? There is a tradeoff here between achieving a low enough level of detail to obtain practical results on the one hand, and remaining at a high enough level of detail to assure adequate memory for running the simulations on the other hand. The issue of the dependency of results on a single plant model must be addressed as well. Three model decisions are representative of the tradeoff:

(a) It was necessary to measure line throughput in order to determine an AGV strategy's effect on the achievement of the master production schedule. Typically, line throughput measurement is no problem in a detailed line simulation, with setup times and machine processing times. For our purposes, however, this would require too much detailed information for modeling all 15 lines in addition to the AGV system. We obtained data concerning the number of AGV trips to particular stands on each line, as well as throughput data for the line. Empirical functions were designed from this data which express throughput as a multiple linear regression function of AGV trips to particular P/D stands on the line.

(b) Initially, the number of individual sources of randomness in the simulation which the memory requirements could handle was unknown. As a step in the modeling process, it was decided to reduce the number of attributes for individual P/D stand entities to one: spatial position. This attribute was a must for our analysis purposes. Orders were generated for a particular P/D stand from a line by creating a 'general' order distribution for each line for time between orders. A second (initially uniform) distribution was used to determine which P/D stand on the line the order originated from. This reduced the number of distributions from which sampling was required for the order process from 296 to 30.

(c) It was decided not to model individual part numbers in the simulation. The Acustar plant contains thousands of different parts, each of which would have to be tracked separately for the simulation. Backorders for parts are randomly generated and applied to an arbitrarily-chosen order for AGV delivery to a particular stand on a line. This decision is in keeping with that of not representing each production line at a detailed level (i.e. machining times and setup times for machines). The interest is in AGV trips to the line, not necessarily in what is being delivered at a low level of detail.

In each case above, the model decision resulted in a loss of information and a degradation of the accuracy of simulation results, from the standpoint of the particular plant being modeled. On the other hand, the results may be useful and they can be generalized to similar plants. Such decisions must be carefully considered; they represent the real expertise in setting up the experimental approach.

### 5.5.2 Simulation experimentation

A series of simulation experiments was planned in order to test the effect of various combinations of AGV dispatch and routing strategies on typical shop floor scenarios (Romero and Romero, 1993). The experimental results are used to tie particular strategy combinations to particular shop floor conditions. This knowledge is incorporated into the intelligent AGV controller (Romero, I.S., 1993).

Figure 5.2 depicts each of the 39 alternatives which are considered in the simulation. Each alternative represents a combination of dispatch and routing strategies employed in the experiment. For each of the alternatives, a factorial design experiment is performed. The following six factors are incorporated:

1. volume of orders generated from production line stations (randomly dispersed): high or normal;
2. volume of orders generated in a specified area (creation of congestion in a randomly-selected portion of the plant): high or normal;
3. volume of backordered parts (randomly-selected jobs): high or low;
4. line downtime (randomly-selected lines): high or low;
5. amount of 'expedite' requirements for finished goods (randomly-selected lines): high or low;
6. AGV failure to deliver: high or low.

A factorial design (Law and Kelton, 1991) was developed to allow one factor at a time to vary to the factor level which 'stressed' the manufacturing system. (For example, factor 1 was set to 'high' while all other factors were set to 'low' or 'normal'.) Three responses were

| Dispatch Strategies | Routing Strategies | | | |
|---|---|---|---|---|
| | Shortest Path | Generalized Assignment | Min-Max | Optimization by Column Generation |
| First In - First Out | • | • | • | • |
| Least Utilized Vehicle | • | • | • | • |
| Longest Idle Vehicle | • | • | • | • |
| Nearest Vehicle | • | • | • | • |
| Random Vehicle | • | • | • | • |
| Earliest Pickup | • | • | • | • |
| Extracted Rule [Erlang] | • | • | • | • |
| Random Work Station | • | • | • | • |
| Shortest Travel Time/Distance | • | • | • | • |
| Vehicle Looks for Work | • | • | • | |

**Fig. 5.2** Strategy alternatives.

considered: AGV utilization, average time between order creation and order completion, and throughput. Comparison of alternatives with this design allows for the determination of which AGV strategies perform best under certain plant conditions. Multiple, simultaneous plant conditions were not considered at this stage because of the added complexity to the experimental design.

## 5.6   MODELS, GENERALITY, AND AI

### 5.6.1   Overview

The models we are building in this phase of the research are distinct from the simulation models, which emphasized the accurate representation of specific plant operations. In this section we concentrate on the reasoning about the plant and the simulation models. There are two kinds of reasoning. The first is reasoning tightly bound to the Acustar plant. This reasoning is gathered both from persons involved in scheduling and the empirical simulation results. The second sort of reasoning is more general and concerns a variety of issues that are applicable to the Acustar plant and other plants as well. A clear example concerns the selection and use of dispatching and routing strategies that are not available at the facility. However, it should be noted that the reasoning about these strategies is tied to the plant in the sense that knowledge of strategy application is based on empirical simulation results.

### 5.6.2   The domain and the general problem

As noted above, some events on the shop floor create manufacturing problems requiring a great deal of reasoning for resolution.

1. A machine breaks down unexpectedly on a critical line. It is desired to use the material handling subsystem in a nonstandard way to aid in overcoming production delays caused by the machine downtime. By focusing attention on this problem and identifying available AGV resources when the breakdown occurs, work-in-progress inventory (WIP) can be routed to a similar machine on another line with less strenuous production requirements and routed back to the critical line for continued processing. This is an example of the use of the material-handling subsystem to aid in achieving the goals of a particular production line in order to meet the more global goals of the master production schedule.
2. A supplier notifies production management that the delivery of parts to the plant will be delayed. When the backordered parts

arrive at the plant, a peak demand for AGV deliveries of the part to the line will occur. By focusing attention on this problem before the parts arrive, AGV deliveries to other lines can be performed 'early', allowing more AGV resources to be used to expedite backordered part delivery. This planning represents a decision to temporarily violate JIT principles in order to achieve the goals of the master production schedule for critical production lines.

3. Less dramatically, there are shop floor conditions that call for the selection of a dispatch or routing strategy other than the 'default' strategy. This may be necessary when certain areas of the shop floor are congested, there is a high level of demand placed on the AGV system, or there is some obstructed path. Additionally, there may be reasons to depart from the default strategy when the P/D stands are in a particular configuration, or when a certain assembly is being produced.

The monitoring of any physical system is knowledge-intensive and time-dependent. Incoming low-level data must be filtered, interpreted, and abstracted for reasoning purposes. During this process, much of the data is converted into a higher-level, qualitative representation. Given these representations, higher-level models encode knowledge of system interdependencies, and enable the reasoner to recognize significant combinations of data values occurring concurrently and to plan accordingly. A reasoner must be able to recognize a significant change on the shop floor, and correspondingly change the AGV strategy. The foregoing cases suggest the need for a selective attention mechanism. This mechanism provides the resources for focusing on a current shop floor problem in a current context while allowing the rest of the system to proceed as usual.

Achievement of the global goals of manufacturing requires an understanding of interacting schedules and tradeoffs. Past manufacturing research has produced many analytic techniques for generating *individual* schedules. Typically, however, interactions with other schedules (with the exception of the master production schedule) are handled by making assumptions. The assumptions can be very limiting, rèsulting in an unrealistic schedule. Figure 5.3 depicts various schedules in manufacturing with their bounding assumptions.

Figure 5.3 can be understood in terms of the reasoning of a collection of agents. Agents embody knowledge and have resources for communication and negotiation. The latter characteristic separates them from a traditional knowledge-based system. Each agent has its own information and knowledge which guides its actions. Each agent informs other agents of its results or requests resources from other agents.

For any particular agent, the sensors, databases, and actions of other

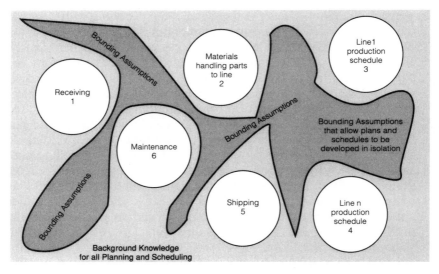

**Fig. 5.3** Various schedules in manufacturing.

agents constitute the external world. The agent's internal world is constituted by the knowledge and understanding that it has of its particular task. The spectrum of organizational models for agents ranges from strict hierarchies (agent results reported to a higher level where authority exists to make decisions) to planes (agent results reported to all other agents; all have equal authority). Intelligent focus of attention attempts to avoid the excesses of these poles. Unlike the focus of attention in a hierarchical system, agents do take account of the rest of the world. Unlike the focus of attention in planar systems, agents are constrained in the actions that they may take. In brief, intelligent focus of attention requires that agents be aware of their own resources, goals and constraints in the effort to solve a problem that potentially affects the entire collection of agents.

A scheduler agent views the plant as a closed, idealized world, in which cause-and-effect analysis across organizational units is ignored. Bounding assumptions keep the plans and schedules in relative isolation, and embody assumptions about the abilities of neighboring subsystems. The application of rigorous analytical OR techniques is normally tightly constrained by bounding assumptions. Problems occur when the boundaries are violated. Further, collections of agents (or schedulers) working on the same global goals may be unable to achieve local goals, or the converse.

Agent conflicts are produced from the closed-world assumption that any knowledge not explicitly represented in the agent nor inferable from such knowledge is irrelevant to the domain. The closed world assumption can be mitigated by providing higher-order representations and reasoning to moderate conflicts among lower-level agents. The closed world is opened to some degree by allowing more heuristic styles of reasoning to moderate conflicts and the violations of the boundaries. Heuristic styles of reasoning do not provide the analytical detail of the OR techniques, but encompass a greater range of events. The inclusion of heuristic techniques is also subject to the closed world assumption, but in combination with the OR techniques a 'larger' closed world is created.

Given incoming information from the shop floor, an intelligent material handling system may choose to change dispatching or routing strategy. Strategies for scheduling are contained in agents and two types of commitment are available. The first type provides an adequate schedule or analysis based on agent knowledge of its assumptions, used to determine validity of the analysis given the current state. The second type secures adequate resources to meet the demands of the schedule. Collectively, agents handle interacting schedules and commit limited resources to accomplishment global and local goals.

Several researchers are examining dynamic scheduling for manufacturing and for other domains. Zweben *et al.* (1992) have performed research in rescheduling with iterative repair for the scheduling of shuttle operations. Xiong, *et al.* (1992) performed work in intelligent backtracking techniques for job shop scheduling. Shaw and Whinston (1989) examined FMS scheduling via a goal-directed inferencing method, in which the primary form of interaction among subproblems was the sharing of machine resources. Pan *et al.* (1992) describe the development of an intelligent material handling system which is integrated with other CIM functions. Wu and Wysk (1990) have developed an inference structure for control and scheduling in manufacturing which provides both forward and goal-directed inference, simulation capabilities, and other modes of analysis. Ringer (1992) employs time phased abstractions for combining predictive and reactive scheduling. Hadavi *et al.* (1992) have developed a recursive architecture for real-time distributed scheduling.

Figure 5.4 depicts one-way communication from the master production schedule agent (MPSA) to other schedulers. The MPSA dictates requirements to lower-level agents, and is bounded by the assumption that supporting schedulers can satisfy its dictates. Figure 5.5 shows interagent communication which allows for cooperation in dynamic scheduling. Changes in particular manufacturing parameter values, or 'cues,' serve as triggers for dynamic rescheduling (Interrante and Rochowiak, 1993). Collectively, the 'cues' for dynamic scheduling comprise an information pool from which agents are informed and by which they communicate. The effect of such communication is to overcome the bounding assumptions and to reduce isolation in schedule

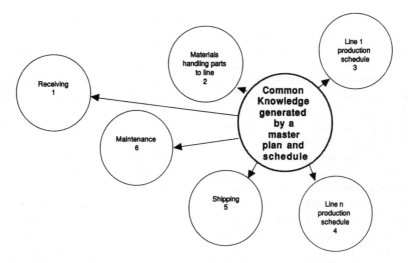

**Fig. 5.4**  One-way communication from the MPSA to other schedulers.

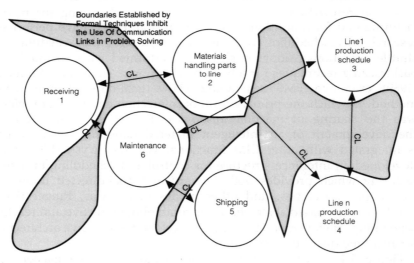

**Fig. 5.5**  Interagent communication.

development. However, there is also a need for selectively attending to the various cues.

### 5.6.3  Selective attention

Intelligent focus of attention assumes that there are sufficiently accurate data, suitable higher-level representations, and mechanisms for converting the lower-level data into higher-level representations. Selective attention is a mechanism for accomplishing intelligent focusing of attention. Agents with selective attention select information about changing events in the external world, focus their reasoning on the effect of that information on their goals, project new courses of action, and evaluate courses of action in terms of a global goal. Selective attention is both reactive and proactive and forms the basis of a tightly integrated feedback loop between the current state of the system and the reasoning processes of the system (Interrante 1991). Other researchers have pursued the related notions of selective perception (Simmons *et al.*, 1992) and active vision (Swain 1991; Ballard 1991).

Selective attention can be applied to the plant when it is represented as a complex, interacting set of subsystems. For example, an increase in MPSA requirements for a particular product causes a demand for increased throughput on a particular line or lines. This in turn requires heavier utilization of AGVs to deliver raw materials. Suppliers must increase the amount or rate of shipment of the raw materials. Such relationships are well understood by those with manufacturing experience, and some of them have been captured in computer models (e.g. simulation models). What is the impact of the increased need for AGVs on other manufacturing processes? Since we view resources as being committed to particular activities during the planning and scheduling process, when a change requires a corresponding change in planned resource usage, the model must reflect the fact that limited resources such as AGVs are no longer available to meet previous commitments. The cause-effect relationships which define the commitment and release of resources and the corresponding effect on global manufacturing goals must be understood and modeled. The degree to which changes in commitment propagate to other subsystems (degree of localization of change) must be determined.

Table 5.1 depicts typical schedules, resources, constraints, and goals. Figure 5.6 shows a simplified view of schedules with accompanying resources and schedule interactions. MPSA is the main driver for other schedules.

At any given moment, a particular situation on the shop floor may critically affect the achievement of production goals. Scheduling efforts and available (limited) resources should be directed to such critical focal points, with a minimum impact on other operations. As conditions on the shop floor change, the focus areas change. Further, the use of selective attention allows the system to adapt control strategies to the introduction of new products and to quickly adapt to real-time information in order to support just-in-time manufacture.

### 5.6.4 General issues of knowledge representation

A knowledge representation technique or scheme is a collection of formal conventions (syntax) and interpretations (semantics). These are

**Table 5.1**  Typical schedules, resources, constraints and goals

| scheduling | resources | goals | constraints |
|---|---|---|---|
| Receiving | Trained | Min. scrap and | Policies, rules – |
| Shipping | maintenance | rework | company, |
| Scheduling jobs | people and | Max. throughput | union, laws, |
| on machines | operations | Control inventory | etc., OSHA, |
| and shifts | support (SW) | (accuracy) | free trade zone |
| Materials handling | People – | MH load leveling | Absenteeism |
| scheduling to | motivated | (part orders) | Training |
| machines to | Parts – raw | Meet production | AGV track |
| FG – among | materials – | goals (min. | Limitations on |
| machines | subassemblies | stockouts) | resources |
| Layout changes/ | Machines | Min. personnel | |
| capital | MH equipment – | Constant | |
| improvements | (totes), trucks, | production rate | |
| Maintenance | conveyors | Min. backorders | |
| Plant shut-downs | cranes, | Min. inventories – | |
| Inventory counts | computers, | WIP & FC | |
| Personnel – shift | broad-band, | Min. power, | |
| changes, breaks, | etc. | space, etc. | |
| etc. | Finished goods | Max. utilization of | |
| Master production | Utilities | non- | |
| schedule | Space | consumable | |
| Process planning | Packing materials | resources | |
| (high level) | | Min. downtime | |
| | | (on line or MH | |
| | | system) | |
| | | Min. line changes | |
| | | Max. profit | |

basic parts of any knowledge-based system where there is a knowledge base that holds the conventions and an interpreter that adds, deletes, and compares them. It is often useful to separate the knowledge base into two parts: the atomic part (facts) and the connective part (rules).

The following table is adapted from Tanimoto's summary of knowledge representation schemes (Tanimoto, 1990). In the table the organization column refers to the degree to which the representation scheme allows or prohibits the inference mechanism from accessing some of the knowledge that is represented. For example, in a concept hierarchy the mechanism cannot access all levels of the hierarchy at the same time.

Combinations of the above schemes can represent knowledge at various levels of abstraction. For example, knowledge at one level can be represented in a collection of rules, and at another level the knowledge in various collections of rules can be represented in a frame-like structure.

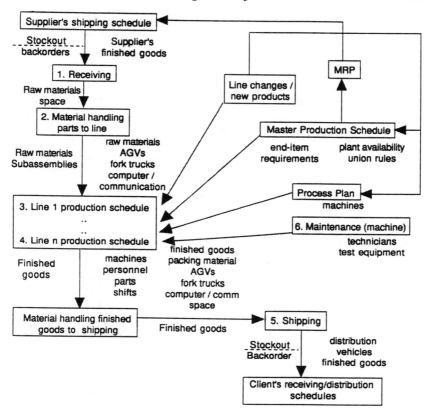

**Fig. 5.6** A simplified view of schedules.

A modified rule representation is used as one portion of our knowledge representation scheme. This representation has the virtue of being well-understood and it is easily implemented. The basic structure of the rule is: If ⟨shop floor condition⟩ then ⟨use strategy SSS on data DDD⟩. At a higher level we have adopted agents as a representational style.

### 5.6.5 Selective attention: inside a computational agent

Selective attention requires the development of future expectations and contingency plans (Fig. 5.7). Initially, a complete reading (snapshot) of shop floor sensor data is provided to the agent. Domain-dependent knowledge is used to filter and abstract the data into a form useful for problem-solving. Abstracted data is input to a high-level domain model which includes specialized mini-planner agents (e.g. AGV strategy agents).

**Table 5.2**  Summary of knowledge representation schemes (from Tanimoto)

| scheme | relations | inference mechanism | organization | comments |
|---|---|---|---|---|
| Propositional Logic | Boolean truth values | Modus ponens and others; resolution. | Weak | Does not model the content of the statements. Can be made to work very fast. |
| Concept Hierarchy | IS-A | Graph search and related techniques. | Strong | There is only one relation in the hierarchy. New relations such as HAS-A might be added. |
| Predicate Logic | Any predicate | Much the same as propositional logic. | Weak | Needs unification and Skolemization to be added. Difficult to organize knowledge into chunks. It can be very expressive and awkward to control. May need a better presentational and analytic set of constructs. |
| Production Rules | If __ then __ | Rule activation, agenda manipulation. | Weak | Awkward for non-procedural, declarative knowledge. Can support different directions in reasoning. Might be used in conjuction with other schemes. |
| Relational Databases | n-ary | Selection, projection, join. | Fair | Awkward for control information. |
| Sematic Nets | Binary or ternary | None, possible to use associational links of various kinds. | Weak to fair for partitoned nets | Highly dependent on the target domain. |
| Constraints | Any predicates | Propagation and satisfaction. | Weak | Highly dependent on the target domain. |

**Table 5.2**  *Continued*

| scheme | relations | inference mechanism | organization | comments |
|---|---|---|---|---|
| Frames | Binary or ternary | None, possible to use associational links of various kinds. | Strong | Could be considered a style rather than a scheme. Can be extended into a weak scheme on which stronger schemes can be based. |

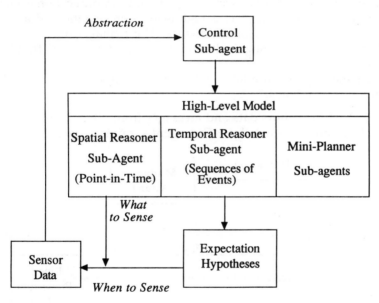

**Fig. 5.7**  Future expectations and contingency plans.

The high-level model contains two specialized reasoners (Interrante, 1991). The temporal and spatial reasoners vary in importance with the domain and reasoning tasks. The spatial reasoner represents a point-in-time spatial analysis of a physical system. This reasoner is separated from the temporal reasoner since time and space cannot be analyzed simultaneously. The agent plans with temporal events and 'switches' to the spatial reasoner for a particular point-in-time when necessary.

The selective attention mechanism develops future expectations on a time line and identifies the data needed to verify the expectations. The controller uses this knowledge to initiate active sensing to confirm or refute expectation hypotheses before contingency plans are developed and executed based on expectations. (Interrante, 1991) The selective attention model draws upon knowledge of common or critical situations for the identification of critical data. Each situation represents a system state which must be recognized by the controller when it occurs. Control action must be taken by the agent when such a state occurs. (Interrante, 1991; Interrante and Rochowiak, 1991)

### 5.6.6   Selective attention: a collection of computational agents

Selective attention also applies to a collection of agents as in Fig. 5.8. These agents share a common master production schedule and there is an agent for each manufacturing subsystem. Communication is enabled through a manager that has a high-level representation of all of the agents and their special characteristics. The manager agent is capable of higher-level reasoning including the identification of trends and tradeoffs.

The general procedure for a collection of agents that exhibit selective attention is as follows.

An agent **monitors** data and messages from other agents according to its internal knowledge. The agent's primary **goal** is given by

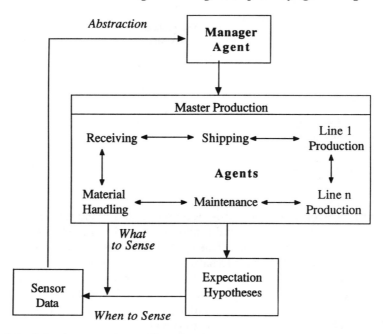

**Fig. 5.8**   Selective attention: collection of agents.

the MPSA. The agent first attempts to satisfy the goal in terms of classical OR techniques. Otherwise, heuristic techniques are used. Upon success, the agent notifies the manager. Otherwise, the agent notifies the human agent and waits for further instructions. An agent may **detect** one of two types of problems. Either the agent cannot satisfy its goal, or the agent receives a request that breaks one of its bounding assumptions.

**Suppose** the agent is **failing** to meet its goal. The agent attempts to fix the problem. If it cannot fix the problem, the agent searches for a solution that might affect other agents. If it finds such a solution, the agent locks itself. Locking means that it cannot engage in any other modifications until this problem is solved. The agent directs the manager to notify the affected agents and report if they can accept the change. The agent continues to poll the data stream. If the problem disappears, it directs the manager to tell the affected agents to discontinue work on the problem. If the manager agent indicates that all of the affected agents agree, then the new schedule is promulgated and the master is notified. If no solution is found in a set period of time, all of the relevant human agents are notified and the system waits.

If there are no pending requests the manager processes the current requests. If there are other requests, the manager determines if any of them can be satisfied while the current request is being processed. If there is more than one such request, then a priority system is used until a single request remains. The request is then sent to the relevant agents. When the manager is notified of a result, it passes the result to the appropriate agents.

## 5.7 CONCLUSION

Selective attention is being employed in an agent-based structure to allow a reasoner to more effectively accomplish automation in an AGV material-handling system. Moreover, interactions between the AGV system and other manufacturing subsystems can be modeled via schedule changes, and selective attention can be useful in such modeling. In this way, the effect of dynamic changes in one schedule on the global production goals can be determined.

Benefits of this research include:

1. an increased responsiveness to volatile shop floor conditions;
2. incorporation of both mathematical and heuristic manufacturing analysis techniques within a semi-formal descriptive language and a distributed system of agents;
3. the modeling of schedule interactions to achieve cooperation in dynamic scheduling;

4. agent commitment to the satisfaction of assumptions before scheduling occurs;
5. resource commitment to particular critical situations in time via selective attention;
6. the identification and use of manufacturing parameters as 'cues' to trigger rescheduling analyses;
7. dynamic adjustment of AGV control strategy to satisfy global production goals.

Several important questions must be answered in the course of this research:

1. Can we recognize significant shop floor changes and react accordingly in real time?
2. Can we successfully change AGV strategies in mid-production?
3. Can we model the 'ripple effect' of changes in interacting schedules?

The approach of employing empirical results as knowledge for a reasoner based on AI representations and OR analyses and a semiformal plant description can be exercised in the pursuit of answers to these questions.

## ACKNOWLEDGMENTS

This material is based upon work supported by the National Science Foundation under Grant No. DDM-9113983. The Government has certain rights in this material.

## REFERENCES

Allen, J.F. (1984) Towards a general theory of action and time. *Artificial Intelligence*, **23**, pp. 123–54.
Badiru, A.B. (1988) *Project Management in Manufacturing and High Technology Operations*, John Wiley, New York.
Ballard, D. (1991) Animate vision. *Artificial Intelligence*, **48**, pp. 57–86.
Bartholdi, J.J. and Platzman, L.K. (1989) Decentralized control of automated guided vehicles on a simple loop. *IIE Transactions*, **21**(1), pp. 76–81.
Beasley, J.E. (1984) Fixed Routes. *Operational Research Society*, **35**(4), pp. 49–55.
Bond, A.H. and Gasser, L. (eds) (1988) *Readings in Distributed Artificial Intelligence*, Morgan Kaufmann.
Claassen, A. (1993) Dynamic scheduling.
Cohen, P.H. and Joshi, S.B. (1990) *Advances in integrated product design and manufacturing*. Winter Annual Meeting of the American Society of Mechanical Engineers, Nov. Dallas, Texas
Collins, H.M. (1990) *Artificial Experts: Social Knowledge and Intelligent Machines*, MIT Press, Cambridge, Massachusetts.
Descotte, Y. and Latombe, J. (1985) Making compromises among antagonist constraints in a planner. *Artificial Intelligence*, **27**(2), pp. 183–217.

Desrochers, M., Desrosiers, J. and Solomon, M. (1992) A new optimization algorithm for the vehicle routing problem with time windows. *Operations Research*, **40**(2), Apr.

Desrochers, M., Lenstra, J.K., Savelsbergh, M.W.P. and Soumis, F. (1987) Vehicle routing with time windows: optimization and approximation. *Vehicle Routing: Methods and Studies*.

Egbelu, P.J. (1987) Pull versus push strategy for automated guided vehicle load movement in a batch manufacturing system. *Jnl. of Man Sys*, **6**(3), pp. 359–74.

Egbelu, P.J. and Tanchoco, J.M.A. (1984) Characterization of AGV dispatching rules. *Int Jnl. of Prod. Res.*, **22**(2), pp. 359–74.

Egbelu, P.J. Positioning of Automated Guided Vehicles in a Loop Layout to Improve Response Time.

Ehn, Pelle (1989) *Work-Oriented Design of Computer Artifacts*, Lawrence Erlbaum, Hillsdale, New Jersey.

Fisher, M.L. and Jaikumar, R. (1981) A generalized assignment heuristic for vehicle routing. *Networks*, **11**.

Floyd, C. (1987) Outline of a paradigm change in software engineering, in *Computers and Democracy – a Scandinavian Challenge* (eds G. Bjerknes *et al.*), Avebury, Aldershot, UK.

Goetschalckx, M. and Jacobs-Blecha, C. (1989) The vehicle routing problem with backhauls. *Eur Jnl. of Op Res*, **42**.

Hadavi, K. *et al.* (1992) An architecture for real time distributed scheduling, in *Artificial Intelligence Applications in Manufacturing*, AAAI Press/MIT Press, Menlo Park, California, pp. 215–34.

Halpern, J. (1986) *Theoretical Aspects of Reasoning about Knowledge*, Morgan Kaufmann.

Hirschhorn, L. (1984) *Beyond Mechanization.*, MIT Press, Menlo Park, California.

Hodgson, T.J., King, R.E. and Monteith, S.K. (1987) Developing control rules for an AGVS using Markov decision processes. *Material Flow*, **4**, pp. 85–96.

Hodgson, T.J., King, R.E. and Wilson, C.M. (1991) *Analysis and control of multiple vehicle AGV systems*.

Interrante, L.D. (1991) *A model for selective attention in monitoring and control reasoning tasks*. Proceedings of the 1991 IEEE Conference on Systems, Man, and Cybernetics, Oct. 13–16, Charlottesville, Virginia, pp. 1931–5.

Interrante, L.D. (1991a) *Representing temporal, spatial and causal knowledge for monitoring and control in an intelligent simulation training system*. Proceedings of the AI, Simulation and Planning in High Autonomy Systems Conference, Apr. 1–2, Cocoa Beach, Florida, pp. 263–9.

Interrante, L.D. and Biegel, J.E. (1989) *Reasoning in time and space: issues in interfacing a graphic computer simulation with an expert system for an intelligent simulation training system*. Proceedings of the Winter Simulation Conference, Dec. 4–6, Washington, DC, pp. 1102–6.

Interrante, L.D. and Rochowiak, D. (1991) *ELICITing monitoring and control knowledge for an intelligent simulation training system*. Proceedings of the 1991 AAAI Workshop on Knowledge Acquisition, July, Anaheim, California.

Interrante, L.D. and Rochowiak, D. (1992). *Dynamic focus of attention for manufacturing design, operations and control*. Proceedings of AAAI-92 SIGMAN Workshop on Knowledge-based Production Planning, Scheduling, and Control, San Jose, California.

Interrante, L. *et al.* (1992) *Intelligent dynamic control of automated guided vehicles*. 1992 Proceedings of the International Material Handling Research Colloquium, May 1992, Milwaukee, Wisconsin, pp. 215–28.

Interrante, L., Rochowiak, D., Rogers, J. and Shields III, L. (1993) *Intelligent control of AGVs: dynamic scheduling via selective attention.* Proceedings of the 1993 NSF Design and Manufacturing Grantees Conference, Jan., Charlotte, North Carolina.

Interrante, L. and Rochowiak, D. (1993) *Active Rescheduling for Dynamic Manufacturing Systems.* Proceedings of the 1993 AAAI SIGMAN Workshop on Intelligent Manufacturing Technology, July, Washington, DC.

Jaikumar, R. (1986) Postindustrial manufacturing *Harvard Business Review,* Nov.–Dec., pp. 69–76.

Kaspi, M. and Tanchoco, J.M.A. (1990) Optimal flow path design of unidirectional AGV systems. *Int Jnl. of Prod Res,* **28**(6), pp. 1023–30.

Kim, C.W. and Tanchoco, J.M.A. (1991) Conflict-free shortest-time bidirectional AGV routing. *Int Jnl. of Prod Res.*

Lathon, R. (1993) Strategy-Agent Discrimination Net System.

Law, A.M. and Kelton, W.D. (1991) *Simulation Modeling and Analysis,* McGraw Hill, New York.

Luger, G. and Stubblefield, W. (1993) *Artificial Intelligence: Structures and Strategies for Complex Problem Solving,* Benjamin/Cummings, Redwood City, California.

Mahadevan, B. and Narendran, T.T. (1990) Design of an AGV based material handling system for a flexible manufacturing system. *Int Jnl. of Prod Res,* **28**(9), pp. 1611–22.

Majchrzak, A. (1988) *The Human Side of Factory Automation,* Jossey-Bass, San Francisco.

Milacic, V. (1988) *Intelligent Manufacturing Systems I,* Elsevier.

Mital, A. (1988) *Recent Developments in Production Research,* Elsevier.

Moodie, C.L. *et al.* (1988) State transition table as a data structure for adding intelligence to material flow control in computer integrated manufacturing, in *Artificial Intelligence: Manufacturing Theory and Practice.*

Newton, D. (1985) Simulation model calculates how many AGVs are needed, *IE,* February.

Nof, S. and Moodie, C. (1988) *Advanced Information Technologies for Industrial Material Flow Systems,* Springer Verlag.

Pan, L. *et al.* (1992) Using material handling in the development of integrated manufacturing. *Industrial Engineering,* Mar., pp. 43–8.

Rathmill, K. (1988) *Control and Programming In Advanced Manufacturing,* Springer Verlag.

Ringer, M. (1992) *Time phased abstractions for combining predictive and reactive scheduling methods.* Proceedings of the 1992 AAAI SIGMAN Workshop on Knowledge-Based Production Planning, Scheduling and Control, July 12–16, San Jose, California, pp. 147–61.

Rochowiak, D. (1992) *Local correctness, knowledge promotion, and the looseness of knowledge.* Proceedings of AAAI '92 Workshop on Communicating Scientific and Technical Knowledge, San Jose, California, pp. 89–96.

Rochowiak, D. and Interrante, L. (1991) *Heterogeneous knowledge based systems and situational awareness.* Proceedings of the AAAI Workshop on Cooperation Among Heterogeneous Intelligent Agents, July 15–18, Anaheim, California.

Rochowiak, D. and Interrante, L. (1993) *Concurrent engineering and design: person-centered and computer-assisted.* Proceedings of the AAAI Workshop on Collaborative Design, August, Washington, D.C.

Romero, I.S. (1993) NSF grant project shop floor scenarios.

Romero, N.R. and Romero, I.S. (1993) *Combination of AGV dispatch and routing strategies.*

Romero, N.R. (1993) NSF grant project system description.

Romero, N.R. (1993) Scheduling agent for AGV strategies.

Shaw, M. and Whinston, A. (1989) An artificial intelligence approach to the scheduling of flexible manufacturing systems. *IIE Trans*, **21**(2), June, pp. 170–83.

Shoureshi, R. (1987) Intelligent Control. *ASME Publication G00382*.

Simmons, R. *et al.* (1992) *Working Notes of the AAAI Control of Selective Perception Spring Symposium*, Mar 25–27, Palo Alto, California.

Sinriech, D. and Tanchoco, J.M.A. (1991) Intersection graph method for AGV flow path design. *Int Jnl. of Prod Res*, **29**(9), pp. 1725–32.

Swain, M. (1991) Proceedings of the NSF Active Vision Workshop, August, Chicago, Illinois.

Talavage, J., and Ruiz-Mier, S. (1988) Knowledge representation for dynamic manufacturing Environments, in *Artificial Intelligence, Simulation, and Modeling*.

Tanimoto, S. (1990) *The Elements of Artificial Intelligence Using Common Lisp*, Computer Science Press, New York.

Tsatsoulis, C., and Kashyap, R.L. (1988) Planning and its application to manufacturing. *Artificial Intelligence: Manufacturing Theory and Practice*, Institute of Industrial Engineers.

Walton, R., and Susman, G. (1987) People policies for new machines. *Harvard Business Review*, March-April, pp. 98–106.

Wilson, C.M., Hodgson, T.J., and King, R.E. (1991) A note on incorporation of improved information in the control of multiple vehicle AGV systems.

Winograd, T. and Flores, F. (1986) *Understanding Computers and Cognition – a New Foundation for Design*, Ablex, Norwood, New Jersey.

Wu, S. and Wysk, R. (1990) An inference structure for the control and scheduling of manufacturing systems. *Comp and Ind Eng*, **18**(3), pp. 247–62.

Xiong, Y., Sadeh, N. and Sycara, K. (1992) *Intelligent backtracking techniques for job shop scheduling*. Proceedings of the 1992 AAAI SIGMAN Workshop on Knowledge-Based Production Planning, Scheduling, and Control, July 12–16, San Jose, California, pp. 147–61.

Zuboff, S. *In the Age of the Smart Machine*, Basic Books, New York.

Zweben, M. *et al.* (1992) *Rescheduling with iterative repair*. Proceedings of the 1992 AAAI SIGMAN Workshop on Knowledge-Based Production Planning, Scheduling, and Control, July 12–16, San Jose, California, pp. 162–76.

# CHAPTER 6

# An integrated planning and control system for scheduling in flexible manufacturing systems

*Ai-Mei Chang*

## 6.1 INTRODUCTION

In recent years there has been a renewed interest in planning, scheduling and control problems with the evolving overlap of artificial intelligence, operations research, and control theory disciplines (Parunak and Judd, 1990). Although planning and scheduling problems have been researched in depth from both the artificial intelligence viewpoint (Smith, Fox, and Ow, 1986) and the operations research viewpoint (Graves, 1981), applications integrating both viewpoints have evolved only recently (Kempf, Russell, Sidhu *et al.*, 1991). Increasing competitive pressures amidst unstable markets have rendered the manufacturing environment dynamic and unpredictable with incomplete and uncertain information. Production planners are increasingly called upon to plan under time pressure with partial information. Under such circumstances, integrated planning, scheduling and control systems using techniques from AI, OR and control theory can help planners to react to changing environments effectively and efficiently.

In this paper, we propose an architecture for an integrated planning and control system for scheduling in a flexible manufacturing environment. The architecture we propose has evolved from previous research efforts and experiences in using AI and OR with scheduling problems. It seeks to combine the positive aspects of both while minimizing the disadvantages. The central component of the integrated system is the knowledge system which embeds rules generated by domain experts (such as shop foreman, production planner, operations researcher) and rules generated through simulation runs. Consistency in the knowledge system is maintained using an assumption-based truth maintenance system (ATMS), (de Kleer, 1986). Rules that are learned using simulation methods are incorporated into the knowledge system while the ATMS ensures that consistency is maintained. In addition to reasoning knowledge (rules) that are stored in the knowledge system, there is a current status module in the knowledge system which keeps track of the status of each resource in the system at any point (descriptive knowledge). The scheduling decisions are made in a reactive fashion in real-time using the rule-base and current status module. In this paper, we also discuss the implementation of a prototype system using Smalltalk/V and Prolog/V. In what follows we give a short description of previous research and formulate our research in the light of it.

Although operations researchers have worked on planning, scheduling, and control (PSC) problems earlier (Graves, 1981), advances in AI-based planners and schedulers for manufacturing (Kempf, 1989) have provided a significant impetus to research in integrated PSC systems.

There are three types of AI-based planning and scheduling systems that have been proposed (Kempf *et al.*, 1991):

1. fully automated systems using expert system technology;
2. those automated systems using 'deeper' methods;
3. those systems that are interactive.

Many knowledge-based systems have been developed based on rules derived from domain experts (Farhoodi, 1990). In many instances, the rule base is complemented by the use of discrete-event simulation to generate schedules (Jain, Barber, and Osterfield, 1990; Norrie, Fauvel, and Gaines, 1990). The possible problems with this approach are that in many instances the knowledge engineering process is complex because of the large amount of knowledge that must be acquired from the domain experts; second, it is hard for human experts to develop a feeling for the global effect of their decisions. Thus their focus tends to be myopic and short-term. In some cases, it is not possible for a domain expert to visualize all constraints. Finally, there is the question of reliability of the expert.

The 'deeper' methods of AI-based planners and schedulers concern methods of constraint satisfaction. A good example of such a system is ISIS – a knowledge-based system for factory scheduling developed by Fox and Smith (1984). ISIS incorporates all relevant constraints in the construction of a schedule, which are then used in conducting a constraint-directed search for an acceptable schedule (thereby effectively controlling the combinatorics of the underlying search space). Such systems incorporate very specific knowledge about the problem domain and generate schedules (Sadeh and Fox, 1989; Subhash and Salgame, 1989). While the deeper methods can provide a superior performance to ordinary expert systems, it has been argued that interactive systems can better the performance especially when constraints cannot be properly represented or when the constraints change rapidly over time. Interactive systems allow the human to build plans and schedules by methods that they naturally use but are difficult to represent as algorithms. They can incorporate intuition and guidance. Thus, the third approach to designing planners and schedulers is to provide intelligent help to the human planner (Clark and Farhoodi, 1987; O'Grady and Lee, 1989). While this approach has a high likelihood of acceptance on the shop floor, there are risks involved. The scheduling problem may be too large for the human scheduler to comprehend, and the pace of human interaction may waste valuable time which may be better spent in searching for a good alternative (Kempf *et al.*, 1991).

The integrated planning and control system that we propose combines the knowledge of the domain experts, results from operations research, and knowledge derived from simulation runs using an interactive approach. Our approach recognizes the shortcomings of deriving knowledge from domain experts. In practice, when plans and scheduling rules are formulated by domain experts, they make some simplifying assumptions about (1) various system parameters such as demand distribution, machine breakdown frequencies, etc. and (2) factors that constrain the plans and schedules. When domain experts input their plans into the system, the assumption-based truth maintenance system (ATMS) checks the existing knowledge base to identify **contexts** under which such a plan would be valid. These contexts essentially are a collection of conditions and assumptions under which the plan can be believed to lead to the intended optimal or 'satisficing' result. The system checks the assumptions and conditions input by the expert with its knowledge base to determine any conflicts or violations with the many contexts it has for a plan. If there is no conflict, the expert is allowed to implement the plan. In the case of a conflict, the system interacts with the expert in identifying the 'offending' elements, which leads to a revision of the plan. In a similar manner consistency is maintained by the ATMS when new rules are added to the system. The simulator is invoked in instances where a scheduling rule (expert-generated or otherwise) is not available.

By complementing domain experts' knowledge with results from operations research and simulation, we avoid depending on domain experts for complex problems, where they could make myopic decisions. At the same time, using the interactive approach we let planners use their intuition in generating/changing schedules in addition to choosing the planning and scheduling criteria. An additional feature of our system is that it allows for new rules to be added and 'refined' over a period of time to render the system more suitable for the planning and scheduling application.

In the next section we discuss our architecture for designing an integrated planning, scheduling, and control system. We provide detailed descriptions of the components of the system, the knowledge acquisition process, and the assumption-based truth maintenance system that preserves the consistency in the knowledge base. In section 6.3, the implementation of the system using Smalltalk/V and Prolog/V is discussed. We conclude in section 6.4 with directions for future research.

## 6.2 SYSTEM ARCHITECTURE

### 6.2.1 Problem definition and scope

In the planning and scheduling literature, flexible manufacturing systems (FMSs) have been defined in different ways from alternative viewpoints (Kusiak, 1985; Stecke, 1985; Rachamadugu and Stecke, 1988; Gupta and Goyal, 1989; and Gershwin *et al.*, 1984). For our purposes, we define FMSs as a collection of machines which provide for a moderate to high level of automation as far as the part **volume** that is generated, and a moderate level of flexibility as far as the part **variety** it can handle. While our proposed architecture is general (that is, it does not pre-suppose a particular configuration of the system), we do assume that the design problem has already been solved and the system parameters such as machine capacity, fixtures, product range, and aggregate process plans are fixed. The PSC process is viewed as a hierarchical process where in the first stage, machine grouping and resource allocation decisions are made, followed by on-line scheduling decisions including dispatching rules and contingency plans at the second stage and followed by control problems of output control and maintenance at the third stage. Our model focuses more on the second and third stage decisions which control production over the course of each day by optimal routing of jobs and efficient use of resources subject to various constraints. Resource availabilities, resource capabilities (i.e. capacity limits, throughput limits, which operations can be run on a given machine), precedence, machine setup, changeover times, and buffer space are typical physical constraints. Organizational goals like machine utilization, due dates, processing priorities, cost and WIP levels are also viewed as constraints.

When the top level decisions are already made and input to the system, our system focuses on the dispatching rules necessary for on-line scheduling. These dispatching rules primarily help the foreman on the shop floor to make appropriate decisions. The system expects the criteria for optimality to be input by the user. The decisions related to these dispatching rules include pre-release decisions, part-releasing decisions and part-routing decisions (Gupta *et al.*, 1989). We assume that the pre-release decisions such as setting the due date, determining priorities and lot sizes are already made. Part release decisions consisting of which part to select next and when it is to be released are advised by the system. The part routing decisions are also advised by the system. This task of providing advice on scheduling is accomplished by a rule-based expert system that forms the core of our proposed system. The expert system is a repository of expertise acquired from domain experts, simulation results and other analytical research.

Thus, to summarize our scope, the proposed architecture describes the design of a **generic**, object-oriented model and rule-based expert system for on-line scheduling and control in an FMS. The generic design is broad enough to encompass the basic features of most FMSs and at the same time is flexible enough to allow for easy modification and tailoring to specific applications.

### 6.2.2  System components

The system consists of six major modules as shown in Fig. 6.1. Each module performs a specific function and has distinct information input and output points. Each module is a distinct stand-alone unit which can perform its separate function. The interconnections of each module, which indicate the flow of information between modules, facilitate the processing of specific inputs from the user or master scheduler or domain expert to lead to a scheduling decision and ultimately real-time and on-line scheduling and control of the FMS. An overview of each module is given below while the specific details are discussed in the implementation section.

**Current status module:** this module is designed to provide a snapshot of the various entities in the FMS. The module is a repository of descriptive knowledge of the status of the FMS at any point in time. It maintains information on all jobs that have entered the system including due date, number of operations with processing times, operations already completed, future schedule, expected finish time, etc. It maintains similar information on machines: sequence of jobs in each machine, processing times, machine breakdowns, schedule for tool changeovers, etc. The current status module is updated in response to any scheduling event and accurately reflects the FMS status, which may be required as input for generating a schedule either through the expert system or simulator and for truth maintenance purposes. The entities represented in the current status module could include all elements of the FMS including material handling devices, material transportation devices, machining devices and tools.

**Information interface:** this module provides the interface between the user (or the domain expert) and the knowledge system consisting of the simulator/scheduler, the expert system, and the truth maintenance system.

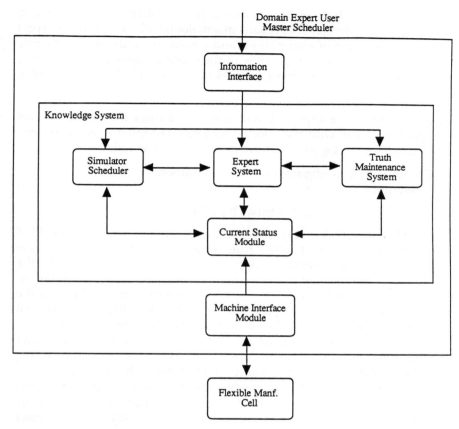

**Fig. 6.1** System architecture.

The input could be of two types:

1. descriptive information including part pre-release information, priorities, policies, managerial objectives and device maintenance
   information covering scheduled maintenance and breakdowns;
2. reasoning knowledge in the form of new dispatching rules for
   scheduling.

The kinds of actions initiated as a response to these two types of
input are quite different. In the first case, the expert system takes into
account the information on new parts, part pre-release information,
constraints and objectives along with the current status of the system
in suggesting a revised scheduling rule to be followed. The detailed
schedule corresponding to this rule is enumerated by the scheduler. In
the second case, the new scheduling rule is checked by the truth

maintenance system as to whether it is consistent with the existing reasoning knowledge. This results in a dialogue with the domain expert through elicitation of assumptions and conditions under which the rule is valid. If the rule is consistent with the existing reasoning knowledge, it can be enumerated by the scheduler (and also added to the reasoning knowledge base).

**Expert system:** the expert system consists of rules that relate various dispatching rules with the specific management criteria, assumptions, and conditions. The general form of the rules is as follows:

$IF\{$criteria $= X,$ configuration $= \ldots,$ conditions $= \ldots,$
     assumptions $= \ldots\}$                                      (6.1)

$THEN\{$Dispatching Rule $= \ldots\}$

The rules should be very specific to the configuration of the FMS and should be derived either from simulation studies of the system with actual or representative data, or from domain experts with the necessary experience with the system and the particular configuration, or from analytical OR studies. Thus based on the current status of the FMS, input data and the parameters of the jobs, the expert system advises an optimal or satisficing dispatching rule. The actual development of the schedule based on this rule is carried out by the scheduler which has appropriate algorithms to elaborate the schedule. If the expert system is unable to come up with a **best** scheduling rule, then the set of rules selected as viable alternatives are referred to the simulator for further analysis through simulation. If this presents problems during real-time execution, the user can intervene to select an appropriate rule.

**Simulator/scheduler:** this module performs the dual functions of a scheduler and a simulator. The scheduler provides details of the start time and finish time for each job, their sequence of operations and routes and for each machine the sequence of jobs scheduled on it using algorithms for different dispatching rules. Taking into account all the physical constraints, the scheduler uses the dispatching rule and the routing rule provided by the expert system along with the algorithms to flesh out the detailed schedules. The same algorithms are used to compare the suitability of the different dispatching rules for a given scenario using the simulator. This provides an enhanced decision-making capability in the event the expert system is unable to generate one best rule. Each candidate alternative would be tried out by the simulator to test the efficacy of the rules and to identify the best rule for a given scenario. This task, however, cannot be performed if there is a time constraint for real-time execution. In such cases, the simulations are run at a later stage and the results are used to develop additional rules to be added to the rule-base.

**Truth maintenance system:** the truth maintenance system plays an important role in maintaining consistency in the knowledge system as new reasoning knowledge is added to the system. Also, when the user initiates a rule to be implemented, the truth maintenance system first ensures that the rule does not contradict the existing knowledge base, thus preventing execution of wrong policies. If the conflict is detected, then the truth maintenance system aids in identifying or localizing the offending elements that contribute to the conflict. The function of truth maintenance is accomplished through the use of truth maintenance rules. For each dispatching rule that exists in the expert system, the truth maintenance system maintains **contexts** under which the rule is valid. The contexts refer to conditions and assumptions which accompany the rule. On receiving a new rule, the system checks whether the input is in conflict with any of the contexts using the truth maintenance rules. The details of the system are further elaborated in the next section.

**Machine interface module:** this module provides an interface between the system and the FMS. It helps in executing the schedules in real-time, updates the current status module and ensures that it truly reflects the status of the FMS.

It is clear from the above discussion that the expert system and the truth maintenance system form the core of our proposed architecture. In the next section we present our scheme for knowledge representation, knowledge revision and truth maintenance.

## 6.3 KNOWLEDGE REPRESENTATION USING ATMS

Before providing a description of the knowledge representation scheme and how consistency in the knowledge base is maintained, it is necessary to discuss what this knowledge is and how it is acquired. The expert system contains inference rules that are used to identify the optimal dispatching rules in a given situation or scenario. The parameters used in defining such a scenario include the following (Schriber and Stecke, 1986): FMS configuration, primary resources, secondary resources (buffer space, loading/unloading station), geometric configuration, secondary time requirements (transfer times, fixtures, tool changeover time), operating discontinuities (machine breakdown, scheduled maintenance), job characteristics (due dates, precedence constraints, fixture requirements) and demand characteristics (distribution, mean, and variance). While some of these parameters are fixed (such as FMS configuration, primary resources, geometric configuration, etc.), others change during the short-term planning horizon. Given a set of parameters and a set of managerial objectives, the inference rules provide the 'best' or a 'good' dispatching rule to use for on-line scheduling in the short-term. Thus, the form of the inference rules is as given in Equation (6.1) where some of the parameters of the scenario are facts (e.g. FMS configuration, primary capacities), while some are assumed (e.g. demand characteristics), albeit assumptions which have a high likelihood of being true.

Since the inference rules form the core of the system, it is important that they are obtained error-free. In order that the inference rules are useful, it is also necessary that they be very specific to the FMS configuration for which the scheduling system is being designed. The OR literature on suitability of different dispatching rules for the various scenarios can provide a good starting source, but since there is no basic framework to compare these results (Gupta, Gupta, and Rector, 1989, p. 375), its utility may be limited. Simulation studies based on the specific FMS configuration, covering the whole range of values expected in practice for all the scenario parameters, would provide the best source for the inference rules (see Schriber and Stecke, 1986 for a thorough study using the shortest processing time (SPT) dispatching rule).

It is not necessary that the inference rule-base be exhaustive to begin with. One of the advantages of our architecture is that reasoning knowledge can be added to the system over time as different situations are encountered and the simulator is called upon to analyze and provide good heuristics. Rules from domain experts can also be incorporated into the rule-base in a similar manner.

We organize the inference rules using an assumption-based truth maintenance system (ATMS) (de Kleer, 1986), which is a knowledge representation scheme for storing reasoning knowledge about various propositions (results) that are derived from the simulation studies, OR studies, or domain experts. The propositions are the THEN part of the IF-THEN rules and are of the form: 'SPT would be an optimal rule under this scenario', where the scenario is defined as the parameter values in the IF part of the rule. Associated with each proposition are assumption/premise sets called environments or contexts, that contain the minimal assumptions/premises (the IF part) under which the proposition can be believed. Thus, a proposition 'SPT is an optimal rule' has contexts or minimal conditions under which it can be believed. Thus the ATMS uses these inference rules as its elementary units of knowledge and draws conclusions by combining rules to form explanations. If a new rule that is added to the system conflicts with a proposition that has been arrived at based on previously held contexts, the ATMS calls for a revision of the appropriate contexts in such a way that the result is most plausible and in agreement with the new rule.

Formally, the ATMS consists of the following parts: an inference rule base, $\{K\}$, consisting of causal and logical rules, a premise set, $\{I\}$, consisting of concrete facts and empirical information and an assumption set, $\{Z\}$, containing assumptions about uncertain events (e.g. demand characteristics). Based on these, the propositions, $\{P\}$, are derived. The ATMS maintains for each propositional node in the system sets of contexts called label $L(P)$, under which the corresponding proposition can be proven or explained. The label $L(P)$ can yield only three possible truth values for the proposition $P$: **believed, disbelieved**, and **unknown**. If any context in $L(P)$ is believed, then $P$ is believed; if any context in $L(\neg P)$ is believed, then $P$ is disbelieved; if we can confirm neither $L(P)$ nor $L(\neg P)$, then $P$ is unknown.

Figure 6.2 provides a graphical representation, called a rule-structure, of a two-rule ATMS in our application context. Both rules relate to the proposition $(P_1)$ that 'SPT performs well under the scenario'. The fixed parameters of the FMS are denoted as premises $(I_1, I_2$ and $I_3)$ and the other parameters are denoted as assumptions $(Z_3$ through $Z_5)$. The assumptions $Z_1$ and $Z_2$ relate to managerial objectives of minimizing WIP inventory and minimizing makespan, respectively. The rule $K_1$ states that IF $\{$Criteria = Minimize WIP, conditions = $I_3 \ldots I_5$, assumptions = $Z_3 \ldots Z_5\}$ THEN $P_1$. Rule $K_2$ can be expressed in a similar form. The label $L(P_1)$ consisting of two separate contexts under which $P_1$ can be true, is given below:

$$L(P_1) = \{(Z_1, Z_3, Z_4, Z_5, I_1, I_2, I_3), (Z_2, Z_3, Z_4, Z_5, I_1, I_2, I_3)\} \quad (6.2)$$

If either of the two contexts holds true, $P_1$ is true.

The ATMS can perform three useful functions in organizing and revising the reasoning knowledge base:

- **Producing explanations.** If a proposition $P$ has to be true, the ATMS can retrace the justification paths and identify the argument of proof justifying that proposition, as well as the assumptions upon which it is founded. This is a useful function as the user of the system can readily recall all the contexts under which a particular dispatching rule can prove useful and understand the context under which the particular dispatching rule was chosen.
- **Identifying inconsistencies.** When a new rule is entered into the system, the ATMS can check if the new information conflicts with the already existing knowledge base and identify the particular elements (assumptions or premises) that lead to the conflict. Such conflicts need to be resolved by the domain expert which may lead to modifications of the contexts of existing rules or creation of new contexts or modifications in the new rule. This truth maintenance function prevents inconsistencies in the knowledge base.

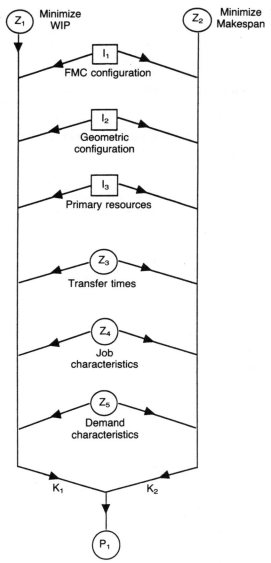

SPT rule performs the best under this scenario

**Fig. 6.2**   A graphical representation of a two-rule ATMS.

- **Guiding the acquisition of new information.** If a certain proposition is in an unknown state, then the label $L(P)$ provides clues as to the information required to render it 'believed' or 'disbelieved'. That is, if a confirmation of assumption $Z$ is all that is missing from one set in $L(\neg P)$, while the confirmation of $\neg Z$ is missing from some set in

$L(P)$, then a test leading to the confirmation or denial of $Z$ should be devised. That is, the ATMS provides clear guidelines for gathering additional information to render the reasoning base fully specified.

In order to understand how the truth maintenance operation is carried out, it is necessary to know how the rule-structure (as the one in Fig. 6.2) of the rules in the system relate to the structure of a new rule being added to the system with respect to a particular proposition $P$, such as 'SPT is the best rule'. Any new rule that is being entered into the system has a particular **context** to it, specifying the conditions under which a proposition $P$ is valid. This context can be related to the contexts or label sets of $P$ in the system in three different ways: **disjoint**, **overlap**, and **inconsistency**.

If the contexts have null intersection, that is, if there are no common assumptions or premises between them, their relationship is disjoint. Disjoint contexts provide different perspectives or explanation to the same proposition. If the assumptions/premises do not contradict each other (that is, $Z$ in one context and *not* $Z$ in the other), the new rule is directly added to the system as it is augmenting an incomplete section of the knowledge base. If the intersection of the contexts is not null and is a strict subset of each of the label sets, then the relationship is called overlap. In this case, there may be a few common assumptions and/or premises between the two contexts; however one context is not a subset of the other. Since the contexts contain the minimal assumptions/premises under which a proposition is valid, any contradiction between the assumptions/premises of contexts (e.g. $Z$ and *not* $Z$) leading to the same $P$, need to be examined and resolved before both contexts can exist in the rule-base. The truth maintenance rules will detect conflicts of this kind and will demand resolution before a new rule can be added.

If there exist common premises and assumptions which lead to proposition $P$ in one context and proposition *not* $P$ in the other, the two contexts are said to be **inconsistent** with respect to proposition $P$. A relationship that is not inconsistent is called consistent. Whereas disjoint and overlap are mutually exclusive, inconsistency is not mutually exclusive with these relationships. Since in our application the rules are concerned with which rule is best/good under circumstances and not which rule is not good, we will not focus much on these type of inconsistencies. In case of any contradictions, the truth maintenance module will identify the offending elements which would need resolution by the domain expert or by further simulation/analytical studies.

We elaborate on the specifics of the system components and the details of the truth maintenance operations in the context of our implementation of a prototype discussed in the next section.

## 6.4  IMPLEMENTATION OF THE SYSTEM

The development of a prototype implementing the proposed system architecture has been accomplished using the object-oriented language Smalltalk/V and Prolog/V. The Smalltalk/V environment was chosen primarily due to the availability of Prolog/V as a sub-class of the available class structure. This provides for a powerful environment, possessing Prolog's logical and declarative power for building the expert system and truth maintenance modules within the flexible class structured environment of Smalltalk/V, which is ideal for rapid proto-typing and incremental development. The prototype has been de-veloped with a generic FMS in mind but can be easily expanded and tailored to a specific FMS. We first provide a short description of object-oriented programming concepts and then discuss the proto-typing of the various system components using Smalltalk/V and Prolog/V.

### 6.4.1  Object-oriented programming

Object-oriented programming (OOP) is recognized as a more 'natural' approach which encapsulates both data and the algorithms that specify that data in a single object, as against conventional programs which separate data and procedures that operate on the data. Thus, in object-oriented systems there are interacting objects, operations on which take place through a well-defined interface. The six central concepts in this programming language are: objects, messages, methods, classes, inheritance and instances.

An object is considered a single programming entity encapsulating data and the algorithm that specifies the object's behavior. From a programming viewpoint, they may be viewed as elements of an OOP system that send and receive messages which control an object's behavior. The same message could have different meanings for dif-ferent objects. For example, a message 'divide' to an object 'number' would half the number, whereas the same message to an object 'sphere' may result in two hemispheres. How an object would react to a message depends on the code it has in it to perform a given command. Thus, since the data and the procedure that operates on it constitute an object, we can use messages that have different connotations for dif-ferent objects. Addition of any new object would involve its creation as well as creating its ability to react to any message.

While messages are commands passed from one object to another, a method is a procedure or function in a procedural language. It is the code in an object that tells the object how to react to a message that has the same name as the method. On receiving a message the method executes the instructions within it. It may even trigger another method in another object or send a message back indicating the completion of the instruction. Methods are the primary means through which the fields within an object may be manipulated or modified. If a message is sent to an object that does not have a method of that name defined, then the system issues an error message to that effect.

Classes (or object types) are templates defined by system designers that describe the properties and behaviors of a set of common objects. Objects in a class react to one or more messages in the same manner. Each object within a class has the same number and type of fields and only differs from the other objects in the class in the data that exists in the fields. All objects share the same methods, typically by pointing to a common method table or dictionary. A class, like everything else in an object-oriented world, is itself an object. One class may be defined in terms of another class, leading to the concept of subclasses and inheritance. Subclasses inherit the characteristics of higher-level ancestor classes. All objects in OOP have at least one ancestor class. The closer an object is to the root class, the fewer ancestors it has.

A method of a subclass may be redefined so as to override the characteristics it inherits from its ancestor class. This makes it possible to considerably reduce the amount of code that an application programmer needs to write. The only code necessary is that which explains how a method differs from the parent method. All other methods that have not been over-ridden will respond in the same way as objects in the parent class. Application specific objects are called instances. In general, instances are unique objects that cannot have any sub-kind links of their own.

Object-oriented systems have been found particularly suitable for the design of diagnostic and scheduling problems (Elorantes *et al.*, 1989). The OOP approach is very natural and allows an easier transition from conceptualization to implementation (Lazarev, 1988). OOP languages lead to rapid prototyping and encourage an exploratory 'design-prototype-refine' approach to applications development. Moreover, the notion of objects communicating through messages is fundamental to reusable software components, which reduces programming effort tremendously.

### 6.4.2  Current status module

For the design of our prototype we assume a hypothetical problem modeled after the configuration and parameters described in Schriber and Stecke (1986). The current status module is an object-oriented model built around commonly found features in most flexible manufacturing systems. The entities involved fall easily into a hierarchical class structure, as shown in Fig. 6.3, with each sub-class inheriting the properties of the higher class. The major sub-classes include material

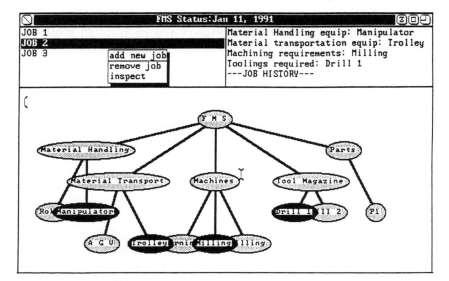

**Fig. 6.3**  Class hierarchy of the FMS objects.

handling devices, material transportation devices, machining devices and tools. Further subclassification into specific devices are made for each category, with instances of each provided as well. One of the advantages of the OOP framework is that it allows for a flexible design, making it easy to add new classes or sub-classes or any new methods to the existing classes. Starting with the basic framework shown in Fig. 6.3 as the foundation we can build any specific structure as needed. Similarly specific methods can easily be incorporated.

The specific problem for the current status module was to devise a method for obtaining a description of the FMS status at any point in time. The various elements of the FMS and their attributes would need to be stored and retrieved. Jobs entering the system would need to be added to the job queue and changes in the FMS elements would need to be updated. Thus, in the present prototyping of the system the entire FMS class structure can be viewed in a window together with pop-up panes in an animated logical diagram on the screen. The FMS class structure in Fig. 6.3 serves as an event-driven model as well. An event results in a change in the FMS objects and can affect one or more classes. For example, addition of a new job to the job list can potentially affect the schedules of the machines class, material handling and material transportation devices class, tools class and the various sub-classes of each. Changes of this form follow the tree defined information flow. However, there may also be certain events which affect parallel sub-trees. For example, a breakdown in any machine may lead to

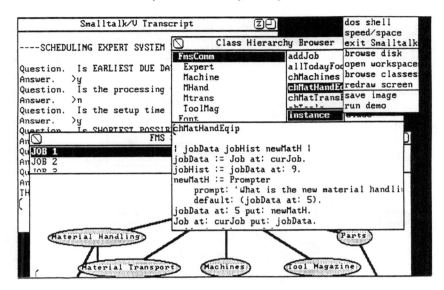

**Fig. 6.4**  Current status module window panes.

schedule changes in material handling, material transportation and tooling classes and their sub-classes. Hence the logical tree structure has an underlying graphical structure with each node connected to the other. Although inheritance follows the class hierarchy, messages need not flow in a hierarchical fashion.

The current status module is accessed through a main FmsComm Window Pane. The top half is defined as a text pane and the bottom half as a graph pane (screen-print in Fig. 6.4). The text pane is divided into two adjacent panes, the first being a list pane and the second being a description pane for selecting the job. The graph pane represents the FMS class structure. Depending on the job selected via the cursor, the appropriate current job information is pulled up on the descriptor pane and at the same time, the relevant nodes on the graphical FMS structure are highlighted. Further selection and pop-up menus have been defined to access other windows and initiate processes. By means of these panes, other windows for accessing the machine, material handling, material transportation and tools data dictionaries can be opened (Figs 6.5 and 6.6).

The dictionary class facilitates the development of an underlying database for storage of current status information (descriptive knowledge). The necessary means to add, delete, change and query this database have been designed and built into the prototype. Static presentation graphics are used to provide Gantt charts, routing charts and other graphical aids for scheduling and control purposes. Statistics

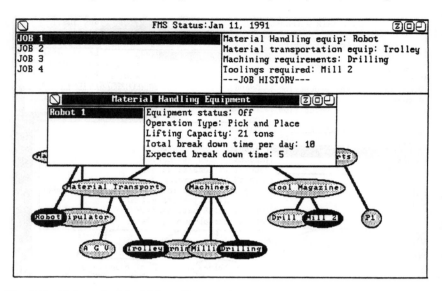

**Fig. 6.5** Status of material handling equipment.

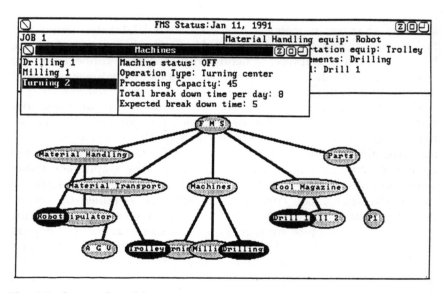

**Fig. 6.6** Status of machines.

on the system performance are also collected and displayed with the presentation graphics. The current status module can easily be expanded and tailored to suit any existing FMS by simply creating and adding to the existing structure the appropriate sub-classes, instances, and methods. Suitable enhancements could be made to the graphical interface as well.

### 6.4.3 The expert system

Since the prototype is a generic one, the scheduling rules used in the expert system are basically a set of high-level heuristics implemented more for expository purposes than for real use. The rules have been derived from the survey article on FMS scheduling by Gupta, Gupta, and Rector (1989). As discussed previously, in an actual application context, these rules should be configuration specific and derived from simulation studies to provide useful expertise for scheduling. Prolog/V has been used as the tool for developing the expert system. Prolog is known to be a powerful declarative programming language ideally suited for rule-based expert systems development. Since Prolog/V is embedded in the Smalltalk/V environment, its capabilities are substantially augmented through features such as inheritance, window manipulation, I/O interface and graphics interface.

In the prototype, the Prolog/V environment is represented by a group of classes. Each class is broken into methods. Each method contains clauses with the same name which are edited and saved as a group because they are actually compiled into one Smalltalk method. The name of a clause is the name of its leftmost relationship which is in turn the name of its leftmost predicate. Prolog/V is implemented as a subclass of Logic, which itself is a subclass of class Object. As in Smalltalk/V, a Prolog/V subclass inherits the predicates implemented in its superclass. Communication between parallel classes is possible. Each Prolog/V class can be viewed as an expert system and these can consult each other.

In the Prolog/V environment, a dynamic knowledge base is created by the 'database' predicate. This knowledge base contains premises, assumptions and propositions, which can be dynamically asserted or retracted. The knowledge base is global in the sense that once created, it can be accessed by an instance of any Prolog class. Variables, structures, lists and rules are similar in implementation to those in standard Prolog, with changes only in the syntax.

In the prototype, simple rule-trees were developed for the four major management objectives of scheduling:

1. minimum work center and system utilization;
2. processing times;
3. due dates;
4. production rate.

An example of the rule tree is presented in Fig. 6.7. A description of the dispatching rules used in the rule base is given in Table 6.1.

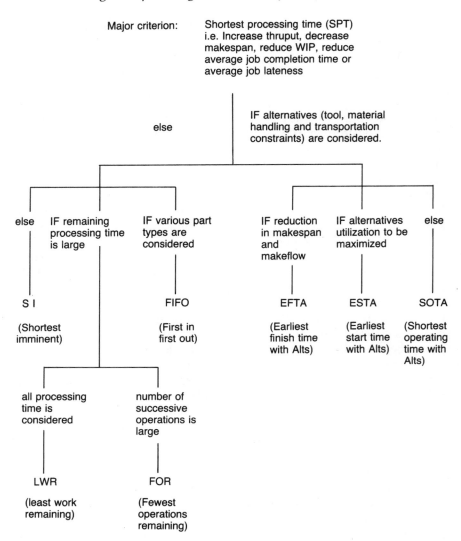

Major criterion:     Shortest processing time (SPT)
i.e. Increase thruput, decrease
makespan, reduce WIP, reduce
average job completion time or
average job lateness

else

IF alternatives (tool, material
handling and transportation
constraints) are considered.

else     IF remaining        IF various part                    IF reduction     IF alternatives     else
        processing time      types are                          in makespan      utilization to be
        is large             considered                         and              maximized
                                                                 makeflow

S I                          FIFO                                EFTA             ESTA             SOTA

(Shortest                    (First in                          (Earliest        (Earliest        (Shortest
imminent)                    first out)                         finish time      start time       operating
                                                                 with Alts)       with Alts)       time with
                                                                                                   Alts)

all processing      number of
time is             successive
considered          operations is
                    large

LWR                 FOR

(least work         (Fewest
remaining)          operations
                    remaining)

**Fig. 6.7**   Example of the expert system decision tree.

The development interface for the expert system is provided through the logic browser window, which may be opened from any text pane. It is similar to the browser window used for adding and viewing Smalltalk methods and classes. The user interface for the expert system was developed by integrating Prolog/V classes with Smalltalk/V. Communication with Smalltalk/V is primarily through the 'consult' and 'is'

**Table 6.1** Explanation of the dispatching rules

| acronym | explanation |
|---------|-------------|
| SI | Select the job with the 'shortest imminent' operation time |
| FIFO | Select the job with 'first-in, first-out' |
| LWR | Select the job with 'least work remaining' |
| FOR | Select the job with 'fewest operations remaining' |
| SOTA | Select the job with 'shortest operation time with alternatives considered' |
| ESTA | Select the job with 'earliest starting time with alternatives considered' |
| EFTA | Select the job with 'eariliest finishing time with alternatives considered' |
| CR | Select the job with the 'highest critical ratio |
| SPJL & NEP | (i) Load the part which has the smallest proportion of job launched (ii) as each pallet is loaded, reload it with like product, if possible. |

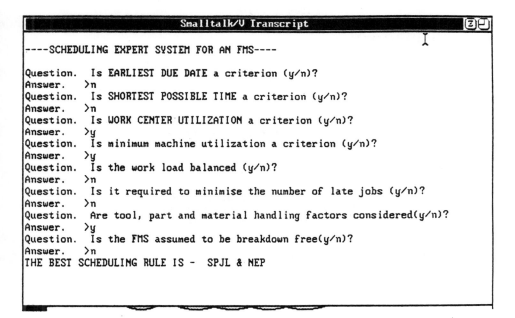

```
┌─────────────────────────────────────────────────────────────┐
│                    Smalltalk/V Transcript              [z][◻]│
├─────────────────────────────────────────────────────────────┤
│----SCHEDULING EXPERT SYSTEM FOR AN FMS----                   │
│                                                               │
│Question.  Is EARLIEST DUE DATE a criterion (y/n)?             │
│Answer.    >n                                                  │
│Question.  Is SHORTEST POSSIBLE TIME a criterion (y/n)?        │
│Answer.    >n                                                  │
│Question.  Is WORK CENTER UTILIZATION a criterion (y/n)?       │
│Answer.    >y                                                  │
│Question.  Is minimum machine utilization a criterion (y/n)?   │
│Answer.    >y                                                  │
│Question.  Is the work load balanced (y/n)?                    │
│Answer.    >n                                                  │
│Question.  Is it required to minimise the number of late jobs (y/n)? │
│Answer.    >n                                                  │
│Question.  Are tool, part and material handling factors considered(y/n)? │
│Answer.    >y                                                  │
│Question.  Is the FMS assumed to be breakdown free(y/n)?       │
│Answer.    >n                                                  │
│THE BEST SCHEDULING RULE IS -  SPJL & NEP                      │
│                                                               │
└─────────────────────────────────────────────────────────────┘
```

**Fig. 6.8** Expert system interface.

predicates. These two predicates provide the bridge between the two worlds. The expert system module (and the truth maintenance module) can be called from within the current status window. The system poses a series of questions which appear in the transcript

```
"scheduling rules"


scheduled('ST') :
              criterion (#edd),
               constraint (#setuptimelarge),
                constraint (#processingtimelarge),not (constraint
                (#manyoperationsremaining)),
                   rules('ST').

scheduled('CR') :
              criterion (#edd),
                constraint (#processingtimelarge),not (constraint
                (#manyoperationsremaining)),
                   rules('CR').

scheduled('ST/O') :
              criterion (#edd),
                constraint ( #setuptimelarge),
                 constraint (#processingtimelarge),
                    constraint (#manyoperationsremaining),
                       rules ('ST/O').

scheduled('LWR') :
              criterion (#spt), not(constraint(#alternatives)),
                constraint (#processingtimelarge),
                 constraint (#allprocessingtimeconsidered),
                   rules ('LWR').

scheduled('FOR') :
              criterion(#spt), not(constraint(#alternatives)),
                constraint (#processingtimelarge),
                 constraint (#largenoofsuccessiveoperations),
                   rules ('FOR').

scheduled('SI') :
              criterion(#spt), not(constraint(#alternatives)),
               not(constraint (#processingtimelarge)),
               not(constraint(#variousparts)),
                   rules ('SI').
```

**Fig. 6.9** Examples of scheduling rates.

window, with answer and selection panes provided for replies (Fig.
6.8). The questions are designed to obtain relevant information regarding the criterion to be used and other constraints. On the basis of the
answers, the expert system provides the best scheduling rule(s) to be
used for the situation. This information is also relayed to the scheduler
for fleshing out the details. Figure 6.9 presents a view of some of the
rules that make up the expertise.

### 6.4.4 Truth maintenance module

The truth maintenance module is invoked whenever new rules, either derived from a domain expert or from the simulator, is added to the system. This module is also implemented using Prolog/V as a group of classes separate from the expert system. This module essentially consists of a knowledge base consisting of lists of propositions, assumptions/premises and context sets and a set of rules known as truth maintenance rules. Truth maintenance rules help in detecting contradictions. They are of the form:

IF $\{Z$ AND *not* $Z\}$ THEN contradiction.

The truth maintenance operation can be best explained by using an example. Let the expert system contain rules $K_1$ and $K_2$ as shown in Fig. 6.2 along with another rule $K_3$. The form of the rule base can be represented as below:

| Rule base |
| --- |
| $K_1$: $\quad I_1 \wedge I_2 \wedge I_3 \wedge Z_1 \ldots \wedge Z_5 \rightarrow P_1$ |
| $K_2$: $\quad I_1 \wedge I_2 \wedge I_3 \wedge Z_2 \ldots \wedge Z_5 \rightarrow P_1$ |
| $K_3$: $\quad Z_3 \wedge I_4 \wedge I_5 \rightarrow P_2$ |

The above rules lead to appropriate proposition nodes, assumption/premise nodes, and contexts in the truth maintenance module.

| Proposition nodes | | | |
| --- | --- | --- | --- |
| Prop-ID | Justification | Context-pointer | Status |
| $P_1$ | Rules $K_1$, $K_2$ | 21, 22 | Believed |
| $P_2$ | Rule $K_3$ | 23 | Believed |
| ... | ... | ... | ... |
| $P_8$ | External | Domain expert | Believed |

In the above representation, each proposition is associated with a rule(s) that explains it, a context-pointer which indicates where its context resides and a status field which indicates whether the proposition is currently believed or disbelieved. $P_8$ is a reference to a new proposition that has been added to the system which does not have its rule yet. The domain expert will provide its context. Till then the proposition will not be evaluated by the truth maintenance rules.

| Assumption/premise nodes | |
|---|---|
| Assump-ID | Status |
| $I_1$ | Believed |
| $I_2$ | Believed |
| . . . | . . . |
| $Z_1$ | Believed |
| . . . | . . . |
| $Z_3$ | Believed |
| . . . | . . . |

The assumption/premise nodes provide the status of each of the elements. The context sets shown below provide the contexts for each proposition. The status of each could be either compatible or incompatible depending on whether the individual elements comprising a context are all believed at the same time.

| Context set | | |
|---|---|---|
| ID | Context | Status |
| 21 | $\{Z_1, Z_3, Z_4, Z_5, I_1, I_2, I_3\}$ | Compatible |
| 22 | $\{Z_2, Z_3, Z_4, Z_5, I_1, I_2, I_3\}$ | Compatible |
| 23 | $\{Z_3, I_4, I_5\}$ | Compatible |
| . . . | . . . | . . . |

If a new rule is added to the system which deals with a proposition already existing in the system, the truth maintenance rules will check if contradicting assumptions/premises arise within the contexts explaining the proposition. If so, the status of the contexts are set to incompatible and the system would highlight the offending assumptions/premises. The intervention of the domain expert/analyst is needed at this time to resolve the conflict.

In the Prolog/V implementation, the truth maintenance rules are implemented in the same manner as the expert system rules. The proposition nodes and assumption/premise nodes are subclasses of the node class. The determination of the status of these is accomplished through appropriate methods. In the present implementation, the truth maintenance rules are not exhaustive. They can be expanded upon easily depending on the specifics of the FMS and scheduling rules.

The remaining modules, the simulator and the machine interface module, are yet to be developed in the prototype. For simulation purposes, an object-oriented language such as SIMSCRIPT would be suitable. A key advantage of the object-oriented approach is that new objects can be incorporated into the simulation module with minimum disruption to the system that has already been developed. The development of the machine interface module would depend on the specific FMS to which the system is tailored.

### 6.4.5 Practical considerations in implementation

**Scheduling problem.** The combinatorial explosiveness of the scheduling problem makes it a particularly difficult problem to solve. The number of possible schedules grows exponentially along each dimension such as operations, machines, tools and jobs. The possibilities are further increased by the alternate machines that can perform the same operations, machines that can perform several different operations and other such factors which actually render the FMS flexible. In addition to these, the schedules are also functions of the machine configurations. Thus, for the development of a practical and useful scheduling system it is essential to identify, represent and efficiently use available constraints to bound the space of possible schedules. The first step in that process is to develop the scheduling system **specific** to an FMS application. Once the system configuration is known, the schedule possibilities can be further pruned by eliminating schedules that are known to perform poorly. Such knowledge can be obtained either from the domain expert (who may know this through his/her experience with the system) or from application-specific simulation studies. Although heuristics tend to yield acceptable results, while developing the rule-base it is important to focus attention on those variables that can make the most difference on the criteria of interest. Rather than concentrating on low-level heuristics, it is advantageous to develop high-level heuristics based on (human) domain expertise. Reasoning at a high level of abstraction can help reduce the effective space search and also help maintain the right focus (Rickel, 1988).

A high level of abstraction is a necessity in our prototype because of the nature of the rules in the expert system and in the truth maintenance system. The rules are basically propositions which lay out minimal conditions under which a proposition such as 'SPT rule performs the best' can be believed. If the size of the minimal conditions is large, then the performance of the system in terms of system response in answering queries and in checking whether truth is maintained in the knowledge base can deteriorate quickly, resulting in poor on-line response. The size of the truth maintenance operations can also grow quickly although algorithms are available which can limit the growth to the order of $O(n^2)$. Only by simplifying the problem in the above manner can the combinatorial aspect of the problem be controlled.

**Building the system.** One of the key advantages of using an object-oriented approach is that it allows the system to be developed in stages proceeding from the basic rudimentary system to the most sophisticated and advanced system. Developing and refining the procedural knowledge in any system development process can be a time-consuming process. However, by using Smalltalk/V we found that many methods were simple to implement and could be developed by modifying existing Smalltalk methods. Here, the advantages of using an object-oriented approach were well evidenced. We could start out by first developing interim versions of methods which were rebuilt later to suit a specific purpose. The methods that were written in different contexts could be recycled and modified to later requirements quite easily. The use of Smalltalk/V also allowed us to develop the system module by module. Once the rudimentary structure of the system was laid out, it was easy to enhance the functionalities of each module.

**Knowledge acquisition.** This is an important step in the development process. For the prototype we elicited knowledge from scheduling literature and various simulation studies and scheduling experts. In a practical shop-floor application, the knowledge could come more from domain experts, further augmented by simulation studies. It is not uncommon to find this knowledge coming in different forms, some as rules of thumb, some in the form of Lotus spreadsheets, etc. In addition, there could be conflicting information and data coming from different sources. It is essential that the knowledge input to the system be validated and confirmed from another source before being implemented as rules.

**Users.** It is essential to decide in advance who the users of the system are going to be before implementing the system – are these domain experts, master schedulers, or the shop-floor personnel? The computer literacy level of the users will decide the final form and look of the user interface and the manner and the frequency with which they will interact with the system. Our system has been designed for use by production schedulers. Even so, on a mock test we found that there were many disparities between the schedulers' expectations of what a scheduling system should do and what our system was actually capable of doing. This only proves that an actual implementation and use of the system is a much more complex process than just the development of the system.

## 6.5 CONCLUSION

Our prototype design demonstrates to a significant degree how an environment possessing the combined benefits of Smalltalk/V and Prolog/V can be used for developing generic solutions for complex dynamic systems. However, the design of the prototype is only a start in the process of validating the proposed system architecture and its applicability in practice. While the use of object-oriented systems for developing scheduling systems have become fairly routine (Steffen, 1986), our contribution is mainly in providing a system for consistently updating the knowledge in the expert system as more scheduling experience is gained with an FMS system. Updating the knowledge base is no easy task and the utility of expert systems in providing error-free support depends to a high degree on how consistent the knowledge base is. This aspect assumes added importance when one considers that almost all of the available literature on on-line scheduling in FMS provided very system-specific results (Gupta, Gupta, and Rector, 1989). Thus in a specific situation, one cannot depend much on available analytical results, using instead simulation studies and domain experts' knowledge to characterize the system behavior. In such cases, it is essential to methodically assimilate new information and results with the existing rules in the system. Our architecture using ATMS provides the necessary means for such knowledge revision.

## ACKNOWLEDGMENT

I acknowledge the help and programming efforts of P.K. Kannan, Amit Vyas, Belinda Wong and Joe Latimer during the various phases of this project. Part of the work was conducted under the support of University of Arizona Small Grant No: 428180.

## REFERENCES

Clark, M. and Farhoodi, F. (1987) *Artificial Intelligence in Production Scheduling.* Proceedings of the ESPRIT Technical Conference.
de Kleer, J. (1986) An Assumption-Based Truth Maintenance System. *Artif Int,* **28**, pp. 127–162.
Elorantes, E., Hammainen, H., Alasuvanto, J. *et al.* (1989) An Evaluation of Expert Systems Design Tools. *Knowledge-Based Systems in Manufacturing* (ed. A. Kusiak), Taylor and Francis, London, pp. 331–79.
Farhoodi, F. (1990) A Knowledge-Based Approach to Dynamic Job-Shop Scheduling. *Int Jnl. of Comp Integ. Manuf,* **3**(2), pp. 84–95.
Fox, M.S. and Smith, S.F. (1984) ISIS: A Knowledge-Based System for Factory Scheduling. *Expert Systems,* **1**, pp. 25–49.
Gershwin, G.B., Hilderbrandt R.R., Suri R. *et al.* (1984) *A Control Theory's*

*Perspective on Recent Trends in Manufacturing Systems*. Proceedings of 23rd Conference on Decision and Control, Las Vegas, December, pp. 209–25.

Graves, S.C. (1981) A Review of Production Scheduling. *Op. Res.*, **29**(4).

Gupta, Y.P. and Goyal, S.K. (1989) Flexibility of Manufacturing Systems: Concepts and Measurements. *Eur Jnl. of Op Res*, **43**, pp. 1–7.

Gupta, Y.P., Gupta, M.C. and Rector, C.R. (1989) A Review of Scheduling Rules in Flexible Manufacturing Systems. *Int Jnl. of Comp Integ Manuf*, **2**(6), pp. 356–77.

Jain, S., Barber, K. and Osterfield, B. (1990) Expert Simulation for On-Line Scheduling. *Communications of the ACM*, **33**(10), pp. 54–60.

Kempf, K. (1989) *Manufacturing Process Planning and Production Scheduling: where we are and where we need to be*. Proceedings of the IEEE Fifth Conference on Artificial Intelligence Applications, Miami Beach, FL, pp. 13–19.

Kempf, K., Russell, B., Sidhu, S. *et al.* (1991) AI-based Schedulers in Manufacturing Practice: Report of a Panel Discussion. *AI Mag*, Special Issue, pp. 46–55.

Kusiak, A. (1985) Flexible Manufacturing Systems: A Structural Approach. *Int Jnl. of Prod Res*, **23**, pp. 1057–73.

Lazarev, G.L. (1988) Prolog/V: Prolog in Smalltalk/V Environment. *Dr. Dobb's Journal*, Nov., pp. 68–102.

Norrie, D.H., Fauvel, O.R. and Gaines, B.R. (1990) Object-Oriented Management Planning Systems for Advanced Manufacturing. *Int Jnl. of Comp Integ Manuf*, **3**(6), pp. 373–78.

O'Grady, P.J. and Lee, K.H. (1989) An Intelligent Cell Control System for Automated Manufacturing in *Knowledge Based Systems in Manufacturing*, (ed. A. Kusiak), Taylor and Francis, London.

Parunak, H.V. and Judd, R. (1990) Sharpening the Focus on Intelligent Control. *Inter Jnl. of Comp Integ Manuf*, **3**(1), pp. 1–5.

Pearl, J. (1988) *Probabilistic Reasoning in Intelligent Systems*, Morgan Kaufmann Publishers, Inc., San Mateo, CA 94303.

Rachamadugu, R. and Stecke, K.E. (1988) Classification and Review of FMS Scheduling Procedures. Working Paper No. 481c, University of Michigan, Ann Arbor, MI.

Rickel, J. (1988) Issues in the Design of Scheduling Systems. *Exp Sys and Intell Manuf*, **1**, pp. 70–90.

Sadeh, N. and Fox, M.S. (1989) *Focus of attention in an activity-based scheduler*, Proceedings of NASA Conference on Space Telerobotics, NASA, Bethesda, MD.

Schriber, T.J. and Stecke, K.E. (1986) *Machine Utilizations and Production Rates Achieved by using Balanced Aggregate FMS Production Ratios in a Simulated Setting*. Proceedings of the Second ORSA/TIMS Conference on FMS.

Smalltalk/V Windows (1991) *Tutorial and Programming Handbook*, Digitalk, Inc., Los Angeles, CA 90045.

Smith, S.F., Fox, M.S. and Ow, P.S. (1986) Constructing and Maintaining Detailed Production Plans: Investigations into the Development of Knowledge Based Factory Scheduling Systems. *AI Mag*, Fall, pp. 45–61.

Stecke, K.E. (1985) Design, Planning, Scheduling, and Control of Flexible Manufacturing Systems. *Annals of Op Res*, **3**, pp. 3–12.

Steffen, M.S. (1986) *A Survey of Artificial Intelligence based Scheduling Systems*. Proceedings of the Fall Industrial Engineering Conference, pp. 395–405.

Subhash, C. and Salgame, R. (1989) A Knowledge-Based Approach to Dynamic Scheduling in *Knowledge-Based Systems in Manufacturing* (ed. A. Kusiak), Taylor and Francis, London.

# CHAPTER 7

# Autonomous control for open manufacturing systems

*Grace Y. Lin and James J. Solberg*

## 7.1  INTRODUCTION

### 7.1.1  Characteristics of manufacturing systems

The manufacturing environment is undergoing fundamental changes. Transfer lines, numerical control, computerized numerical control machines, automated material handling systems, automatic tool change systems, computer integrated manufacturing (CIM), and flexible manufacturing systems (FMS) have come into being. Recently, the intelligent manufacturing system (Solberg *et al.*, 1985) concept was raised in the hope that it will increase product quality and productivity and achieve a high degree of customization. We can expect to see these trends continue because they are driven by competitive forces. It is economically advantageous for a company to adopt any technology that works effectively in the direction of greater integration, greater flexibility and greater responsibility. However, the manufacturing environment is a large, complex system of interrelated activities. Success of these systems depends on the development of computer-based decision making and efficient control structures to manage its activities.

Solberg and Heim (1989) have demonstrated the vast quantity of manufacturing information embedded in a factory. A complete catalogue of information about all items in a factory could take in the neighborhood of $10^{15}$ bytes, without even considering the relationship between items, control methods and procedures and different time scales. The decision making based on this information is therefore inherently complicated.

Uncertainty is another characteristic of a manufacturing environment. Physical manufacturing systems are large, very complex mechanical systems. They are subject to forces, wear and tear and a variety of hostile conditions. Failures and unforeseen exceptions will always occur despite any effort to prevent them. The reliability problem will remain even if better forecasting models are available. New products, engineering changes to current products, management objective changes, technological changes, organizational changes, customer-driven changes, responses to competition, new equipment or software, delays in receiving raw casting or semi-finished parts, etc. all contribute to the uncertainty of the manufacturing environment.

### 7.1.2  Manufacturing control and scheduling

Manufacturing control and scheduling is the allocation of available resources such as machines, tools, fixtures and automatic guided vehicles to specific manufacturing operations such that the part orders received by the factory are produced under certain production performance criteria. It is an important issue because it has a direct effect on the smooth functioning and the overall cost of manufacturing goods. Each step of production adds value to the items being produced, but each point of delay or storage adds cost. Statistics for discrete manufacturing show that a major fraction of the cost of a typical product consists of non-value-adding factors, and work-in-process. However, despite decades of academic effort into the study of controlling and scheduling of manufacturing systems, many real-world manufacturing practices still rely on *ad hoc* procedures and manual record keeping with almost arbitrary decision making. The main reason for this is that the real manufacturing world is so complex and uncertain that the sheer dimensionality of available alternatives is almost overwhelming. Traditional academic approaches suffer either from over-simplified assumptions made to get clean results or are too complicated for real-world applications. Also, few of them take into account uncertainty; their views are basically deterministic. Global competition and advances in technology force fundamental changes in the manufacturing environment itself and the rate of change keeps increasing. A practical control/scheduling strategy which can handle complexity, cope with the changing environment and accommodate changing multi-objectives with the aid of advanced production technologies and facilities is vital to the success of this transformation.

### 7.1.3  Open manufacturing systems

A modern factory floor often consists of various machine tools linked by various material handling systems from different makers with various degrees of automation and flexibility. Computers or micro-computers are embedded in equipment to control the operation of the equipment, but they can also be used for data storage, communication or decision-making when equipped with appropriate software. This environment naturally forms an interconnected and interdependent

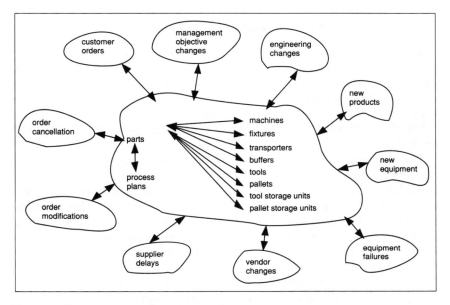

**Fig. 7.1** Dynamic manufacturing environment.

system where both information and decision making are often logically and physically distributed. Unanticipated events such as machine failures, arrival of high priority tasks, product requirement changes, delay of the delivery of raw material and management objective changes can occur anytime and affect the control and physical flows of the environment (Fig. 7.1). Neither centralized decision making nor global data storage is workable because of the complexity and uncertainty of the manufacturing environment. For example, the introduction of a new job to a manufacturing facility would entail either a readjustment of the global scheduler or an adaptive change in local control, whereby each machine would process the job according to some control strategies. In the first alternative, unforeseeable communication delays, or the breakdown of the scheduler, lead to suboptimal system behavior and loss of control; the second approach naturally allows for local decisions to be made in order to adjust to new situations.

These evolving systems are termed **open systems** (not to be confused with 'open system architecture,' which deals with the problems of integrating hardware or software from different suppliers) in the recent computer science literature. More specifically, according to Hewitt (1985, 1988), an open system has the following fundamental characteristics: concurrency, asynchrony, decentralized control, inconsistent information, arm-length relationships and continuous operation. We thus call the modern manufacturing systems open manufacturing systems.

### 7.1.4   New concepts in control and scheduling

Recognizing the nature of the manufacturing environment and supported by advanced computer technology, researchers have started to explore alternative control and scheduling architectures. Cooperative systems, heterarchical structures, object-oriented programming concepts, real time negotiation of resource assignments, opportunistic scheduling concepts and automatic learning have been proposed (Solberg, 1989).

The cooperative system view recognizes that very complex systems are beyond direct control. Instead, they operate through the cooperative behavior of many interacting subsystems which may have their own independent interests, values and modes of operation. The resulting behavior of the entire system is collectively determined. A heterarchical structure provides flexible and efficient interactions among all entities, while chaotic behavior can be avoided by defining some well-formed, simple rules. Object-oriented programming is the environment in which all things are represented as (software) objects and the actions are carried out by messages sent or received between objects. Internally, an object may carry out its function in ways which are 'hidden' from other objects in the systems. These properties ensure that objects will be portable, modular and maintainable. Real-time negotiation and opportunistic scheduling, as opposed to preplanning, blends well with the above concepts. Automatic learning provides the ability to improve performance or even to keep up with environmental changes. This is a very desirable feature in manufacturing software.

### 7.1.5   Integrated flow control approach

A successful model that is capable of handling the complexity and uncertainty of the real-world manufacturing systems however, requires the integration of all the above approaches. Also, massive communication flow may become the bottleneck of the system and myopic decision making because of loss of global information may occur. The success of the implementation relies on the organizational principles that will ensure harmonious operation and effective completion of work. In this chapter, we describe a highly adaptive integrated flow control framework which uses the above concepts and is based on a price and objective mechanism. We envision a manufacturing control and scheduling system that is different from the traditional approach. Instead of machines processing jobs according to a plan established by a global controller, we see a population of 'intelligent' entities operating in cooperation to achieve individual goals. As in a marketplace, with machines operating as vendors, a job enters the marketplace with some currency and a 'shopping list' of process requirements and tries to fulfill the requirement according to some constraints by 'bargaining' with the vendors. No attempt is made to control the aggregate behavior; the collective outcome issues naturally from the relatively simple and independent behavior of the units.

An object-based architecture is used to keep the framework separate from system details so that changes can be made without affecting the functionality of other components. Negotiation offers a reasonable way to use local knowledge in achieving reasonable performance and maintain system robustness, modularity and stability. An opportunistic approach with the aid of advances in computer technology allows coordination of intelligent entities through negotiation to achieve individual goals. A price mechanism is used to serve as an 'invisible hand' to monitor the instant local and global states to ensure a system's smooth operation in a distributed environment. System flexibility and effectiveness are also explored in the following areas: an input part process representation scheme that allows alternative routing, highly adaptive control methodologies that incorporate changes in the objectives and environment easily, distributed decision making that explores concurrency and robustness and distributed information flow that minimizes the information traffic. Two distributed control strategies under the framework are also presented and their performances analyzed and compared.

## 7.2 BACKGROUND AND DISCUSSION

In this section, traditional scheduling paradigms and information management strategies are reviewed and some negotiation-based systems are discussed.

### 7.2.1 Traditional scheduling paradigms

Some significant efforts have been made in studying the control/scheduling problem for manufacturing systems. According to Solberg (1989), they can be categorized as the optimization paradigm, the data processing paradigm and the control paradigm.

#### *Optimization paradigm*

Many people view production scheduling as fundamentally associated with optimization. The goal of their efforts is to formulate and solve the problems as optimization problems, such as mathematical programming, dynamic programming, integer programming, graph-theoretic methods, queuing network analysis, or the use of heuristic rules. Hence the tasks relate to setting up appropriate expressions for the objective functions and the constraints. Graves (1981) has provided an excellent survey of these techniques. A new contribution to the field might be either a new formulation or a new algorithm to solve a class of formal mathematical problems. However, many of the simplified problems, much less realistic ones, would never be solvable in a reasonable time. Also, the simplified assumptions which form the basis of these mathematical models, usually cannot catch the essential aspects of real life production scheduling problems.

Discrete event simulation and artificial intelligence are important variations of the optimization paradigm. Simulation is employed both as a practical tool in industrial applications and as an experimental tool for those who want to compare heuristic planning and scheduling algorithms. For an overview of simulation used for manufacturing control see Grant, 1987. Well-designed simulation can capture as much of the complexity of a real problem with which one wants to contend; however, the model is only descriptive and cannot directly provide 'optimum' results. Artificial intelligence is another variation of the optimization paradigm (Steffen, 1986). Expert systems in particular, offer attractive alternatives to formal optimization because the rules can incorporate all sorts of qualitative constraints (Wysk, Wu, and Yang, 1986 and Kusiak and Chen, 1988). However, it only generates feasible solutions with no measure of optimality. Also, both the simulation and artificial intelligence approaches have unacceptable computational inefficiencies if the models they deal with are realistic enough.

### Data processing paradigm

There is an entirely different community of people who see scheduling problems as data management issues. Focusing upon the enormous quantity of information that must be stored and retrieved, they develop computer-based methods to organize and manipulate the data. Production scheduling software is viewed as a special case of the broader database issues of an enterprise. It is carried out through a comparatively straightforward collection of requirement data stemming from accepted orders like material requirement planning (MRP) systems. Much of the research work conducted under this paradigm is commercial in nature. The more basic work on such issues as relational databases, structured analysis and concurrency control can be found in computer science literature such as Date (1986).

These investigators address a number of issues that are ignored by the optimization community such as data dependencies, redundant or contradictory information and efficient data storage structure. In contrast to seeking the optimum algorithm, they employ matching and sorting algorithms.

*Control paradigm*

A third group of researchers sees production scheduling problems as control issues (Gershwin *et al.*, 1986). They focus upon the command and feedback loops that are necessary to keep a production system operating 'under control' or as the designer intended. The problem is perceived basically as one of following a desired trajectory in time or maintaining equilibrium in the presence of disturbances. When formal mathematics is used, it tends to be continuous differential and integral equations, although integer variables and discrete behavior may be accounted for as nonlinearities. The people and literature adopting this view are commonly associated with electrical, mechanical, or chemical engineering.

There were a few examples of attempts to apply formal control theory to scheduling in the 1960s; the informal application of control theory concepts has been widely regarded as the proper way to approach certain manufacturing system design issues. However, the paradigm focuses on the monitoring and tuning aspects of systems.

### 7.2.2  Manufacturing information management strategies

Manufacturing is a complex and dynamic enterprise. The most obvious aspect of manufacturing information is its vast quantity. To successfully control a manufacturing system, one needs a consciously chosen strategy to handle the enormous amount of information. Solberg and Heim (1989) classify manufacturing information management strategies into five different approaches: subsystem isolation, total integration, hierarchical decomposition, heterarchical decomposition and hybrid methods (Fig. 7.2). We summarize these approaches below.

*Subsystem isolation*

Subsystem isolation, which can also be described as an absence of control strategy, is to select a portion of the problem and to manage that portion by ignoring everything else. In practice, people often do so unconsciously. Whenever a company purchases software packages such as an MRP system, automatical guided vehicle (AGV) system, etc. it is almost always following a strategy of subsystem isolation. This approach often causes system under-performance.

On the other hand, building software between two pre-existing systems, neither of which is designed with interface considerations in mind, is often just as difficult as starting over. Moreover, the number of potential interface pairs grows as the square of the number of the units that have to communicate. Thus, the subsystem isolation strategy, while popular, leaves formidable barriers to achieving real system coordination.

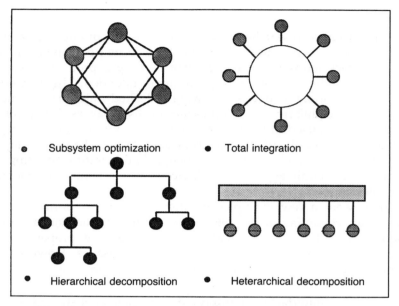

Subsystem optimization        Total integration

Hierarchical decomposition        Heterarchical decomposition

**Fig. 7.2**    Manufacturing information management strategies.

### Total integration

Total integration represents the opposite extreme. Many people have advocated this approach to ensure consistency throughout the entire manufacturing enterprise. They attempt to build a monolithic manufacturing control system to deal with all manufacturing information. However, the complexity of the manufacturing environment makes this approach impossible. Even if it were possible, the dynamism of the manufacturing system would soon render portions of the totally integrated system ineffective or worse.

A more structured approach, based upon decomposition and abstraction, offers better opportunities for coping with the complexity.

### Hierarchical decomposition

Hierarchical decomposition (tree structure) offers the advantage of a simple, consistent organizational structure which supports the notion of decomposition and abstraction. It has been regarded as a natural way to manage complex manufacturing systems (McLean *et al.*, 1983). At the lowest levels, physical devices carry out simple actions under command of software modules at the next higher level. A unit in an intermediate level is responsible for receiving status information from lower level units and issuing commands to them while passing status information upwards and carrying out commands from above. Finally, a single unit at the top of the hierarchy bears ultimate responsibility for fulfilling the goals of the whole system.

Under hierarchical decomposition, the system is partitioned into modules such that each subsystem does a small and comprehensible amount of processing. However, all communication must go through each channel of the hierarchy. In cases where tight control is desirable (e.g. coordinating a robot with a lathe for part loading), the hierarchy makes eminently good sense. In other cases where the associations are loose (e.g. machines in different departments of a factory), the advantages can be outweighed by the burden of overhead. Also, it requires a relatively complete design prior to implementation and, once completed, is not easily changed. Another disadvantage is that a crash at some level can cause the entire system below to come to an abrupt halt.

## Heterarchical decomposition

Recently, researchers such as Duffie and Piper (1986), Hatvany (1984), and Vamos (1983) have proposed the use of heterarchical decomposition to eliminate the deficiency of the hierarchical decomposition. The concept emphasizes modularity by encapsulating information into relatively autonomous 'objects.' Instead of the master-slave relationship in the hierarchical decomposition, heterarchical decomposition adapts a communications structure that allows components of a system to exchange information as peers. Heterarchically decomposed systems function as independent cooperating processes or agents without a centralized or explicit direct control. Control decisions are reached through mutual agreement and information is exchanged freely among the participants. Hence effectiveness, flexibility and robustness can be achieved. However, massive communication flow may become the bottleneck of the system and myopic decision making due to loss of global information can occur. Reducing communication flow, avoiding system deadlock and deriving good control strategies are keys to the success of the implementation of a heterarchical control structure.

## Hybrid control

Both the hierarchical control structure and the heterarchical control structure have strengths and weaknesses subject to the context in which they are applied. The hierarchical approach is appropriate for tightly coupled systems with strong interactions, while the heterarchical approach offers advantages where coupling is more loose. Therefore, combining the two strategies appropriately should provide the most flexible design methodology and opportunity for incorporating changes that are certain to occur (Naylor and Volz, 1988). However, at the same time, the simplicity of a consistent philosophy is lost when the approaches are mixed. Careful and intelligent design are required to ensure the success of its implementation.

### 7.2.3 Negotiation-based systems

Research in negotiation-based systems is concerned with how autonomous agents can interact effectively. In 1972, Farber and Larson proposed the use of a bidding scheme for computing resource allocation in a distributed computing system. Smith and Davis refined the idea into the contract net (Smith, 1980 and Smith and Davis, 1981). They introduced a general negotiation protocol for communication and control among cooperating agents and defined it in the context of a distributed sensor system (Davis and Smith, 1983). Based on the contract net protocol, Malone *et al.* (1983), Stankovic and Sidhu (1984), Ramaritham and Stankovic (1984), Ferguson *et al.* (1988) and Waldspurger *et al.* (1989) developed prototype systems and methodologies for computer task allocation in a distributed environment. Shaw (1984, 1987), Parunak *et al.* (1985, 1986), Duffie and Piper (1986), Maley (1987), Maley and Solberg (1987), Ow *et al.* (1988) and Upton (1988) use it in the manufacturing control domain.

In this chapter, we first introduce a price mechanism into a negotiation-based distributed manufacturing control and scheduling system. It is used as a mechanism to monitor instant system state and explicit and implicit operation costs, to react to part priority and as a bridge for system entities interaction during the decentralized negotiation process. Multiple resource matching and the construction of negotiation functions to ensure the efficient implementation of a negotiation metaphor are explored. Realistic manufacturing environment setting is considered and concrete implementation architecture under a heterogeneous control structure will be presented.

### 7.3 INTEGRATED SHOP FLOOR CONTROL FRAMEWORK USING AUTONOMOUS AGENTS

In this section, we describe an adaptive control and scheduling system framework (Lin and Solberg, 1992). The framework is based on a distributed information processing, distributed decision making and heterarchical market-like model. That is, each functional unit of parts and resources is equipped with an intelligent (software) agent which acts as the representative of the entity. The agents communicate and negotiate with other agents in real time through message passing and a bidding protocol to achieve mutual agreements for task sharing (Fig.

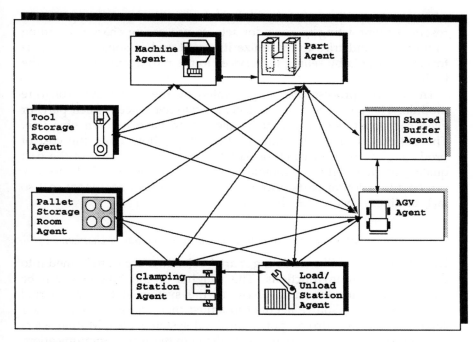

**Fig. 7.3** Communication among system agents.

7.3). The agents can be considered as information filters which acquire and recognize factors pertaining to decision making and make decisions based on these current factors. The intelligence of the agents comes from the specific information about their clients, task specific algorithms, knowledge in its knowledge base and the experience accumulated in the system (learning). No information aggregation and no advanced planning are required.

### 7.3.1  System modeling

In this section, we discuss the system modeling framework which includes the market-like model, a generic bid construction mechanism based on a combination of price and objective mechanism and a critical resource-centered resource unification scheme.

#### *Market-like model*

The model we present here closely resembles an open marketplace: a part agent enters the system with some (fictitious) currency and a flexible process plan (Lin and Solberg, 1989, 1991). It tries to fulfill the processing requirement to achieve its unique set of weighted objectives (time, cost, quality, etc.) by 'bargaining' with resource agents. A resource agent acting as a vendor sets its processing charge according to its status and tries to maximize its profit and/or some system performance criteria by selling its services to the part agent that offers the highest price.

The flexible process plan is generated by the CAD/CAM software module. The part's objective reflects a customer's need. Each part, or even each operation of a part, is allowed to have its own objective. The objective can be the minimization of the flow time, the cost of producing the part (or batch of parts), or the maximization of the part quality, or the weighted combination of the above factors. Part currency can be a function of parts priority, direct cost (plus operational cost so far), tardiness cost, or finished goods price.

### *A combination of price and objective bid construction mechanism*

Under our market-like modeling, parts' objectives are transformed into a set of evaluation functions. The weights of the functions can be adjusted dynamically on the basis of the system state and external conditions. The resource agents can adjust their charge prices by their capability and the current global and local system states. Mutual selection and mutual agreement are made through two-way communication.

Figure 7.4 depicts the combined price and objective mechanism. Research in scheduling to date either seeks to optimize the system

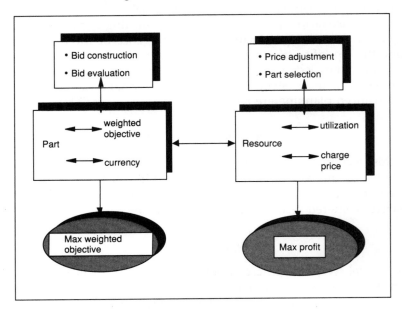

**Fig. 7.4**  The combination of price and objective mechanism.

objective such as mean flow time, throughput, or tardiness and, hence, neglects either the heterogeneous job objectives or job priority. The underlying assumption of the research is that the system has a single criterion to meet. Our combined price and objective mechanism responds to heterogeneous job objectives, job priority and other factors important for decision making and provides a dynamic environment for the agents to cooperate. It allows parts to devote their currency flexibly to the most important and appropriate resources while re- sources can adjust the charge prices dynamically based on their status to optimize their utilization. It also models the real-world constraints into the limited currency that a part agent can spend. It is generic in the sense that all system factors can be transformed into the weighted evaluation functions and prices and control/scheduling algorithms and methodologies can be modeled easily by this mechanism.

The incorporation of a price system is also a distinct feature of our system compared to other negotiation-based manufacturing control and scheduling systems. Based on microeconomics theories (Henderson and Quandt, 1980), the price serves as an invisible hand to guide the negotiation to improve system performance. Under well-defined agent utilities and rules, the control/scheduling system is expected to function based on the equilibrium of the objectives and prices. For example, if the system contains a powerful machine which can complete some operations much faster than other machines, then it will attract many part agents to send their parts to this machine. Very soon, the queue will build up and the expected waiting time will increase. Accordingly, the prices that the machine charges will be increased and the objective values of the interested parts will be decreased. These factors will drive parts to other machines and diminish the demand for this machine. The result is that the system will adjust itself smoothly based on the flow and load of the system, and the overall performance of the system results from the equilibrium of the price system and orchestration of all components involved.

Different price systems and rules will result in different control strategies and system performance. Prices and objectives can be set by an expert, an evaluation scheme, or an expert system and are independent components of the framework. This provides an environ- ment for exercising self-learning.

### Multiple-way and multiple-step negotiation

The decision of the model is made through multiple-way and multiple- step negotiation. It is a multiple way negotiation since a part usually needs the simultaneous service of several resources such as machine, tool, fixture, and 'just-in-time' service of the transporter. Therefore, a part agent will negotiate with different types of resources at a time. It is also a multiple step negotiation since both parts and resources have opportunities to deny or accept an offer during the negotiation process based on individual concerns. Final agreement is made only when parts and resources all commit themselves through multiple-step negotiation.

### Critical-resource-centered resource unification scheme

An important issue during the negotiation process involves ensuring that all needed resources for a task are available at the right place and at the right time. This is done by using a critical resource-based, dynamic resource unification scheme. That is, both part and resources have the ability to activate the bidding procedure; however, the critical one is the one to drive the bidding procedure. And, since a part has the detailed processing requirement information, the final resources matching decision is done by the part. More specifically, knowing the available resources and their charge price through task announcement and bid collection, a part agent will find a best matching for all required resources to maximize its weighted objective function and minimize the resource charge price. It will make arrangements for transporter, machine, tool, fixture, etc. to be available for the operation at the right place and time. On the other hand, the resource will select parts to serve to maximize its profit; the decision of the most critical resource will play the most important role in the part resources matching.

For example:

1. If a required tool is a bottleneck resource, i.e. with a high charge price and a long waiting time, then if an alternative tool exists, the alternative tool will be chosen for the service of the operation even if the precision is lower. However, if that is the only tool which can perform the operation, the part agent will have to take the time slot and pay the price that the tool agent sets and try to reserve other resources for that period. The tool agent, realizing the criticality by its high processing charge and long queue, will try to sequence the tasks to maximize profit using the algorithms in the intelligence file.
2. If an AGV controller is incorporated into the system and it is a critical resource, then the part agent will try to arrange other required resources according to the AGV schedule given by the AGV controller.

### Summary

Under the critical-resource-centered resource unification scheme, both parts and resources are aggressively participating in the negotiation procedure to achieve their individual goals through mutual selection;

the most critical resource will have the biggest impact on decision making. Since a part only needs one machine and one tool, etc. at a time, this decision-making scheme can decrease information flow, simplify a decision-making procedure and avoid some deadlocks. This approach is very different from the subcontractor approach. In the subcontractor approach, a hierarchical scheme is used. For instance, a part first searches for a machine. Therefore, premature commitment due to the lack of second or third level resources, or conflicts due to several machines searching for the same second or third level resource for the same part, often occurs. The original system conditions that led to the current decisions may have been changed by the time a sub-contract is obtained. Thus, renegotiating is needed, and this causes system inefficiency.

The use of a price system and the resource unification scheme also resolves bottleneck problems. Bottleneck resources have been recognized as an important factor in system performance, and they play the central role in several scheduling systems (Fox, 1983 and Adams et al., 1988). The central theme is that the throughput of a bottleneck resource determines the throughput of the system. However, a bottleneck resource does not necessarily span the entire scheduling horizon and tends to shift before the system is entirely scheduled if advanced scheduling is used. Under our modeling, the bottleneck resources, which can be a machine, tool, transporter, etc. are monitored throughout the negotiating and processing process. The system will not only fully utilize the bottlenecks, but will also direct jobs to other machines (if other options exist) and thus minimize the bottleneck to utilize system flexibility and enhance system performance.

It is believed that a part-centered scheduling procedure performs better in a lightly loaded system, and a machine-centered scheduling procedure performs better in a busy system. However, in a distributed environment, global states such as the up-to-date system load and critical resources are not available to the local processors. This makes it difficult, if not impossible, to apply the above approach. It is also worth pointing out that the overall system load does not necessarily reflect the load of the critical resources. The use of our price mechanism and the cooperating part and critical resources decision-making process address this issue in a rather sophisticated manner. The price fluctuates according to the system state and thus reacts rapidly to the change of system load. In a lightly loaded system, a part may get many low price bids. It can choose resources that fit it best. In a heavily loaded system, critical resources will get many requests and therefore get chances to increase the service charge, and explicitly (through part selection) and implicitly (through price adjustment) select which part to serve.

To summarize, the presented model not only has the adaptivity offered by market-like negotiation-based modeling, but it also allows heterogeneous job objectives, admits job priorities, recognizes multiple resource types, and allows multiple step negotiation between parts and resources to ensure system effectiveness.

### 7.3.2  System architecture

The control modules in the control architecture include part agents, resource agents, intelligence agents, monitor agents, communication agents and database management agents (Fig. 7.5). The part and re-source agents negotiate with each other to manage the operations of part entities and the functioning of the resources. The intelligence agents provide different bidding algorithms and strategies; the monitor agents are used to supplement the system status. The database management agents manipulate inter-agent information; the communi-cation agents carry out all communication between entities. In this section, we describe the organization and responsibilities of each module in detail (Lin and Solberg, 1992).

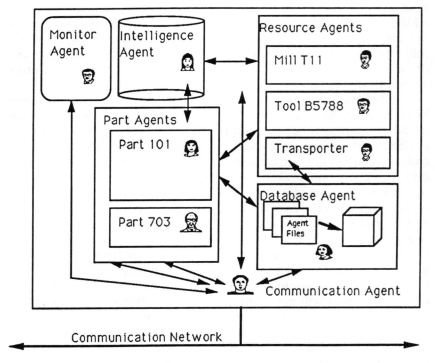

**Fig. 7.5**  Control components.

*Part agents*

When a part (or a functional unit of parts) enters the system, a part agent is created with a list of processing requirements, currency and objectives. It manages the part throughout the period when the part is in the system. The agent is an active process that resides in a control computer. The control computer can be a very small one which is attached to the part or pallet and only responsible for the particular part, or it can be any fixed or non-fixed computer in the system that runs the part process. In the latter case, a new agent process is created in the new control computer, but it inherits all the knowledge of the current agent. In all cases, the part agent of a part identifies the part by its part ID which is unique for each part in the system.

The part agent has a local database called an **agent data file**, a local knowledge base called **agent intelligence file**, and an inference engine. The agent data file records information of the part such as part identity number, part process plan or a link to the dynamic process planning system, current part position, part history, part objective (cost, time, due day, quality requirement, etc.), weights of the objectives and the amount of currency that the part possesses. It also records the current 'state' of the system. The part intelligence file stores the current state of the bid in progress and built-in algorithms for the part agent to make decisions during the negotiation process. More complicated or intelligent algorithms and algorithms with different characteristics are stored in a distributed knowledge base that is manipulated by part intelligence agents which will be described in more detail in a later section. The inference engine has the ability to invoke a negotiation process and can perform reasoning processes and enhance the agent's capability by acquiring alternative algorithms from the distributed knowledge base through a simple bidding protocol.

The negotiation process for parts to select resources includes task announcement, bid collection, bid evaluation, task offer submission and task commitment. All built-in algorithms for the process may be adjusted according to the system states by an expert or by rules in the knowledge base. Alternatively, a learning module can be attached to select the appropriate algorithms. Part agents can obtain the specialized algorithms by matching the pre-conditions and post-conditions of the algorithms using a simple bidding procedure similar to the bidding of resources: the part agent announces the requirements of the algorithm using the bid announcement protocol, the part intelligence agents with matching algorithms then submit bids from which the part agent chooses one. The part agent then incorporates the newly acquired knowledge into its inference engine. Under this model, the bidding procedures are black-boxes where different algorithms can be inserted or removed. This allows the characteristics of the procedures to be changed dynamically.

*Resource agents*

A resource is either a single resource such as a machine, a tool, a fixture, a transporter, or a collection of resources that is viewed as a functional unit or controlled by a single controller such as an AGV controller or tool magazine controller. Each resource is represented by an agent who resides in a control module. Similar to the part agent, each resouce agent has associated with it an agent data file, agent knowledge file, agent's inference engines and the ability to participate in the negotiation process to sell the processing time. The resource agent file records resource information such as price, position, equipment status, queue status, remaining tool life, machine capability, reliability, precision and records of all pending bids that have been submitted. The resource intelligence file stores built-in algorithms for the resource agent to use during the negotiation process and to set up a reasonable selling price. More detailed algorithms with different characteristics are stored in a distributed knowledge base and manipulated by resource intelligence agents. Similar to the parts inference engine, the resource inference engine controls the invocation of negotiation procedure and the selection of algorithms to suit the changing environment. The negotiation procedures for resource management include availability announcement, task announcement monitoring, bid construction, bid submission, task offer collection, task offer evaluation and task offer acceptance.

*Intelligence agents*

Intelligence agents are responsible for the manipulation of knowledge bases. The intelligence knowledge base contains a collection of algorithms used by parts or resource agents to make decisions. These decisions include when to announce a bid, how to construct bids, next operation(s) selection, bid evaluation, bid award, etc. For each function there may be several associated algorithms with different characteristics. For example, it can be either a fast brute-force approach without look ahead (to be used when quick response is required or when the computing power is limited), or a more sophisticated algorithm with global considerations (to be used when the computing power is sufficient). Some may even have the capability to learn from past experience. There are pre-conditions, post-conditions, advantages and shortcomings associated with each algorithm.

After the part or resource agent sends out an algorithm request bid which specifies the objectives, the intelligence agent will match the objectives with the pre-condition and post-condition of the algorithms in the knowledge base and choose one or more to send to the part agent. In certain implementations, the intelligence agent may actually run the algorithms for the part. In this case, the part sends the parameters to the intelligence agent and the latter sends back the results (decisions). A link can be created between the part agent and the knowledge base for later communication if necessary. Note that the algorithm selection procedure or the built-in algorithm overwritten procedure is controlled by part or resource built-in inference engines. And the intelligence of agents can be enhanced by an input of new algorithms, a learning module built in the part or resource intelligence file and experience accumulated from monitor agents.

The intelligence agents can be distributed in control computers which are independent from the part and resource agents. They communicate with part or resource agents through simple bidding protocols and provide them with the needed knowledge or algorithms. The distributed approach makes the system robust; even if some of the computers are down, the remaining system can still function smoothly. Changes can be made and algorithms can be added or removed from the system easily without having to modify other parts of the system or disturbing the operations of the manufacturing systems.

### Monitor agents

In a truly distributed environment, all entities in the environment make decisions independently on the basis of the local information and the very limited global knowledge they possess. Although accurate global information may improve the effectiveness and efficiency of the decision-making process, the centralized nature of the global information is exactly what makes the distributed decision making attractive and thus global information should be avoided. Our solution to this conflict is to let the entities make decisions independently but provide certain timely global information to the entities upon request. This task is accomplished by distributing a few monitor agents in the system. The monitor agents watch the information flow around their assigned region and collect and analyze the information to provide timely and accurate global information. They are also responsible for other global-oriented tasks such as setting up job priority, price and currency, maintaining consistency between part currency and resource charge price and collecting statistics, scheduling maintenance, searching non-active jobs and resources and interfacing with other manufacturing components and agents within the control component. It can provide part, resource and intelligence agent knowledge accumulated from history and change the agent's default control algorithms.

### Communication agents

A communication agent resides in each control computer. The duties of the communication agent include distribution of incoming messages, sending of outgoing messages and arranging for arrival and departure of the agents. The existence of the communication agents makes the underlying network architecture transparent to system entities. This also allows the system to be ported easily to different manufacturing environments.

### Database management agents

There is also a database management agent in each of the control computers which maintains and manipulates information for the system entities in the same control computer. Its duties include the update, retrieval and manipulation of part and resource information, the creation, transaction recording and destruction of part files, etc. When global knowledge about the manufacturing environment is needed, the database agent contacts the monitoring agent to obtain the most up-to-date global information. The global information will be buffered in the local database but has a lifetime that depends on the nature of the data. The consistency issue of the global information is solved by consulting monitoring agents and the communication agents periodically.

### 7.3.3  Communication

Although the model calls for heterarchical architecture, the underlying physical network does not need to be so. The basic network requirement for such a system is a communication network that supports reliable data transmission. The basic network functions are read, write, broadcast and multicast. Any network protocol can be used as long as it guarantees the proper arrival and ordering of the messages at the application level.

The bidding communication protocols needed are bid construction and bookkeeping. The bid construction protocol includes bid announcement, acknowledgment, submission and award. The bookkeeping protocol includes task arrival, file disposal, and information propagation.

The package types are task arrival, bid announcement, bid evaluation, bid submission, bid acknowledgment, bid award, bid rejection, file disposal and information update.

For some manufacturing environments, high speed and high load capacity may be required for quick real-time control.

### 7.3.4  Features of the system

The system architecture we described above has the following features: robustness – resource failures will not disrupt operation of the system; adaptiveness – changes in the environment, objectives, or the system itself can be incorporated quickly and smoothly; and deadlock prevention – no parts will be stuck in the system indefinitely. Furthermore, other resources such as tools and fixtures are integrated in the framework. The modular approach allows the control and scheduling system to be used in both simulation and real-time control.

## Achieving robustness

When an error such as machine failure, tool failure, fixture failure or part failure occurs, the corresponding agent withdraws itself from the communication network or stops the bidding operation and may call for an error recovery module for diagnosis. The rest of the system continues operation without disruption caused by the error.

To assure robustness when a computer goes down, each resource agent can keep a copy of the part file which it is currently serving. When the computer in which the part agent resides goes down, the current operation agent can detect the failure and act as the new part agent or help to create a new part agent for it. If the current operation resource computer is also down, the previous operation agents can be invoked. An age mechanism is used to maintain the files. The age is increased by one when a new operation is finished and moved to another machine. The part agent file is removed from the machine when it becomes too 'old.' The definition of the 'old age to die' is system-dependent. For a reliable system, the old age can be set to a small number such as two or less. When the resource agent computer is down, a new agent can be created residing in another control computer based on the knowledge of the resource stored in other control computers. The resource and the related parts can either keep the commitment and continue the operation or restart the bidding process to look for a new part or resources.

## Change incorporation

When change occurs, only the corresponding agents are affected. The deletion of a physical component only causes the related agents to resubmit bids. For the addition of a new component, the only thing needed to be done is to attach a control module to it and connect it to the network. If the objective or the price of a part changes, or the data file of the resources needs to be changed, the corresponding agents can be found by tracing them through their paths or by broadcasting the change information with the corresponding part numbers or resource numbers to identify the affected entities. Algorithms can be added or deleted from the knowledge base easily by informing the intelligence agents. For most changes, continued operation of the system is guaranteed.

*Deadlock prevention*

To avoid a part being stuck in the system for a long time, an age factor can be added in the part objectives, and price functions and the corresponding weights will be increased when the part becomes old in the system. This will lead to the essential acceptance of the part to the resources.

*Integration of machine tools*

We separate the control flow, database, knowledge base and network communication protocol in the proposed framework. In this way the system can be implemented in an existing computer-aided manufacturing environment and be adapted easily to run in future factories. The integration of various machine tools from different vendors (which is an important requirement of a framework for scheduling and controlling the coordination of part flow and information flow) can be done by the unified communication protocol.

### 7.3.5    The implementation of an integrated shop floor control system

An integrated shop floor control and scheduling system, called a flexible routing adaptive control system (FRACS), has been developed based on the proposed architecture. The system contains a control system and a simulation system. The control system consists of a collection of autonomous agents who negotiate with each other to reach job processing decisions using one of the several supported multi-stage negotiation protocols. The simulation system is coupled with a very high level programming language that contains constructs to model the manufacturing environment and flexible processing process plans. The modeling of the environment includes the definition of machines, transporters, tools, tool magazines, tool storage rooms, pallets/fixtures, pallet storage rooms, machine breakdowns, factory layout geometries, initial system states and other factors common in manufacturing environments. By employing the language as the front end to the simulation system, very realistic modeling of manufacturing systems is possible. The integration of the two systems is based on three interface modules: the control interface, the monitoring interface and the communication interface.

The control interface provides a set of functions for controlling the activities of part entities and resources in the simulation system or in the controlled job shop. The monitoring interface module supports a set of queries for obtaining states of part entities, resources or the global states of the environment.

The same interfaces can be used for the control and scheduling of both the simulation system and the actual shop floor. By integrating the control framework with the FRACS simulation system, we created a sophisticated test bed for experimenting with different control and negotiation strategies. By decoupling the control and scheduling system from the simulation system and linking it to the real-time control and monitoring interface of a manufacturing environment, a real-time manufacturing control system can be created. This arrangement allows us to test the proposed concepts in the controlled simulation system and also allows the control system to be easily ported to real-time control of manufacturing systems.

All communication between agents in the controlled system or between the control system and the controlled units and sensors goes through a communication interface module which is an abstraction of the underlying communication network and provides utilities for sending and receiving messages, broadcasting messages and transmitting commands and bids, etc.

The simulation system and the control system are both event driven. The simulation system is driven by the time-dependent events generated by the shop simulator. And the control system is driven by the events caused by receiving messages from other agents or from interface modules.

The modular description of the framework is shown in Fig. 7.6.

## 7.4  PRICE-BASED NEGOTIATION SCHEMES

In this section, we present two part-machine negotiation schemes: part-initiated, machine-centered negotiation scheme and machine-initiated, part-centered negotiation scheme under the integrated shop flow control model described in the last section.

The generic control framework is as follows: initially, the unit service charges of the resources are labeled according to the unit processing costs and capabilities of the resources. Parts enter the system carrying a flexible process plan and a budget for each operation. The budget of the operation for each resource is given based on the knowledge of the price of the resource and the priority of the part. The negotiation between parts and machines then follows according to the control scheme to achieve each entity's goal under the price constraints. Note that high priority jobs are allocated with more currency so that they have more purchasing power to buy critical resources to achieve their operation objectives. Since a budget is set aside for each operation, a part will not run out of funds before completing its processing. Moreover, to prevent a low priority part staying in the system too long, an 'extra fund' is allocated for each part to use in case some critical

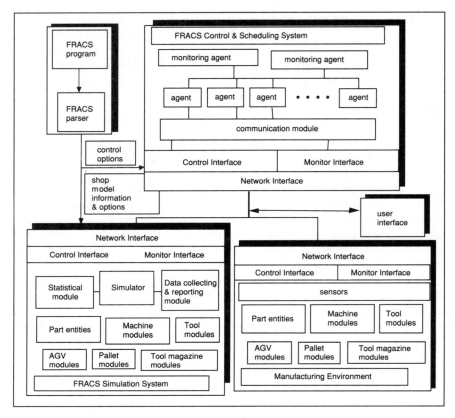

**Fig. 7.6**  An integrated control framework.

events occur. When a machine breaks down, the part which is being processed is considered as defective and is discarded. Other parts which have committed to this machine will start their negotiation again to look for new machines to serve their operations. The part-initiated, machine-centered negotiation scheme is initiated by parts. A part which is ready (or about to be ready) announces its ready operations and corresponding offer prices to machines. Machines which have open slots select parts based on the prices that parts offer. The part then selects machines that best match its objective. The machine-initiated, part-centered negotiation scheme lets the machines announce their availabilities and current processing charges, then parts select the most attractive machines to process one of their ready operations. The machines then choose the best bids from parts.

The negotiation schemes are each presented in three groups of interacting procedures of parts, machines and transporters, respectively. The procedures are event-driven so the traditional way of static al-

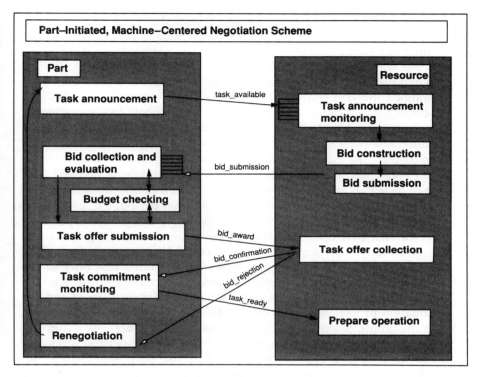

**Fig. 7.7**  Part-initiated, resource-centered negotiation scheme.

gorithm description is not appropriate. Therefore, we list the procedures in modules and briefly explain the interaction of controls among the modules. Events for the two schemes and the flow of control are illustrated in Figs 7.7 and 7.8 respectively.

### 7.4.1  Part-initiated, machine-centered negotiation scheme

Under this scheme, parts will initiate the part-machine negotiation process by sending task announcements to a selected set of machines. The machines monitor task announcements and select a subset of tasks that best match their objectives to submit bids. The parts who receive the bids select the 'best' bid to process the operations. Below we list and explain the modules in groups of parts, machines and AGVs.

**Part procedures**
**Task announcement.** Task announcement is invoked if one of the following conditions occurs

1. a part arrives;
2. the current operation is near completion;

3. the machine that the part is on has broken down;
4. no satisfiable reply is received after a prescribed time interval of the previous announcement.

When the task announcement module is activated, it announces the operations that are ready to be processed, the corresponding budget, processing specification and processing requirements to the machines. It also inquires about the AGV availability and transmission time by sending a request_AGV_information message to AGV controllers. It then invokes the bid collection and evaluation module to collect bids.

**Bid collection and evaluation.** When this module is invoked, the part collects bids submitted by machines as well as information sent by AGV controllers. At the end of the bid-collection period, bid_evaluation subroutine is invoked to select an operation and a corresponding machine that best optimizes the objective function of the part (taking the availability, cost and service time of AGV into account). It then invokes the task offer submission module to submit a task offer to the selected machine. If no acceptable reply is received during the bid collection period, the task (re)announcement module is invoked and the negotiation process is started over again. At the same time, if the part is in the chunk or output buffer of a machine, a request_AGV_service message is sent to the AGV controller to obtain transporting service to move itself to the system buffer so that the part won't block the machine. When the number of reannouncements exceeds a prescribed number, the budget checking module is called to adjust the budget to increase its chance of getting affordable bids.

**Budget checking.** When this module is invoked, it calls the extra fund allocation subroutine which allocates a portion of the surplus fund to the new operations if there are surplus funds left. Otherwise, it reports to the monitor agents to see if it can get extra funds for the operations.

**Task offer submission.** After a bid is chosen, the part submits the task offer to the chosen machine confirming the service charges and service starting time. The part then invokes the task commitment monitoring module to wait for responses from machines to close the negotiation.

**Task commitment monitoring.** The part will wait for task offer acceptance messages from the chosen machine in this module. After it obtains commitments from the needed resources, it then sends a request_AGV_service message to the AGV controller to request transporting service (based on the part_AGV_evaluation procedure). Finally, it updates the state information of the part, surplus fund, and prepares to be physically moved to the selected machine for processing. If the task commitment did not arrive in the monitoring period, or a bid_rejection message arrives, the renegotiation module is invoked.

**Renegotiation.** The renegotiation module calls the budget checking module to adjust its current incentive for resources to provide services or requests for more funds if the part does not get any satisfiable bids after several renegotiation attempts. It then restarts the part negotiation process by calling the task announcement module.

## Machine procedures

**Task announcement monitoring.** This module monitors task announcement messages from parts and screens task announcements that the machine is capable of doing. It calls the bid construction module periodically to choose tasks to submit bids.

**Bid construction and submission.** The bid construction module will first call current_charge function to adjust its current charge. It will then select tasks that pay the highest unit processing fees. The number of tasks selected is the number of slots available in the input buffer. It then calls the procedure prepare_bid to compute the submitted charge to the selected tasks. The prepared bids are then submitted by the bid submission module. It will then invoke the task offer collection module to collect task offers.

**Task offer collection.** The task offer collection module waits for task offers to arrive. When it receives an offer, it reserves an in_buffer position and adjusts the offer if necessary. It then sends back a task offer acceptance message to the part.

**Prepare operation.** This module will update the schedule of the resource and prepare serving the chosen parts.

## AGV procedures

**Respond to request_AGV_information.** When an AGV receives the request_AGV_information message from a part, it sends back information that includes the earliest time for an AGV to arrive at the part's location, the speed and the transportation cost and time.

**Respond to request_AGV_service.** The AGV controller assigns the part an AGV which is nearest the location of the part when there is one available. It then sends the AGV information to the selected part and waits for confirmation. Confirmation is needed since there may be more than one AGV controller that picks the same request to serve. It will start the physical movement after it receives a part's confirmation message.

Under the part-initiated negotiation scheme, the resources needed to price their charge carefully. In order to win the best bids to utilize themselves efficiently, they should try to charge as much as possible to parts with a high budget by offering them higher priority. But they also need to limit the charges so that the parts won't go to other resources. On the other hand, once a part is selected, the resource availability for the part is guaranteed at the specified time interval so there is no need for parts to perform price adjustment when processing task offers. Therefore, the parts try to select the best set of resources among bids that they receive to optimize their objectives under the budget constraints.

### 7.4.2 Machine-initiated, part-centered negotiation scheme

In this section, a machine initiated negotiation scheme is presented. Under this scheme, machines periodically announce their availability and concurrent charge price to parts. When parts are ready to schedule the next operation, they select the machines that fit their objectives (time, price, etc.) best, and at the same time, try to adjust the price they offer to win the bids that they submit. Machines on the other hand, select offers from parts based on price to optimize their profit. The negotiation scheme is described below.

**Machine procedures**
**Machine availability announcement.** The machine announces its availability when there is a space available or expected to be available in the

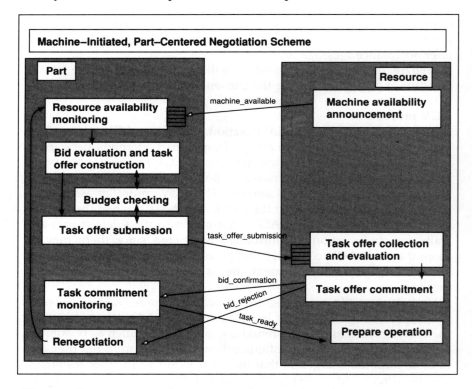

**Fig. 7.8** Resource-initiated, part-centered negotiation scheme.

near future; this is controlled by a timing control module. The availability announcement includes the number of spaces available, current average charge price (over a certain period of time), capabilities and earliest available time. It will also set a timer to invoke the Task offer collection and evaluation module.

**Task offer collection and evaluation.** When this module is invoked, it collects bids for a certain period of time. It then evaluates the task offers, selects parts with best offers to serve and invokes the task offer commitment module to send the task commitment messages. The number of offers that it can accept depends on the space in the input buffer.

**Task offer commitment.** After a task offer is accepted, the machine reserves a slot in the in_buffer for the part if it is not already on the machine. This module then sends the task commitment message to the chosen parts.

**Prepare operation.** This module is the same as the one in the part-initiated scheme.

**Part procedures**
**Resource availability monitoring.** A part starts the bidding process by monitoring the resource availability announcement information in the network a short time before the completion of the current operation or when the part is idle. This module screens resource announcements and passes the up-to-date information of eligible resources to the bid evaluation and task offer construction module.

**Bid evaluation and task offer construction.** This module uses the procedure bid_evaluation as described in the last section to select a machine and a ready operation based on the resource announcement. It then calls the offer_construction procedure to construct a task offer and invokes the task offer submission module to submit the task offers to the selected machine and tool. Task offer is set as the sum of the machine's announced charge price and a ratio of the part's surplus fund. The ratio is a system parameter and is adjustable. If no satisfiable bid is found during the bid evaluation period, it will call the renegotiation module to restart the process.

**Task offer submission.** This module will submit the offer to the chosen machines and invoke the task commitment module. It also starts the bid_confirmation_timeout timer so that the part won't wait for the bid confirmation message indefinitely.

**Task commitment.** The task commitment module monitors the network waiting for bid confirmation message. When the module gets the bid confirmation message, it updates the part schedule, states and fund. It then requests transportation service by sending a request_AGV_service message to transportation controller. If it does not receive the confirmation message from the chosen machine within a predefined time interval, then it will call the negotiation module to select another machine.

**Renegotiation.** This module is the same as the one in the part-initiated scheme.

Under the machine-initiated negotiation scheme, parts first select the resource according to the resource availability announcement information, then machines select parts among those parts which have expressed interest in being served. In this way, parts have more complete information about resources and may find a better set of resources to achieve their objectives. However, there is no guarantee that the machine the part is interested in will pick the part for service. Therefore, the part will have to construct its bid strategically. It should try to submit the most competitive amount of service charge to the resources that are most likely to choose it in order to win the bid. Different pricing and bidding strategies were discussed in Lin (1993); we also show one example of price setting in the next section.

### 7.4.3   Simulation results discussion

The part-initiated and machine-initiated negotiation schemes have been implemented based on the integrated shop flow control system described in section 7.3.5. Several experiments under different manufacturing environments and different price adjustment algorithms have been conducted to test the robustness and effectiveness of the control schemes and the corresponding price systems. The test results show that the control maintains a coordinated information flow and physical part flow and can cope quickly with sudden environmental changes such as machine failures or changes in part objectives. They also show that the price mechanism reflects and reacts to individual part objectives, priorities, machine capabilities and system loads. For example, Figs 7.9–13 show the results of a sample system that consists of three milling machines and two drilling machines (each has a different processing speed), a set of automatic guided vehicles and a central buffer. The initial machine charge price is set as two per unit processing time. In the first test case, the part-initiated negotiation scheme is used. All parts are of a single part type with a flexible processing process plan and each operation has a budget of two per unit processing time. Furthermore, the objective of the parts is to minimize the flow time. This implies that only the time factor is weighted and the part evaluation function that we described in the bid evaluation module of parts is thus defined as the finishing time of the operation on the machine if the bid is accepted.

$$evaluation\_function(o_i, m_j) = T_{finish\_time}(o_i, m_j)$$
$$= \max\{T^{m_j}_{available\_time}, T_{arrival\_time}(entity(o_i), m_j)\} + T^{m_j}_{o_i}, \quad (7.1)$$

where $m_j$ is the machine, $o_i$ is the operation of the part entity that we are considering and $T^{m_j}{}_{o_i}$ is the processing time of operation $i$ on machine $j$. Also, the function $T_{arrival\_time}(entity(o_i), m_j)$ is the earliest time that the entity that owns operation $o_i$ can arrive at the machine $m_j$ and is defined to be $T_{now}$ if the entity is already on the machine. Otherwise, it is defined as:

$$T_{arrival\_time}(entity(o_i), m_j) = \max_{AGV}\left\{T_{arrival}(entity, AGV, m_j)\right\}$$

$$= \max_{AGV}\left\{\max\left\{T^{entity}{}_{ready}, T_{now} + T^{AGV}{}_{arrival}(entity, AGV)\right\}\right.$$

$$\left. + T_{traverse\_time}(entity, AGV, m_j)\right\}, \tag{7.2}$$

where $T^{AGV}{}_{arrival}(entity, AGV)$ is the arrival time of the AGV to the entity's position and is defined to be:

$$T^{AGV}{}_{arrival}(entity, AGV) = \max\{T_{now}, T^{AGV}{}_{ready}\}$$

$$+ \frac{distance(position^{AGV}, position^{entity})}{speed^{AGV}}. \tag{7.3}$$

The extra fund allocation function $extra(o_i, entity)$ is defined to be the ratio of the processing time of $o_i$ in the remaining processing time times an adjustable incentive ratio times the surplus fund, that is:

$$extra(o_i, entity) = iratio * \frac{T_{processing}(o_i)}{T_{processing}(o_i) + T^{srp}{}_{entity}} * Surplus, \tag{7.4}$$

where $iratio$ is the adjustable incentive ratio (initialized to be 0.1), $T_{processing}(o_i)$ is the standard processing time of operation $i$ and $T^{srp}{}_{entity}$ is the shortest remaining processing time of the part entity.

On the machine side, the current charge, $current\_charge(m_j)$, is set to be the average unit processing charge of machine $j$ over a sliding time window. Since the machine may already have some commitments, the time interval is chosen to reflect both past and current price history. More specifically, the time interval contains two parts; the time interval from $(t_{now} - time\_interval)$ to $t_{now}$ represents the prices of tasks that have already been processed by machine $j$, and the time interval from time $t_{now}$ to $(t_{now} + time\_interval)$ represents the prices of the currently committed tasks.

The price of a particular operation is the product of the processing time of the part on the machine and the current_charge of the machine. That is:

$$price(m_j, o_i) = current\_charge(m_j) * T_{processing\_time}(m_j, o_i). \tag{7.5}$$

**Machine Utilizations Over Time**

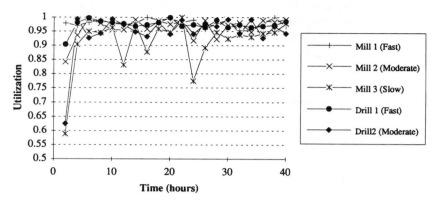

**Fig. 7.9**   Machine utilization comparison.

**Machine Price Changes Over Time**

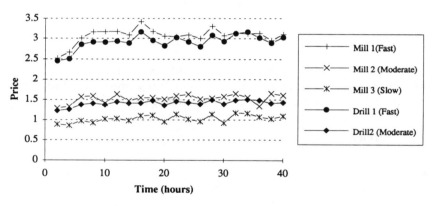

**Fig. 7.10**   Machine price comparison.

Although the average of the recent charge prices is a good indicator for setting a new price, the machine needs to be more aggressive when it is about to become idle. Therefore, in this example, the actual charge of the machine $m_j$ for an operation $o_i$ is set in the function *prepare_bid*($m_j$, $o_i$) as the minimum of the price offered by operation $i$ and *price*($m_j$, $o_i$) if the machine is idle, otherwise it is set as the minimum of the two prices plus three quarters of their difference.

Our test results show that the machine utilization and the price are all proportional to the processing speed of the machine, as shown in Figs 7.9 and 7.10. This means that the parts try to utilize the faster machines as much as possible and, at the same time, slower machines

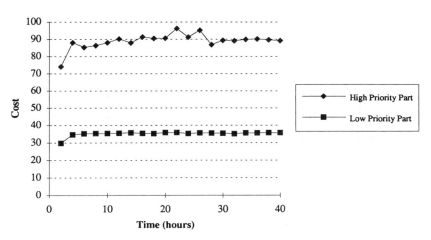

**Fig. 7.11** Part cost comparison.

still win some bids and prevent the faster machines from becoming 'hot spots.'

In the second test case, the machine-initiated negotiation scheme is used and two sets of parts with the same process plan enter the system: one set with minimizing flow time as an objective, a budget of two per unit processing time and an extra fund of 30 units. The other set has minimizing cost as an objective, a budget of 1 unit per unit processing time and no exra fund. For this case, the evaluation function has one additional term which is the product of the weight of the cost and cost of operation in the equation. The results show that the set of parts with cost as the objective did get a lower processing cost but higher flow time and the set with time as the objective got a lower flow time with higher cost and higher throughput, as clearly demonstrated in Figs 7.11, 7.12 and 7.13.

We also compared the performances of the two negotiation schemes that we have discussed by varying factory setup and price-adjustment strategies. The results show that both versions perform well in the sense that they can react well to part priority, machine capability, environmental changes etc. However, the part-initiated negotiation scheme achieves a slightly better system performance under either heavy or light system loads. That is, it can achieve slightly higher machine utilization and slightly lower mean flow time. The machine-initiated scheme reacts better and is more sensitive to the individual part's objective. The higher priority job that has time as the objective can achieve a lower mean flow time under the machine-initiated

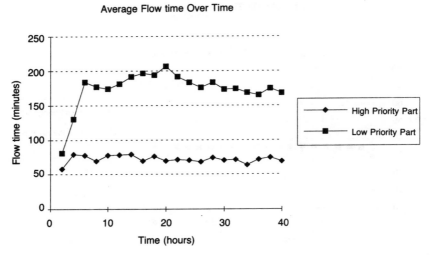

**Fig. 7.12**   Part average flow time comparison.

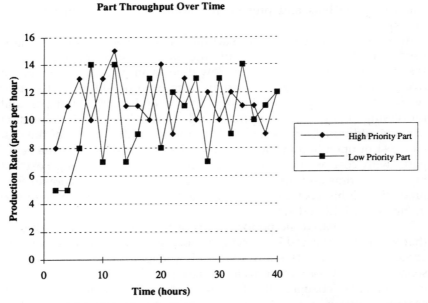

**Fig. 7.13**   Part throughput comparison.

scheme compared to the result of the part-initiated version. For more
test results and discussion, see Lin (1993).

## 7.5 CONCLUSION

In this chapter, an open manufacturing system concept is introduced. Traditional scheduling paradigms, information management strategies and negotiation-based systems are discussed. New concepts in control and scheduling are discussed. We describe a framework for adaptive control and scheduling of open manufacturing systems using autonomous agents. Under the framework, the part agents make decisions based on the objective functions of the part while the resource agents make decisions based on the price evaluation system. The goal of the part is to achieve the objectives with a minimum price paid, and the goal of the resource agent is to maximize the charged price to fulfill the part's requirements. The system functions are based on the equilibrium of the objectives and prices under the generic bid construction model. A resource synchronization scheme is proposed to assure the efficient and simultaneous services of resources to a part. Mutual agreement is made through two-way communication between the parts and resources.

An implementation architecture is designed. The basic control modules are part agents, resource agents, intelligence agents, monitoring agents, communication agents and database management agents. Part and resource agents are responsible for control flow and decision making. They have the ability to make decisions autonomously. Intelligence agents are responsible for the manipulation of knowledge bases. The monitoring agent monitors the system and interfaces with other system components. The communication and database management agents do bookkeeping and manipulation of the database, respectively. The control, decision making, knowledge base and database are all distributed throughout the system.

The framework is designed so that the structures of each control module are the same. Inevitable differences among modules are abstracted into the databases and knowledge bases. Modules communicate through a communication network with the same message passing protocol heterarchically. The amount of the software rewritten and the effect of system uncertainty are reduced to a minimum. The system is highly fault-tolerant. The separation of the part and machine intelligence from the bidding framework also makes it easy to incorporate new algorithms and to include any new parameters deemed important for decision making.

Two distributed control strategies based on the proposed framework are presented. An integrated control simulation system is built to test the flexibility and effectiveness of the proposed framework and control strategies. The results show that the proposed framework and control strategies provide a good foundation for highly adaptive, real-time control of open manufacturing systems.

## ACKNOWLEDGMENT

This work was supported by the Engineering Research Center for Intelligent Manufacturing Systems under a grant from the National Science Foundation, number CDR 8803017, with additional funding from the Defense Advanced Research Projects Agency (DARPA).

## REFERENCES

Adams, Joseph, Egon Balas and Daniel Zawack (1988) The Shifting Bottleneck Procedure for Job Shop Scheduling, *Mgmt. Sci*, **34**(3), Mar.

Date, C.L. (1986) An Introduction to Database Systems, Addison-Wesley Publishing Company.

Davis, Randall and Reid G. Smith (1983) Negotiation as a Metaphor for Distributed Problem Solving, *Art Intell*, **20**, pp. 63–109.

Duffie, Neil A. and Rex S. Piper (1986) Nonhierarchical Control of Manufacturing Systems, *Jnl. of Manuf Sys*, **5**(2), pp. 137–39.

Farber, D.J. and Larson, K.C. (1972) *The Structure of a Distributed Processing System – Software*, Proceedings of the Symposium on Computer Communication Networks and Teletraffic, Polytechnic Institute of Brooklyn, New York, Apr.

Ferguson, Donald, Yechiam Yemini and Christos Nikolaou (1988) *Microeconomic Algorithms for Load Balancing in Distributed Computer Systems*, Proceedings of the International Conference on Distributed Computer Systems, IEEE, pp. 491–99.

Fox, Bob (1983) OPT: an Answer for America, *Inv and Prod*, **2**(4 and 6).

Gershwin, S.B., Hildebrandt, R.R., Suri, R. *et al.* (1986) A Control Perspective on Recent Trends in Manufacturing Systems, *IEEE Cont Sys Mag*, **6**(2), pp. 3–15.

Grant, F.H. (1987) *Simulation and Factory Control: An Overview*, Proceedings of the IXth ICPR, pp. 576–83, Cincinnati, OH.

Graves, Stephen (1981) A Review of Production Scheduling, *Op Res*, **29**(4), pp. 646–75.

Hatvany, J. (1984) *Intelligence and Cooperation in Heterarchical Manufacturing Systems*, Proceedings of the 16th CIRP International Seminar on Manufacturing Systems, pp. 1–4, Tokyo, Japan.

Henderson, James M. and Quandt, R.E. (1980) *Microeconomic Theory – a Mathematical Approach*, 3rd edn., McGraw-Hill Book Company.

Hewitt, C. (1985) The Challenge of Open Systems, *Byte*, pp. 223–42, April.

Hewitt, C. (1988) Offices Are Open Systems, *The Ecology of Computation* (ed. B.A. Huberman), pp. 5–24, Elsevier Science Publishers B.V.

Kusiak, Andrew and Mingyuan Chen (1988) Expert Systems for Planning and Scheduling Manufacturing Systems, *Eur Jnl. of Op Res*, **34**, pp. 113–130.

Lin, Grace Y.J. (1993) Distributed Production Control for Intelligent Manufacturing Systems, PhD thesis, Purdue University, West Lafayette, Indiana, May.

Lin, Grace Y.J. and James J. Solberg (1989) *Flexible Routing Control and Scheduling*, Proceedings of the Third ORSA/TIMS Conference on Flexible Manufacturing Systems, pp. 155–160.

Lin, Grace Y.J. and James J. Solberg (1991) Effectiveness of Flexible Routing Control, *Int Jnl. of Flex Manuf Sys*, **3**(3/4), pp. 189–212, June.

Lin, Grace Y.J. and James J. Solberg (1992) Integrated Shop Floor Control Using Autonomous Agents, *IIE Trans*, **24**(3), pp. 57–71, July.

Maley, James G. (1987) Part Flow Orchestration in Distributed Manufacturing Processing, PhD thesis, Purdue University, West Lafayette, Indiana, Aug.

Maley, James G. and James J. Solberg (1987) *Part Flow Orchestration in CIM*, Proceedings of the International Conference on Production Research, pp. 17–20, Cincinnati, OH, Aug.

Malone, T.W., Fikes, R.E. and Howard, M.T. (1983) *Enterprise: A Market-like Task Scheduler for Distributed Computing Environments*, Center for Information Systems Research, Sloan School of Management, MIT, Oct.

Mclean, Charles, Bloom, H.M. and Hopp, T.H. (1983) *The Virtual Manufacturing Cell*, Industrial Systems Division, Center for Manufacturing Engineering, National Bureau of Standards, Washington, DC.

Naylor, Arch W. and Richard A. Volz (1988) Integration and Flexibility of Software for Integrated Manufacturing Systems, *Design and Analysis of Integrated Manufacturing Systems* (ed. W. Dale Compton), National Academy Press, Washington, DC.

Ow, Peng Si, Stephen F. Smith and Raddy Howie (1988) *A Cooperative Scheduling System*, Proceedings of the Second International Conference on Expert Systems and the Leading Edge in Production Planning and Control, pp. 43–56, May.

Parunak, H.V.D., Irish, B.W., Kindrick, J. *et al.* (1985) *Fractal Actors for Distributed Manufacturing Control*, Proceedings of the Second Conference on Artificial Intelligence Applications, pp. 653–60, Dec.

Parunak, H.V.D., White, J.F., Lozo, P.W. *et al.* (1986) *An Architecture for Heuristic Factory Control*, Proceedings of the American Control Conference, Seattle, Washington.

Ramaritham, K. and Stankovic, John A. (1984) Dynamic Task Scheduling in Hard Real-Time Distributed Systems, *IEEE Software*, pp. 65–75, July.

Shaw, Michael Jeng-Peng (1984) The Design of a Distributed Knowledge-Based System for the Intelligent Manufacturing Information System, PhD thesis, Purdue University, West Lafayette, Indiana, Aug.

Shaw, Michael J.P. (1987) Distributed Planning in Cellular Flexible Manufacturing Systems, *Infor*, **25**(1), pp. 13–25.

Smith, Reid G. (1980) The Contract Net Protocol: High-Level Communication and Control in a Distributed Problem Solver, *IEEE Trans on Comp*, **C-29**(12), pp. 1104–13, Dec.

Smith, Reid G. and Randall Davis (1981) Frameworks for Cooperation in Distributed Problem Solving, *IEEE Trans on Sys, Man and Cybernetics*, **SMC-11**(1), pp. 61–70, Jan.

Solberg, James J., Anderson, D.C., Barash, M.M. *et al.* (1985) *Factories of the Future: Defining the Target*, Computer Integrated Design Manufacturing and Automation Center, Purdue University, West Lafayette, IN, January.

Solberg, James J. and Heim, Joseph A. (1989) Managing Information Complexity in Material Flow Systems, *Advanced Technologies for Industrial Material Flow Systems* (eds S.Y. Nof and C.L. Moodie), pp. 3–20, Springer Verlag, Berlin Heidelberg, 1989.

Solberg, James J. (1989) *Production Planning and Scheduling in CIM*, Proceedings of the 11th World Computer Congress, IFIP, San Francisco, CA, Aug.

Stankovic, John A. and Inderjit S. Sidhu (1984) *An Adaptive Bidding Algorithm for Processed Clusters and Distributed Groups*, Proceedings of the 4th International Conference on Distributed Computing Systems, IEEE Computer Society, pp. 49–59.

Steffen, Mitchell S. (1986) *A Survey of Artificial Intelligence-Based Scheduling Sys-*

*tems,* Proceedings of 1986 Fall IIE Conference, pp. 395–405, Fall.

Upton, David M. (1988) The Operation of Large Computer Controlled Manufacturing Systems, PhD thesis, Purdue University, West Lafayette, Indiana, Dec.

Vamos, T. (1983) Cooperative Systems – An Evolutionary Perspective, *IEEE Cont Sys Mag,* **3**(2), pp. 9–14.

Waldspurger, Carl A., Tad Hoff, Bernardo A. Huberman *et al.* (1989) SPAWN: A Distributed Computational Economy, Dynamics of Computation Group, Xerox Palo Alto Research Center, Palo Alto, CA.

Wysk, R.A., Wu, S.D. and Yang, N. (1986) *A Multi-Pass Expert Control (MPECS) for Flexible Manufacturing Systems,* Symposium on Real Time Optimization in Automatic Manufacturing Facilities, National Bureau of Standards, Gaithersburg, MD, Jan.

# CHAPTER 8

# Applications of Petri net methodology to manufacturing systems

*MengChu Zhou and Anthony D. Robbi*

## 8.1 INTRODUCTION

Computer-integrated engineering systems are event-driven and often asynchronous, exhibiting concurrent, sequential, competitive and co-ordinated activities among their components. They are often complex and large in scale. These systems include integrated manufacturing systems, concurrent distributed systems, computer operating systems, communication networks, and intelligent machines. They fall into a class of systems called discrete event dynamic systems (DEDSs). The relations of events, their evolution over time and their order of appearance are of main interest. Compared with systems modeled as differential or difference equations, e.g. a manufacturing process, one has to deal with qualitative or abrupt changes characterized as events, instead of with quantitative changes in conventional continuous or discrete time systems. An investigation into the properties of and design methods for a DEDS has to start with an appropriate mathematical representation.

Petri nets (PNs) are one of most rigorous and powerful modeling tools for event-driven systems. Various extensions have greatly enhanced their applicability to various types of DEDS, particularly automated manufacturing systems. We propose here the use of a graphical, Petri net-based software tool to form the basis of manufacturing system design. The tool provides for design specification, analysis, simulation and in some cases the actual control codes for system operation.

The purpose of this chapter is to illustrate the applications of PNs to modeling, performance analysis and simulation of automated manufacturing systems. Several manufacturing systems are used to show:

1. How a manufacturing system can be represented by a PN.

2. How a PN can be used for a simulation.
3. How a PN is simulated with the help of computer graphics.
4. How a PN evolves as a manufacturing system becomes more complex.
5. How PN properties can be analyzed and what implications they bring to the modeled manufacturing system.
6. What assumptions need to be made for the analytical derivation of temporal properties.
7. How the production rate and resource utilization can be derived.
8. How a PN methodology can be used for the design of a flexible manufacturing system.

The rest of this chapter relates Petri nets to manufacturing system analysis and design. Related Petri net concepts are described in section 8.2. Section 8.3 describes and models a simple manufacturing system from a demonstration factory floor as an ordinary PN. Based on such a PN model, important system properties are validated. Its timed version is derived for temporal performance. In section 8.4, simulation methods are applied to evaluate the system performance. A recently developed SUN workstation-based software package is briefly discussed and used in this evaluation. To derive the analytical results, several assumptions must be made in terms of job arrival rate, service rate, etc. Section 8.5 discusses the related issues and performs system analysis by using SPNP (a package developed at Duke University is used to derive the performance for stochastic PNs). Finally, design of a flexible manufacturing system is discussed.

## 8.2   PETRI NETS IN MANUFACTURING AUTOMATION

### 8.2.1   Evolution of application of Petri nets to manufacturing systems

While many activities are currently engaged in Petri net applications to the manufacturing automation area, the following foci could be observed since the late 70s:

1. Early interest in Petri nets arises from the need to specify and model manufacturing systems. The activities in this area start with the Petri net representation of simple production lines with buffers, machine shops and automotive production systems, and proceed with modeling of flexible manufacturing systems, automated assembly lines, resource-sharing systems and recently just-in-time and kanban-based manufacturing systems.
2. Early research focused on qualitative analysis of PN models of manufacturing systems. Reachability analysis shows whether a system can reach a certain state. Desired sequences of events are

validated according to the system requirements. Other PN properties are used to derive the DEDS stability, cyclic behavior and freedom from deadlocks.

3. As temporal or quantitative properties become an important consideration, timed PNs are used to derive the cycle time of repetitive and concurrent manufacturing systems. To deal with the stochastic nature of many production operations, stochastic PNs are used to derive the system production rates or throughputs, critical resource utilization and reliability measures. Their underlying models are Markov or semi-Markov processes. The direct construction of Markov chains is avoided thanks to conversion algorithms for stochastic PNs.

4. When a state explosion problem arises or the underlying stochastic models are not amenable to tractable mathematical analysis, simulation must be conducted for analysis of both qualitative and quantitative properties. Fortunately, PN models can be easily utilized to drive a complex discrete event simulation. Several packages based on PNs exist.

5. Programmable logic controllers (PLCs) are commonly used in industrial sequence control of automated systems. They are designed through ladder logic diagrams which are known to be very difficult to debug and modify. It is observed that a PLC can be converted into a PN and vice versa for a subclass of PNs. Early work includes the conversion of a PN into a PLC for implementation. Direct PN controllers without the help of PLCs can also be implemented through either a Petri net interpreter or its corresponding control codes. For most cases, additional information to represent the real-time signals and status needs to be incorporated into such PN models. The advantages of PNs include their relative ease to represent and modify the control logic and their potential for mathematical analysis and graphical simulation to validate a design. It can be proved that the graphical complexity of PNs grows with system complexity less than that of ladder logic diagrams.

6. With a mathematical representation available, designers are able to use PNs for rapid prototyping of a process control system or discrete event control. Virtual factories can be realized through computer graphics using PNs. Stepwise testing and implementation can be achieved by connecting the actual equipment into a PN-based design system reducing design and development time.

7. Petri nets have been combined with other approaches to achieve various purposes in process planning and scheduling, intelligent control, expert system construction, knowledge representation and uncertainty reasoning. For example, the correspondence between a PN and an expert system can be established. This can greatly aid in consistency checking of an expert system.

### 8.2.2   Petri nets

Petri nets are a graphical and mathematical model for representing information and control flow in a event-driven system. They have two nodes, transitions and places. Direct arcs link places to transitions and transitions to places. Tokens are used to describe the state of discrete event systems. A formal definition of a PN can be found in (Murata, 1989) and (Zhou and DiCesare, 1993).

Consider the operation of a pick-place robot for component insertion (these robots are widely applied to assembly of components to printed circuit boards). The robot picks up a component and then places it in a desirable position. Two events are **picking-up** and **placing**. The first one can take place if a component is available and the robot is ready. As shown in Fig. 8.1, two circles called places are used to represent these two conditions, labeled $p_1$ and $p_2$. Putting at least one dot, a token, into each circle represents that each condition is true. Draw a bar called transition $t_1$ to represent event picking-up. Two arcs link $p_1$ and $p_2$ to $t_1$. After the robot picks up a component, a new condition results, i.e. the robot holds the component, represented by another circle called $p_3$. Then event placing depicted by another bar called transition $t_2$ can occur. Arcs from $t_1$ to $p_3$ and from $p_3$ to $t_2$ show the relationships. Once the robot has placed the component, it is ready for the next pick-up operation. Thus the arc from $t_2$ to $p_2$ is created. Since the component is already inserted, no arc is formed from $t_2$ to $p_1$.

Tokens (dots) in a PN appear identical although they may carry different physical interpretations in different places. For example, a token in $p_1$ means one available component and one in $p_2$ the ready robot. The number of tokens in a place defines a local condition. For example, one token in $p_1$ implies one component available while two tokens in it means two components available. The distribution of tokens in all places determines the status of the entire system, called a system state. A state or marking in PN is formally defined as a vector whose components represent the number of tokens in the corresponding places. For example, the initial system state for the above system is $(1,1,0)^\tau$ given the token distribution in Fig. 8.1. The state changes when a transition fires, i.e. an event occurs. This results in a new marking.

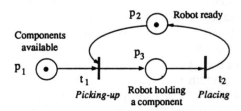

**Fig. 8.1**   A simple PN example: a pick-place robot.

### 8.2.3 Modeling of primitives and concepts with Petri nets

The following primitives as illustrated in Fig. 8.2 can be modeled with places, transitions and arcs.

**Synchronization.** The synchronization between two conditions can be realized through two places linking to a common transition. Even though one place obtains a token, the transition has to wait for the other to get a token. Then the event can take place. Therefore, a transition in a net may have multiple input places. In the context of manufacturing, a raw workpiece's arrival and a machine's availability are two conditions which may lead to occurrence of an event *machine–a–workpiece* or *start–machining*. In assembly systems, several parts must be present before a final product is assembled and in this case several places may be designed as the inputs to the transition representing an assembly operation. Synchronizing various activities to reduce the waiting time and thus to maximize production rate is a challenging problem for production engineers in a flexible manufacturing environment.

**Choice.** When there are two or more machines ready for a coming raw piece, one needs to make a choice as to which machine the piece should be dispatched. This is a dispatching problem. In the context of PNs, a place has two or more outgoing arcs, representing the different possible events to follow, as shown in Fig. 8.2(b). One must be chosen probabilistically, by priority or some other criterion.

**Fork.** The fork structure is used to model an event whose occurrence leads to more than one information and/or control signal. For example, when a machine completes its operation on a raw piece, a machined part is sent out for the next operation and the machine itself becomes available. In other words, two pieces of information are generated due to this event *complete_operation*. Another example is the disassembly process which may lead to different parts for different post-processes.

**Join/merge.** A place may have more than one incoming arc, signifying that several information/control sources lead to a common condition/

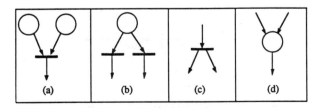

(a)  (b)  (c)  (d)

**Fig. 8.2** Representation of four primitives: (a) synchronization, (b) choice, (c) fork and (d) join/merge.

status represented by a place. The first example is based on manufacturing processes which produce identical parts and send them to a common station for further processing. The second example is a resource shared by several processes. Release from any process resumes the availability of the resource by depositing a token in the common place.

In addition to the above primitives, Petri nets can model the following six structures that are often used in practice, as shown in Fig. 8.3.

**Sequence.** One common structure in event-driven systems is a strict sequence of events. For example, *load_a_part_to_Machine_A*, *process_a_part_on_Machine_A*, and *unload_a_part_from_Machine_A* can be such a sequence. In assembly, several parts may have to follow a strict order to finish an assembly job. With PNs, a series of transitions and places can be used to model these sequences of events as shown in Fig. 8.3(a).

**Concurrency.** In modern manufacturing, many operations have to occur during the same time period to achieve the maximum efficiency

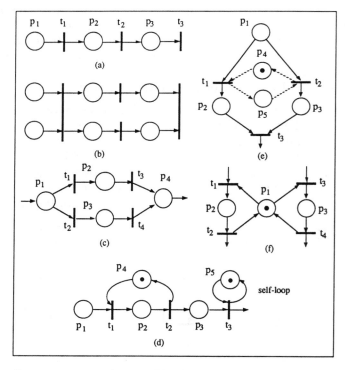

**Fig. 8.3** Representation of six structures: (a) sequence, (b) concurrency, (c) parallel choice, (d) circuit and self-loop, (e) choice-synchronization and (f) mutual exclusion.

and least inventory, which in turn reduces the overall production cost. Pure concurrency of two events can be represented by two non-related transitions. One common case of concurrency, though, can be represented by two or more sequence structures with a common synchronizing transition to start and a fork transition to end as shown in Fig. 8.3(b). If several tokens are allowed in a sequence structure to mark those places, pipeline concurrency has been obtained. Such pipeline concurrency is widely used in manufacturing production lines.

**Parallel choice.** One commonly used choice structure is shown in Fig. 8.3(c). At one point, there exist choices resulting in different paths which merge at a later point. Such a structure can represent the case that the order of installing two parts has no effect on the final product in an assembly process. In flexible manufacturing, a part may be either processed by machine A or B and then C. This case can also lead to such a parallel choice structure.

**Circuit and self-loop.** The above three basic structures do not require any initially marked place. They are especially suitable for modeling manufacturing operations. In practice, many systems or their components exhibit cyclic behavior. Using PNs, a circuit-like structure is adopted as shown in Fig. 8.3(d). A simple circuit may represent the cycle for a robot to perform repetitive activities, e.g. *load_a_part* and *unload_a_part*. A self-loop may represent the availability of a machine and its working status. It is clear that such a simple circuit is suitable for specifying a resource, especially a non-consumable one like a robot or a machine, and its associated repetitive activities.

**Choice-synchronization.** In many instances of manufacturing control it is necessary to enforce particular sequences for a choice structure, particularly when assembly processes are involved. Those choice structures required to enforce sequences are called choice-synchronization structures (Zhou *et al.*, 1992). A typical one is shown in Fig. 8.3(e). Without the enforcement mechanism, places $p_4$ and $p_5$ and dotted arcs, a token residing in place $p_1$ can reach either $p_2$ through $t_1$ or $p_3$ through $t_2$. When tokens are present in both places $p_2$ and $p_3$, the event associated with transition $t_3$ may occur. If there is no restriction on the number of tokens passing through $t_1$ or $t_2$, then the net may be partially deadlocked since either $p_2$ or $p_3$ may never hold a token blocking the event associated with $t_3$. With $p_4$, $p_5$ and related arcs, a strict order can be implemented. If the initial marking is given as shown in Fig. 8.3(e) $p_2$ and $p_3$ will obtain a token alternately with $p_2$ getting it first. Another method to limit the number of tokens passing through sequentially related transitions, e.g. $t_1$ and $t_3$, can be to add an initially marked place to form a circuit structure.

**Mutual exclusion.** A useful structure deals with the resource shared by several processes or machines. For example, a robot may be shared by two machines for loading and unloading. A parallel mutual exclusion example is given in Fig. 8.3(f) in which two parallel processes share the resource in place $p_1$. A sequential mutual exclusion can be obtained if the two processes are sequentially related (Zhou and DiCesare, 1991).

Combinations of the above modeling primitives enable designers to model numerous manufacturing systems with PNs. This is the first step toward the goal to design, synchronize and coordinate production facilities more efficiently.

### 8.2.4   Execution of Petri nets

After the static structure of a PN is defined, the flow of tokens regulated by transition firings describes its dynamics. The event sequence can be studied by executing two rules (Murata, 1989; Zhou and DiCesare, 1993). The enabling rule states that transition $t$ is enabled when all its input places have enough tokens, i.e. the number of tokens in each place equals or is greater than that of the arcs from the place to transition $t$. Transition $t$ being enabled implies that the conditions for its associated event to occur are satisfied. Therefore, it can occur.

The firing rule states that an enabled transition $t$ can fire, or an event can occur. Its firing can be regarded as two separate stages. First, remove the required number of tokens from each of its input places, the quantity determined by the number of arcs between a place and $t$. Second, deposit tokens into each of $t$'s output places; the number of tokens equals the number of arcs from $t$ to the corresponding output place.

The net in Fig. 8.4(a) shows a PN model for an assembly station and its product transfer. Places $p_1$, $p_2$ and $p_3$ model the availability of assembly station, parts and products respectively. Transition $t_1$ models the assembly operation and $t_2$ the transfer. An arc with segment and 4 (meaning four arcs) from $p_2$ to $t_1$ implies that the start of each assembly operation requires four parts, two arcs from $t_1$ to $p_3$ means that each operation produces two products and two arcs from $p_3$ to $t_2$ means that the start of each transfer requires two products. Suppose that initial marking $m_0 = (1,9,0)^\tau$. Since $p_1$ and $p_2$ have enough tokens ($m_0(p_1) = 1$ which equals the number of arcs from $p_1$ to $t_1$ and $m_0(p_2) = 9$ which is greater than the number of arcs from $p_2$ to $t_1$), $t_1$ is enabled. However since $m_0(p_3) = 0$, less than the number of arcs from $p_3$ to $t_2$, $t_2$ is not enabled. Firing $t_1$ takes one token from $p_1$ and four tokens from $p_2$. It deposits one token back to $p_1$ and two tokens to $p_3$. This results in a new marking or state $m_1 = (1,5,2)^\tau$. Note that $p_1$ still has one token after $t_1$'s firing since there is one arc from $p_1$ to $t_1$ and another from $t_1$

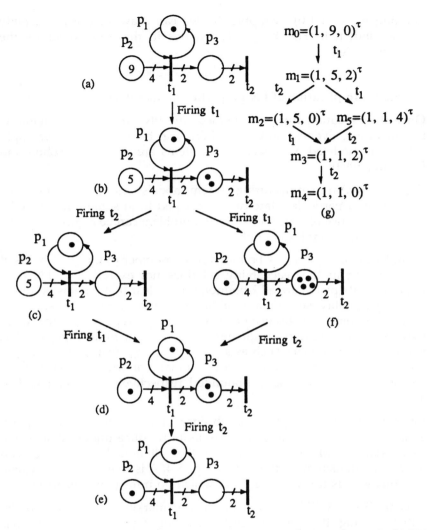

**Fig. 8.4** Example of Petri Net execution: (a) PN with $m_0 = (1,9,0)^\tau$, (b) PN with $m_1 = (1,5,2)^\tau$, (c) PN with $m_2 = (1,5,0)^\tau$, (d) PN with $m_3 = (1,1,2)^\tau$, (e) PN with $m_4 = (1,1,0)^\tau$, (f) PN with $m_5 = (1,1,4)^\tau$, and (g) reachability tree.

back to $p_1$. By checking the conditions to enable $t_1$ and $t_2$, we find that both are enabled. We may pick either to fire first. Firing $t_2$ first leads to $m_2 = (1,5,0)^\tau$. At this marking, only $t_1$ is enabled and its firing leads to $m_3 = (1,1,2)^\tau$. At $m_3$, only $t_2$ is enabled and its firing leads to $m_4 = (1,1,0)^\tau$ at which no transition is enabled.

Alternatively, at $m_1$, we may fire $t_1$ first, yielding $m_5 = (1,1,4)^\tau$. At $m_5$, only $t_2$ is enabled and its firing leads to a previous marking $m_3 = (1,1,2)^\tau$. At $m_3$, only $t_2$ is enabled and its firing leads to the same marking $m_4 = (1,1,0)^\tau$. Graphically, we can draw such a reachability tree as shown in Fig. 8.4(g). The nodes in this tree constitute the reachability set.

### 8.2.5  Implication of PN properties in manufacturing systems

Once the execution rules are defined, the PN properties reachability, boundedness, liveness and reversibility can be analyzed. Their implications in modeled manufacturing systems are as follows (Zhou and DiCesare, 1993):

**Reachability.** In manufacturing system models, one may be interested in whether a particular state can be reached from some other state by firing a sequence of transitions. Such a problem can be computationally intractable for complex enough cases.

**Boundedness.** A place is $k$-bounded if the number of tokens it holds at any marking in the reachability set does not exceed $k$. A PN is $k$-bounded if its places are all $k$-bounded. It is clear that a net is bounded if its reachability set contains a limited number of members. This property implies the absence of capacity overflows. For instance, there may be storage buffers or queues which have a finite capacity. The boundedness of a place such as a buffer or queue ensures that there is no overflow. In Fig. 8.4(a) given the initial marking $m_0 = (1,9,0)^\tau$, $p_2$ and $p_3$ are 9-bounded and 4-bounded respectively according to Fig. 8.4(g).

**Liveness.** Liveness implies the absence of deadlocks. This property guarantees that a system can successfully produce under all specified circumstances. Moreover, it insures that all modeled processes can occur. No transition in the net in Fig. 8.4(a) is live since the system eventually ends up with $(1,1,0)^\tau$ at which no transition is enabled.

**Reversibility.** A PN is reversible if any marking can return to the initial marking. It implies the cyclic behavior of a system. This also has implications for error recovery in the manufacturing context. It means the system can be reinitialized from any reachable state. It is clear that the net in Fig. 8.4(a) is not reversible.

To model cyclic behavior, one often introduces additional arcs to ensure the liveness and reversibility for PN models of manufacturing systems. For example, we might add an arc with weight 4 between $t_2$ and $p_2$, implying that once two products are transferred, four parts will be made available for this system. It indeed models a potentially infinite number of parts by using a limited number of tokens. The resulting PN as shown in Fig. 8.5(a) has the reachability tree shown in Fig. 8.5(b). Then the net is live and reversible. This correctly models

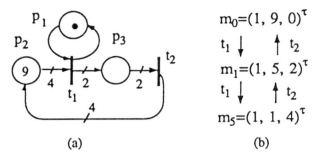

$$m_0 = (1, 9, 0)^\tau$$

$$t_1 \downarrow \quad \uparrow t_2$$

$$m_1 = (1, 5, 2)^\tau$$

$$t_1 \downarrow \quad \uparrow t_2$$

$$m_5 = (1, 1, 4)^\tau$$

(a)         (b)

**Fig. 8.5** Example of bounded, live and reversible PN: (a) PN with $m_0 = (1,9,0)^\tau$, (b) reachability tree.

the cyclic behavior that most discrete manufacturing processes have. Often, boundedness, liveness and reversibility properties are necessary for analytical performance analysis.

The reachability tree of a PN represents all the states and their relationships. If the tree is finite, and the maximum number of tokens in a place over all markings is $k$, then the place is $k$-bounded. If given any two nodes in the reachability tree there is a direct path and all transitions are present, then the net is live. If there is a direct path from any node to the initial marking in the tree, then the net is reversible. The limitation of this enumerative method is the magnitude of the number of the combinatorial states possible. Simulation methods then become a convenient and straightforward yet effective approach for industrial engineers to validate the desired properties of a manufacturing system.

## 8.3 MANUFACTURING SYSTEM MODELING EXAMPLES

The flexible manufacturing system (FMS), a job shop of sorts, presents formidable design problems. Formal discrete system models are useful to predict operational characteristics such as throughput and the utilization of machines. The assembly line, or flow shop, presents a more straightforward problem to the designer, so we discuss first a flow shop model, then an FMS model.

An FMS features diversity in material movement and job types, requiring flexible automation. Flow shop production can be efficiently implemented by hard automation, with material movement handled by conveyors and fixed-movement robots. In an FMS, material flow is mechanized by programmed robots, automatic guided vehicles (AGVs) and people. Since these devices are more costly than their hard automation counterparts, the system design must permit them multiple

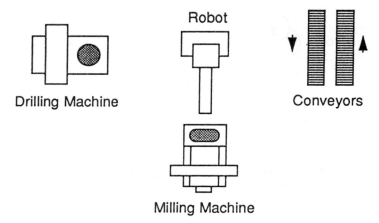

**Fig. 8.6** Fanuc system layout.

roles, allowing competition for their services. System designs which allow for concurrency of activities and competition for resources are difficult to realize. Discrete event system technology can help the designer predict behavior and tune system performance consistent with economic goals.

This section will illustrate Petri net concepts as applied to a Fanuc machining center (located on the 'factory floor' at the Center for Manufacturing Systems at New Jersey Institute of Technology). We will start with a simplified version of the cell and finish with an FMS version. The actual cell is shown in Fig. 8.6. It comprises two workstations, a milling machine WS1, a drilling machine WS2 and a robot R. The operational specifications of this system are:

1. To start a cycle, a raw part and the robot must be available.
2. The robot moves a raw part from the conveyor and loads it at WS1.
3. The milling operation is performed at WS1 while the robot backs off (returns).
4. The robot unloads the part from WS1, loads it to WS2 and returns.
5. The drilling operation is performed at WS2, and simultaneously the robot performs step 2.
6. The robot unloads the finished part from WS2, deposits it on the conveyor and returns.

In steady-state steps 2–6 repeat.

### 8.3.1 Simplified Fanuc system

As described above the Fanuc system is a flow shop, even though there may be competition for resources, the robot, if enough parts

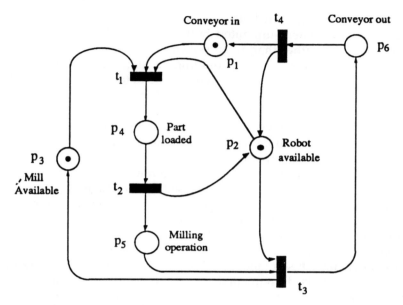

**Fig. 8.7**  Petri net for simplified Fanuc.

become available at the input. Consider first a simpler system without the drill. Its PN is shown in Fig. 8.7 in the situation of step 1 above.

The transition $t_1$ implements step 2 in the specification, $t_2$ implements step 3, $t_3$ implements step 6 and $t_4$ restarts the cycle by presenting the output part as input. Place $p_3$ ensures that no more than one part can occupy WS1 at a time. The logical significance of each place, when occupied by a token, is given in Table 8.1.

The overall net is obviously bounded because the number of incoming arcs for the transitions $t_1$ to $t_4$ equals the number of outgoing arcs (seven). This simple system has only four states, with a transition firing order $t_1:t_2:t_3:t_4$. Note that all four of the primitives shown in Fig. 8.2 appear in this net. The choice situation does not actually arise

**Table 8.1**  Interpretation of places in Fig. 8.7

| *place* | *interpretation* |
| --- | --- |
| $p_1$ | Raw part(s) available |
| $p_2$ | Robot available |
| $p_3$ | WS1 available |
| $p_4$ | Part loaded WS1 |
| $p_5$ | Part milled |
| $p_6$ | Part at output |

because a marking with both $t_1$ and $t_3$ enabled cannot occur. If one wanted to implement unlimited buffering at WS1, then $p_3$ would be connected to $t_2$ rather than $t_1$. In systems with concurrent activity, the throughput can be raised by permitting buffering, as confirmed later.

The dynamic activities of a system can be determined by timed Petri net analysis and simulation. In conventional PNs described earlier, state changes occur instantaneously. Practical systems are constructed from devices which have a finite speed of operation. For example, in the Fanuc system a finite amount of time is required for a robot to complete an activity. Applications of PNs to model such discrete systems require the introduction of time into the model.

Formally, a timed Petri net is a couple (PN, τ) where PN is a classical Petri net and τ is a function assigning a real non-negative number, $\tau_i$, to each transition of the net. This number is the firing time of a transition. Transitions are enabled according to the same rules as apply to classical Petri nets. Firing an enabled transition $t_i$ causes a two step marking change. Instantaneous with the enabling of $t_i$ the marking of its input places is decremented by one token per input arc. After $\tau_i$ time units the marking of its output places is incremented by one token per output arc. A simulation model of a timed net requires a clock reference. In a timed net the transition firing times may be either deterministic or stochastic. It may also contain instantaneous transitions (Marsan *et al.*, 1984). The instantaneous transitions are drawn as solid rectangles as before, and timed transitions are depicted as open rectangles.

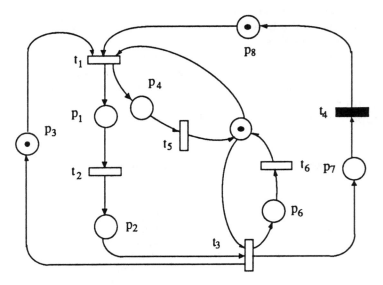

**Fig. 8.8**  Time version of Petri net for simplified Fanuc.

**Table 8.2**   Interpretation of transitions in Fig. 8.8

| transition | interpretation |
|---|---|
| $t_1$ | Robot moving part from input to WS1 |
| $t_2$ | Part being milled |
| $t_3$ | Robot moving part from WS1 to output |
| $t_4$ | Cycle completion |
| $t_5$ | Robot returns to ready position |
| $t_6$ | Robot returns to ready position |

To convert a state net to a timed net is a simple process. Decide which activities require finite time and redraw the corresponding transitions as open. Certain activities may need to be separated in order to simulate correctly the system dynamics. Then the net elements may need reinterpretation. For example the ordinary net of Fig. 8.7 becomes the timed net of Fig. 8.8 with interpretations for enabled transitions in Table 8.2.

The PN representation of this system makes clear that the only complications to this simple situation are the robot return times associated with $t_5$ and $t_6$. The $t_5$ return is overlapped entirely (in a realistic case) by the milling operation $t_2$, but the $t_6$ return adds to the cycle time $\pi$ (time to mill one part). Thus the cycle time $\pi$ is:

$$\pi = \tau_1 + \tau_2 + \tau_3 + \tau_4 + \tau_6 \qquad (8.1)$$

The utilizations of the various devices are computed by summing the associated delays, dividing by $\pi$, and converting the fraction to a percent. For instance, the robot utilization *robot* is:

$$robot = (\tau_1 + \tau_3 + \tau_5 + \tau_6)/\pi * 100 \qquad (8.2)$$

Even in this simple case a PN may help designers see these relationships, particularly if the PN can be drawn on a computer screen and its execution simulated in a step-by-step manner. The situation is more complex if output buffering is permitted, modeled by removing the arc connecting $t_3$ to $p_3$, connecting $t_2$ to $p_3$ instead. This situation is shown in Fig. 8.9 and discussed later.

One of the strengths of the PN is its ability to model conflict, i.e. competition for resources. Rules for conflict handling in timed Petri nets should follow the rules for non-timed nets: conflicts are resolved on the basis of priorities or probabilistically. The resolution procedure is definitive because tokens are removed from input places when transitions are enabled in the timed Petri net tool TPNS (Siddiqi *et al.*, 1993).

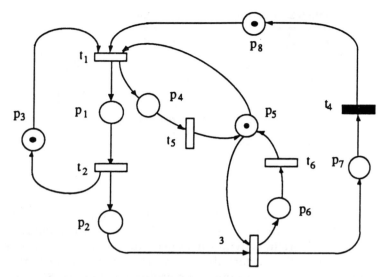

**Fig. 8.9**   Version of Petri net in Fig. 8.8 with output buffer.

### 8.3.2   Fanuc system as an FMS

The Fanuc cell of Fig. 8.6 could have the timed Petri net representation shown in Fig. 8.10. One feature of this cell is that an expensive resource, the robot, is required by the three distinct material movements needed. The place $p_1$ represents both incoming and outgoing conveyor belts. Both workstations are permitted to buffer their output and start new jobs before it is picked up (in $p_5$ and $p_4$). Neither workstation can buffer input. Thus material movement toward a workstation cannot commence until its current task is complete. The 'availability' places $p_7$ and $p_8$ enforce this policy. The robot return operations may have finite times associated with them ($t_6$, $t_7$, and $t_8$), but that time could be made zero if the situation warranted it. The significance of the individual places and transitions is shown in Table 8.3.

## 8.4   MANUFACTURING SYSTEM SIMULATION

The simulation of a manufacturing system has several purposes. One is to debug the design, or its model. Another is to assess its performance. Discrete event simulation languages such as GPSS and Simscript, have been available for decades. Their use is a programming exercise of sorts with all the associated pitfalls. Recognizing this, front ends to such simulation languages which provide a user-friendly interface have been devised. These tools are restricted to performance analysis. We suggest

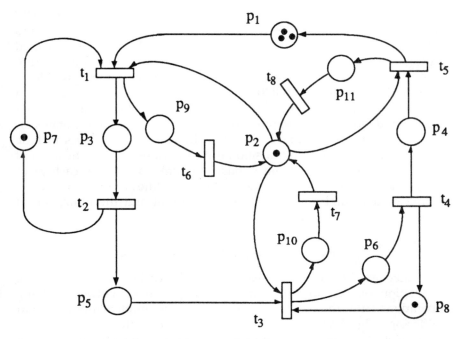

**Fig. 8.10** Timed Petri net for Fanuc with buffers.

**Table 8.3** Functions of places and transitions in Fig. 8.10

| place | interpretation | transition | interpretation | $\tau_i$ |
|---|---|---|---|---|
| $p_1$ | Raw parts | $t_1$ | Robot/part to WS1 | 250 |
| $p_2$ | Robot available | $t_2$ | Milling operation | varies |
| $p_3$ | Part loaded WS1 | $t_3$ | Robot/part to WS2 | 150 |
| $p_4$ | Out buffer WS2 | $t_4$ | Drilling operation | 150 |
| $p_5$ | Out buffer WS1 | $t_5$ | Robot/part to output | 200 |
| $p_6$ | Part loaded WS2 | $t_6$ | Robot returns | 50 |
| $p_7$ | WS1 available | $t_7$ | Robot returns | 50 |
| $p_8$ | WS2 available | $t_8$ | Robot returns | 50 |
| $p_9$ | Robot ready return | | | |
| $p_{10}$ | Robot ready return | | | |
| $p_{11}$ | Robot ready return | | | |

as an alternative that a PN tool can provide a design methodology, simulation, analysis and finally control code generation. Some tools including many of the above features exist. A comprehensive version of such a tool is being constructed at NJIT using C++ in an XWindow environment and will be made available for use.

In this section the use of the timed net simulation aspects of the NJIT tool, TPNS, is illustrated in two cases, the timed nets of Figs 8.8 and 8.9, followed by the Fanuc system of Fig. 8.10. TPNS was implemented in C (Siddiqi *et al.*, 1993). It runs under SunOS in the SunView environment. It is based on an earlier tool (Shukla, 1991) which provides a graphical user interface (GUI) for drawing and editing a PN, and a simulation driver which executes a net (plays the token game) in either a step mode for debugging purposes or a run mode to log the states for successive steps and provide statistics. In TPNS the statistics include the number of times each transition fires, the elapsed time, the amount of time each timed transition is busy, the amount of time each place owns tokens and conflict statistics. These quantities can have manufacturing system implications such as throughput and the utilization of various equipment as illustrated below.

### 8.4.1 Flow shop simulation results

As stated earlier the case of Fig. 8.8 can be easily analyzed and the simulation results confirm that. Permitting overlap between milling operations and the movement of output by implementing output buffering (Fig. 8.9) changes the situation. The throughput should increase and utilization of the robot increase. To simulate limitation of the buffer capacity, a loop like that around $p_1$ could be constructed around $p_2$. We executed the revised net with three tokens in $p_1$ to start, limiting the work-in-process (WIP) to three, and buffer maximum to two because the milling operation is longer than the robot return time.

The performance of the net of Fig. 8.8, $PN_1$, is compared to its buffered output variant defined in Fig. 8.9, $PN_2$, for a range of milling operation times (arbitrary units). All timed transitions are deterministic, with values shown in Table 8.3.

The results are shown in Table 8.4, where the effective cycle $\pi$ is defined as inverse of throughput, *mill* is the % of time WS1 is busy and *robot* is the % of time the robot is busy.

Note the modest improvement in throughput when output buffering

**Table 8.4**  Simplified Fanuc dynamic results

| net | $\tau_2$ | $\pi$ | *mill* | *robot* |
|-----|------|------|------|-------|
| $PN_1$ | 100 | 600 | 17 | 92 |
| $PN_2$ | 100 | 571 | 18 | 96 |
| $PN_1$ | 200 | 700 | 29 | 79 |
| $PN_2$ | 200 | 659 | 32 | 89 |
| $PN_1$ | 500 | 1000 | 50 | 55 |
| $PN_2$ | 500 | 969 | 55 | 60 |

**Table 8.5** Fanuc dynamic results

| case | $\tau_2$ | $\pi$ | mill | drill | robot |
|------|----------|-------|------|-------|-------|
| 1 | 100 | 778 | 13 | 19 | 72 |
| 2 | 100 | 714 | 15 | 21 | 100 |
| 1 | 200 | 774 | 26 | 19 | 97 |
| 2 | 200 | 713 | 29 | 21 | 100 |
| 1 | 500 | 910 | 55 | 16 | 83 |
| 2 | 500 | 851 | 59 | 18 | 83 |

is permitted. We permitted the robot to choose randomly what to do next; the improvement would approximately double if $t_1$ were given priority over $t_3$ when both could fire, i.e. input movement had priority over output. The time parameters chosen here result in a robot-limited situation because we want to be able to compare with the actual Fanuc cell to be discussed next.

### 8.4.2 Fanuc FMS simulation results

The timed Petri net for the Fanuc cell was executed for two cases. The first is the net of Fig. 8.10 with all time delays deterministic, delay values shown in Table 8.3, rightmost column. The second case simulates a situation where the arrival of parts is made stochastic by introducing some random delay, in particular an exponential distribution having a mean value of 50. To accomplish this, a place:transition pair is inserted between $t_5$ (output) and the input buffer $p_1$. To preserve the average loop delay, the deterministic delay at $t_1$ is reduced from 250 to 200. Conflicts for the robot resource are resolved randomly. These two nets were executed by TPNS with the results shown in Table 8.5 for runs of 500 simulation steps with four tokens in $p_1$ to start. The variables are as defined for Table 8.4 with the addition of *drill* representing the % of simulated time that WS2 is busy (WS2 utilization).

In these simulation runs the number of parts produced in 500 steps ranged from 55 (stochastic case 2) to 62 (deterministic case 1). Thus WIP constitutes a very small fraction of the total output used to compute the effective cycle time, and it was ignored. While modern computers produce CPU cycles at modest and diminishing cost, the efficiency of simulations is still a concern, particularly for more complex situations than shown here. (The TPNS tool used here updates the graphic screen step by step as the numbers reported above are collected into a file. During simulation it spends much of its time in I/O. The I/O will become a voluntary feature in future versions.)

In spite of the apparent simplicity of this cell, there are some observations to be made concerning the results in Table 8.5. System balance is improved by increasing the processing time for WS1 from 100 to 200 (with that of WS2 fixed at 150), equivalent throughput with increased utilization of WS1 and other system components. Material movement (robot) tends to be the limiting factor in this case. To make the system more efficient either the robot should be speeded up or the workstations should be given more to do. In all cases the effective cycle time was shortened by less than 10% in the stochastic part arrival case. This is attributable to the shortening of the $t_1$ delay which has more effect on cycle time than the delay added by the stochastic transition $t_9$ because the $t_9$ delay is more often overlapped with other processes.

## 8.5  MANUFACTURING SYSTEM PERFORMANCE

Different from simulation methods, analytical methods can be more efficient in predicating system performance if applicable. On the other hand, many assumptions must be made to achieve such computational efficiency. The accuracy of the results depends on how well the model reflects a realistic system. Therefore, if a manufacturing operation is done in a very deterministic way, then the time for this operation or its corresponding transition firing should be assumed deterministic instead of stochastic. If the variation of randomness is significant, a stochastic variable should be assumed. If the time delay for each transition in a PN model is assumed to be stochastic and exponentially distributed, a stochastic Petri net (SPN) results (Molloy, 1982). Allowing immediate transitions, i.e. with zero time delay, in such SPNs leads to generalized SPN (GSPN) (Marsan *et al.*, 1984). Both models can be automatically converted into their equivalent Markov chains and thus existing mathematical analysis methods are applicable.

Instead of a detailed discussion of these methods for SPNs, the manufacturing system discussed in section 8.3 is used to illustrate the analysis through SPNP software developed by Duke University (Ciardo, 1989).

Assume that all the operations take stochastic time which follows an exponential distribution. Then the rate in its distribution, called transition firing rate, equals the reciprocal of the average delay time of the corresponding event or operation. For example, the machining rate is the reciprocal of the average machining time. Thus if the average time of the related operations in the PN shown in Fig. 8.10 is identical to its deterministic time delay, the firing rates shown in Table 8.6 are obtained. $\tau_2$ is the average milling time equal to 100, 200 and 500 time units respectively. To model the buffers of capacity two, two places marked initially with two tokens are added to the net to limit capacity of places $p_5$ and $p_4$. Note that without the limit, $p_5$ and $p_4$ could hold

**Table 8.6** Transition firing rates for the PN in Fig. 8.10

| $t_1$ | $t_2$ | $t_3$ | $t_4$ | $t_5$ | $t_6$ | $t_7$ | $t_8$ |
|---|---|---|---|---|---|---|---|
| 0.0040 | $1/\tau_2$ | 0.0067 | 0.0067 | 0.0050 | 0.0200 | 0.0200 | 0.0200 |

three tokens due to the stochastic nature of the model. The probability, though, for either to have three tokens is tiny.

To use SPNP for solution, a C file is created to describe this PN. To declare places, transitions and arcs, the following statements are used, taking $p_1$, $p_3$ and $t_1$ in Fig. 8.10 for example:

```
place("p1"); init("p1", 3);
place("p3");
transition("t1"); rateval("t1", 0.0040);
iarc("p1", "t1");
oarc("p3", "t1");
```

The results can be obtained by executing *SPNP filename*. They include the probability for a place to be marked, the average number of tokens in it, probability for a transition to be enabled, and throughput of a transition. Based on such data, we can obtain resource utilization rate, average in-process inventory and production rate or throughput. For example, the system throughput is the throughput of $t_5$ and its reciprocal is the average cycle time for a product. The utilization rate of the milling machine is the probability of place $p_3$ being marked or (1 – Probability $\{p_7$ marked$\}$). Robot's utilization is Probability $\{p_2$ marked$\}$. The average in-process inventory is the sum of average numbers of tokens in places $p_3$, $p_5$, $p_6$, and $p_4$. The results for the net are summarized in Table 8.7.

It is apparent that there exists a difference between the deterministic model and the stochastic PN model due to exponentially distributed time delay in the latter. The stochastic feature evidently permits more overlap of operations resulting in a higher throughput. However, such results are not so far from reality as to be useless. They can be used as a first cut in assessing system performance. For the cases where

**Table 8.7** Fanuc dynamic results for the net in Fig. 8.10

| $1/\tau_2$ | $\pi$ | *mill* | *drill* | *robot* | *in-process inventory* |
|---|---|---|---|---|---|
| 100 | 558 | 21 | 32 | 73 | 1.8 |
| 200 | 613 | 35 | 28 | 75 | 1.7 |
| 500 | 849 | 60 | 19 | 82 | 1.4 |

stochastic models are more suitable than deterministic time ones, refer to Marsan *et al.*, 1984; Zhou and Leu, 1991.

## 8.6  CONCLUSION

This chapter presents Petri nets with their applications to manufacturing systems modeling, analysis and simulation. The use of such a formal methodology to design flexible and complex manufacturing systems can be broken into the following steps:

1. After feasibility studies and preliminary system designs are performed, designers need to select several of the most attractive ones based on the general considerations such as the investment, desired production capacity and the service time in their life cycles. Then these system designs are modeled by Petri nets and their aggregate analysis can be performed. Timed PNs or stochastic PNs might be applied to simplify the analysis. The aggregate analysis should lead to one or two design candidates.
2. While these designs of the manufacturing systems are further detailed, their PN models should be augmented to represent the change through design and synthesis methodologies. More specific performance measures should be analyzed, e.g. resource utilization, production rates, in-process inventory and product cost. Qualitative analysis should be performed at this stage to guarantee deadlock-free, stable and cyclic system with minimum human intervention. At this step, tractable PN models should be applicable by making reasonable assumptions on the operation times (either deterministic or stochastic with exponential distribution). The results could then be used to make a recommendation for simulation studies or to suggest desired modifications to optimize the designs.
3. After more realistic operation data are included, Petri net simulation has to be used to derive more accurate system performance. The results could then be used to make a final recommendation or further modifications.
4. After the final design is selected, manufacturing system control can be partially derived and implemented by using the PN model. The model can also be used as a specification for the system design. Such Petri net-based specification can prove to be valuable for the future maintenance, modification or improvement.

The further reading on Petri nets and their applications are indicated as follows: for PN fundamentals, see Murata, 1989; for PN synthesis and discrete-event control techniques, Zhou and DiCesare, 1993; for PNs in manufacturing systems and robotics Freedman, 1991; DiCesare and Desrochers, 1991; PN's connection with programmable logic controllers, scheduling and expert systems Silva and Velilla, 1982; Silva and Valette, 1990; for their use in intelligent machines Wang and Saridis, 1990; and for their industrial applications Desrochers, 1990.

## ACKNOWLEDGMENT

The authors greatly appreciate the use of SPNP developed by Duke University and Ms. Y. Chen's work to run TPNS at New Jersey Institute of Technology for a portion of simulation results. The support of the Center for Manufacturing Systems at New Jersey Institute of Technology is acknowledged.

## Appendix: GLOSSARY

Deterministic timed Petri net – a timed PN in which timing information is deterministic.

Discrete event dynamic system – event driven systems allowing asynchronous, concurrent and sequential events.

Markov process or chain – a stochastic process whose future state depends only on the current one instead of the history. It can be analytically analyzed.

Petri net (PN) – a graph theoretic modeling tool for discrete event dynamic systems.

Programmable logic controller (PLC) – a logic controller for sequencing the activities in industrial automation systems.

Semi-Markov process – an extension to a Markov process, allowing state transition time to be of general distribution.

Stochastic Petri net – a timed PN in which timing information is probabilistic.

Stochastic process – one to change system states probabilistically.

Timed Petri net – an extension to Petri nets by incorporating timing information into the models.

## REFERENCES

Ciardo, G. (1989) *Manual for the SPNP Package*, Duke University, Feb.

Desrochers, A.A. (1990) *Modeling and Control of Automated Manufacturing Systems*, IEEE Computer Society Press, Washington, DC.

DiCesare, F. and Desrochers, A.A. (1991) Modeling, control, and performance analysis of automated manufacturing systems using Petri nets, *Control and Dynamic Systems* (ed. C.T. Leondes), **47**, pp. 121–72, Academic Press.

Freedman, P. (1991) Time, Petri nets, and robotics, *IEEE Trans. on Robotics and Automation*, 7(4), pp. 417–33.

Marsan, A.M., Balbo, G. and Conte, G. (1984) A class of generalized stochastic Petri nets for the performance evaluation of multiprocessor systems, *ACM Trans on Comp Sys*, **2**(2), 93–122.

Molloy, M.K. (1982) Performance analysis using stochastic Petri nets, *IEEE Trans. on Comp*, **3**(9), pp. 913–17.

Murata, T. (1989) *Petri nets: properties, analysis and application*, Proc. of the IEEE, **77**(4), pp. 541–79.

Siddiqi, J., Chen, Y. and Robbi, A.D. (1993) *A Timed Petri Net Simulation Tool*, Proc. 9th Int. Conf. on CAD/CAM, Robotics, & Factories of the Future.

Silva, M. and Velilla S. (1982) *Programmable logic controller and Petri nets: a comparative study*, in IFAC Conf. on Software for Computer Control, Madrid, Spain, pp. 83–88.

Silva, M. and Valette, R. (1990) Petri nets and flexible manufacturing, in *Advances in Petri Nets 1989* (ed. G. Rozenberg), Springer Verlag, pp. 374–417.

Shukla, A. and Robbi, A.D. (1991) *A Petri net simulation tool*, in Proc. 1991 IEEE Int. Conf. on Systems, Man, and Cybernetics, Charlottesville, VA, pp. 361–66.

Wang, F.-Y. and Saridis, G. (1990) A coordination theory for intelligent machines, *Automatica*, **26**(5), pp. 833–44.

Zhou, M.C. and DiCesare, F. (1991) Parallel and sequential mutual exclusions for Petri net modeling for manufacturing systems, *IEEE Trans. on Robotics and Automation*, **7**(4), pp. 515–27.

Zhou, M.C. and Leu, M.C. (1991) Modeling and performance analysis of a flexible PCB assembly station using Petri nets, *Trans of the ASME, Jnl. of Elect Pack*, **113**(4), pp. 410–16.

Zhou, M.C., DiCesare, F. and Desrochers, A.A. (1992) A hybrid methodology for synthesis of Petri nets for manufacturing systems, *IEEE Trans. on Robotics and Automation*, **8**(3), pp. 350–61.

Zhou, M.C. and DiCesare, F. (1993) *Petri Net Synthesis for Discrete Event Control of Manufacturing Systems*, Kluwer Academic Publishers, Boston, MA.

# CHAPTER 9

# Recent developments in modeling and performance analysis tools for manufacturing systems

*Manjunath Kamath*

## 9.1 INTRODUCTION

This chapter focuses on models for evaluating or predicting the performance of manufacturing systems. Models that predict system performance are useful during several phases – initial and detailed design of a facility, operation of the system, ongoing improvements and facility redesign. Performance prediction involves the task of translating the real system into a mathematical or simulation model and subsequently 'solving' it to yield the desired performance measures. Typical performance measures in the context of a manufacturing system include throughput or production rates, equipment utilizations, work-in-process (WIP) levels and flow times.

Performance measures indicate the effectiveness of a certain configuration of a manufacturing system. The configuration of a manufacturing system is determined by a given set of manufacturing decisions, examples of which are: the products to be produced, the number and types of equipment, part routings, number and types of material handling devices, capacities and locations of the WIP storage areas and operating policies. Performance measures of a manufacturing system are vital in evaluating its design as well as planning its operation, and eventually in determining the competitiveness of the firm (Suri, 1988). Performance prediction or evaluation tools and techniques assist system designers and manufacturing managers in making their decisions (design-related and operations-related) while attempting to achieve their goals with respect to the performance measures described above.

### 9.1.1  Classification of manufacturing system models

*Generative versus evaluative models*

Based on purpose, manufacturing system models can be classified into two main categories: evaluative (or descriptive) and generative (or prescriptive) (Suri, 1985). Generative models (e.g. optimization algorithms) find a set of decisions to meet a given set of performance criteria, while evaluative models (e.g. queueing networks, simulation) predict system performance for a given set of decisions.

Evaluative models provide insight into system behavior and involve the human (the 'decision maker') in the decision-making loop which is not usually the case with generative models. Also, evaluative models form the basis for many generative models used for designing systems. For example, a discrete event simulation model can be used to evaluate the objective function values for the current set of decisions chosen by an optimization algorithm (Glynn, 1989; Nandkeolyar and Christy, 1989).

*Analytical versus simulation models*

Based on the solution methodology, models can be broadly classfied as analytical models or simulation models. Analytical models are based on mathematical relationships, and provide formulas or computational procedures to calculate system performance measures in the case of evaluative models and 'optimal' values of decision variables in the case of generative models. Examples of analytical models are Markov chains, queueing networks and linear programming models. Computer simulation models describe the logical relationships between events such as job arrival, task completion and machine failure that govern the dynamics of the real system. Certain models that fall into neither of these two categories (analytical and simulation) have been labeled as hybrid models (Shanthikumar and Sargent, 1983). Many generative models used for design optimization of manufacturing and assembly systems fall into this hybrid category. For example, in Liu and Sanders (1988), simulation is used to evaluate objective function values and analytical methods are used to update the values of decision variables.

Analytical models often have minimal data requirements (e.g. mean values) and provide results quickly (in seconds to minutes). However, several assumptions have to be made to obtain analytically tractable models, and these assumptions usually affect the accuracy of the re-sults. On the other hand, simulation models can handle considerable complexity and can be made to mimic the real system as accurately as desired. This usually implies long model development time, enormous data requirements and substantial amount of computer time (several hours) to obtain statistically accurate results. The reader may refer to

Buzacott and Shanthikumar (1993) and Suri and Diehl (1987) for more detailed discussions of the analytical and simulation approaches.

### 9.1.2 Role of performance evaluation models

Performance evaluation models of manufacturing systems play a vital role in many stages of decision making, from preliminary system design to operational planning and control. In general, analytical models such as queueing networks provide valuable guidelines at the preliminary system design stage, while simulation provides the detailed modeling capability that is needed for scheduling and control purposes.

*Systems design*

The systems design area has benefited significantly from the application of quantitative design methods based on queueing network models and simulation. The performance measures calculated, for example, equipment utilizations, average queue lengths, flow times, throughput rate, etc. indicate the effectiveness of a certain system configuration. Stated simply, the task of the designer is to choose from several system configurations, one that yields the 'best' system performance. Generally speaking, the performance measures needed for this purpose are the **steady-state** or long-run performance measures.

In a design context, the advantages of the analytical approach are the insight that it provides into how the system performs under various parameter values and its low computational cost compared to that of the simulation approach. However, simulation is the only viable approach when an accurate and detailed performance prediction is required. A systems design approach that combines the speed of analytical methods with the accuracy of simulation is emerging. Analytical models are used in an initial stage to quickly reduce the number of alternatives to a handful of potential designs which can then be evaluated in detail by simulation (Suri and Diehl, 1987; Suri *et al.*, 1986).

*Operations scheduling and control*

The determination of 'optimal' operating rules usually requires the evaluation of the **transient** or time-dependent performance of the system under various operating rules. Very few attempts have been made to analyze the transient behavior using analytical models, as the models soon become mathematically intractable. Discrete event simulation appears to be the only method for deriving accurate estimates of transient performance. Recent research has demonstrated the possibility of achieving significant improvements in the performance of dynamic scheduling systems by using accurate estimates of queue sizes and flow or lead times (Glassey and Petrakian, 1989 and Ramasesh, 1990). The high computational requirement and the low level of reconfigurability of traditional simulation models have often severely limited its application in an operational context. **The recent developments in parallel simulation, object-oriented simulation, and fast simulation offer a strategic opportunity for significantly increasing the applicability of simulation as a tool for on-line decision support.**

### 9.1.3    Scope of this chapter

This chapter describes both analytical and simulation models for predicting or evaluating the performance of manufacturing systems. The focus will be on recent developments and current research activities. Brief summaries of the classical models with several references to archival publications will be presented. Under analytical models, our discussion will be mostly on queueing network models. Furthermore, we will focus on queueing network models with no capacity constraints on queue sizes. A large body of literature exists on queueing network models of systems with finite waiting rooms, primarily of the flow line (tandem) type, and the reader is referred to a recent exhaustive survey by Dallery and Gershwin (1992).

Under simulation modeling, the topics covered include: (i) the object-oriented simulation approach which offers several advantages such as quick reconfiguration, high modeling flexibility and reusability of modeling elements and (ii) the fast simulation approach that is proving to be computationally superior to the traditional event-scheduling approach for certain system configurations. This chapter ends with a brief discussion on the future of systems modelling.

## 9.2    ANALYTICAL MODELS

### 9.2.1    Queueing network models

Queueing network models have emerged as the most widely studied analytical models of modern manufacturing systems. Queueing networks are used to model the flow of parts or jobs through a network of workstations; jobs visit different workstations, in varying sequences, before they are completed. In a queueing network model, customers move from one node to another, competing for server(s) at each node visited with the other customers at that node. The movement of customers within a queueing network is usually described by a routing matrix, the $(i, j)$th element of which gives the probability that a

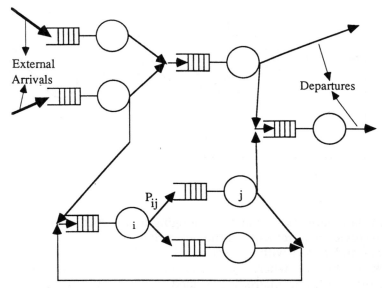

**Fig. 9.1** An open queueing network (OQN).

customer will proceed to the *j*th node after service completion at node *i*. This is often called Markovian routing because the next node probabilistically depends only on the current node and not on past events in the network.

In an **open queueing network** (OQN), customers enter the network from the outside, receive service at one or more nodes and eventually leave the network. A picture of a typical OQN is given in Fig. 9.1. In an OQN, the arrival rate or throughput is an independent variable and the number of customers in the network is a dependent variable. In a **closed queueing network** (CQN), there is a fixed population of customers in the network. In a CQN, neither arrivals nor departures are permitted; instead a fixed number of customers pass repeatedly through the various nodes. A typical CQN is depicted Fig. 9.2. With the closed model, the number of customers in the network is an independent variable, and the throughput rate, which is usually defined as the output rate from some designated node, is a dependent variable. OQNs have been used to model job shops, while CQNs have been used to model flexible manufacturing systems (FMSs). The number of pallets in an FMS determines the number of customers in its CQN model, and the output rate of the load/unload station(s) is the throughput rate.

In **single-class** networks (open or closed), there is only one type of customer – all customers are identical in a probabilistic sense. In a

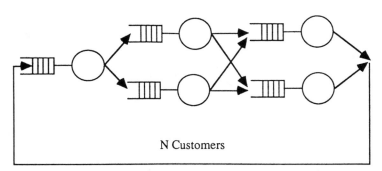

N Customers

**Fig. 9.2**   A closed queueing network (CQN).

multi-class network, customers belonging to a particular class have the same service requirements. If the network is open with respect to some classes and closed with respect to others, then we have a mixed network. In the case where customers do not change classes within the network, a routing matrix is specified for each customer class. In the general case, the probability that a customer of class $r$ who completes service at node $i$ will next require service at node $j$ in class $s$ is specified.

Next, a summary of the key developments in the solution methods for queueing network models will be presented, starting with the pioneering work of Jackson (1957).

*Product-form networks*

**Notation**
{Input or given quantities}

| | |
|---|---|
| $M$: | number of nodes in the queueing network. |
| $m_i$: | number of identical servers at node $i$. |
| $N$: | maximum number of customers allowed in the network, if the network is closed. |
| $\lambda_{0i}$: | mean external arrival rate of customers or jobs to node $i$, if the network is open. |
| $\mu_i\,(\tau_i)$: | service rate (mean service time) of a customer at node $i$. |
| $P_{ij}$: | proportion of those customers completing service at node $i$ that go to node $j$ next. |
| $P_{i0}$: | proportion of those customers completing service at node $i$ that leave the network, if the network is open; $P_{00} = 1$ by definition. |

{Output quantities – all are steady-state measures}

| | |
|---|---|
| $\lambda_i$: | mean total arrival/departure rate of customers or jobs at node $i$. |

$\rho_i$:                     utilization of a machine at node $i$.
$L_{qi}$:                    average number of customers waiting (in the queue) for service at node $i$.
$L_i$:                      average number of customers at node $i$.
$W_{qi}$:                   average time a customer spends waiting per visit before service at node $i$.
$W_i$:                      average time spent by a customer at node $i$ per visit.
$(n_1, n_2, \ldots, n_M)$:  network state vector; $n_i$ is the number of customers at node $i$.

In a landmark publication in 1957, Jackson showed that for a single-class open queueing network with exponential multi-server nodes, FCFS queue discipline at each node, Poisson arrivals to the network and Markovian routing probabilities, the steady-state probability of network state was a product of M terms, with the $i$th term being the probability of finding $n_i$ customers at node $i$.

$$P(n_1, n_2, \ldots, n_M) = P_1(n_1)P_2(n_2) \ldots P_M(n_M). \qquad (9.1)$$

The above result is popularly known as the product-form result and the network, Jackson network. $P_i(n)$ is the steady-state probability of finding $n$ customers in an $M/M/m_i$ queueing system with service rate $\mu_i$ and arrival rate $\lambda_i$. The total arrival rate at node $i$, $\lambda_i$, can be found by solving the following system of linear equations, called the 'traffic-rate equations' – the total arrival rate being the sum of the external arrival rate and the arrival rates from other nodes.

$$\lambda_i = \lambda_{0i} + \sum_{j=1}^{M} \lambda_j P_{ji}, \quad i = 1, 2, \ldots, M \qquad (9.2)$$

If $\rho_i = \lambda_i/(\mu_i m_i) \geqslant 1$, then the $i$th node, and hence the network, is unstable, and the calculations stop. When $\rho_i < 1$ for all $i$, then each node can be treated as an independent node and the familiar machinery of the $M/M/m_i$ queueing system can be used to derive many network performance measures (Buzacott and Shanthikumar, 1993; Viswanadham and Narahari, 1992).

Later, Jackson (1963) showed that the product form result was valid for a broader class of OQNs, where the mean arrival rate of customers depends on the number of customers in the system (state-dependent arrival rate) and the mean service rate at each node depends on the number of customers at that node (load-dependent server). He also examined a case with 'triggered arrivals' where a customer is injected automatically into the network when a departure causes the network population to fall below some threshold value (this extended Jackson queueing network behaves as a CQN) (Conway and Georganas, 1989).

Independently, Gordon and Newell (1967) also showed that CQNs with exponentially distributed service times have a product-form solution. They showed that for a single-class CQN with exponential multi-server nodes, FCFS queue discipline at each node and a stochastic routing matrix, the analytic form of the joint probability distribution of customers at each node has a product form. Their result for a CQN with single-server nodes is shown below:

$$P(n_1, n_2, \ldots, n_M) = \frac{1}{G(N)}\prod_{i=1}^{M} x_i^{n_i} \tag{9.3}$$

where,

(i) $x_i = v_i/\mu_i$, $i = 1, 2, \ldots, M$. $v_i$ is any set of values satisfying the traffic-rate equations representing the conservation of flow:

$$v_i = \sum_{j=1}^{M} P_{ji}v_j. \tag{9.4}$$

Note that since there are no external arrivals to the system and no departures from the system, the traffic-rate equations provide only relative output or throughput rates of the nodes. $v_i$ are often called the visit ratios. In CQN models of manufacturing systems, the visit ratio to the load/unload station, say node 0, is set equal to 1.0, so that the $v_i$ can be interpreted as the average number of visits to node $i$ by each new job (Suri *et al.*, 1993).

(ii) $G(N)$ is a normalization constant defined by requiring the probabilities of all possible states to sum to unity.

$$G(N) = \sum\prod_{i=1}^{M} x_i^{n_i}. \tag{9.5}$$

The computation of $G(N)$ by direct summation over all $\binom{M + N - 1}{N}$ states is expensive as the number of terms increases exponentially with $M$ and $N$.

Because of the difficulty associated with the computation of the normalization constant, the practical value of product-form CQNs remained very limited till Buzen's (1973) discovery of an efficient algorithm to compute $G(N)$. Buzen's algorithm, also known as the convolution algorithm, consists of an efficient recursive scheme ($O(NM)$) for computing $G(N)$. He also showed that many common performance measures (e.g. throughput rate, node utilizations and average number of jobs at a node) are simple functions of the values computed during the recursion. For example, the network throughput rate or the output rate of the designated output node is $G(N - 1)/G(N)$. The development of this efficient algorithm led to a significant increase in the application of CQN models to analyze the performance of computer systems as well as manufacturing systems. Based on Buzen's algorithm, Solberg (1977) developed a manufacturing modeling tool, CAN-Q, for quickly evaluating the performance of FMSs. Solberg's work related to CAN-Q is considered as a pioneering effort by the manufacturing modeling community, because it stimulated several efforts to model flexible/ automated manufacturing systems in the late seventies and early eighties.

In 1975, there was an important publication by Baskett, Chandy, Muntz, and Palacios that greatly extended the class of product-form networks. The BCMP networks retain the product-form of the equilibrium state probability with different classes of customer and new service disciplines. The service discipline and service time distribution combinations that are allowed at a node are: (i) first-come-first-served (FCFS) discipline and the same exponential service time distribution for all classes, and (ii) any one of the following disciplines – processor sharing (PS) (time division) discipline, ample servers (no queueing), and preemptive-resume last-come-first-served (LCFS) discipline and a single-server – together with a distinct service time distribution having a rational Laplace transform for each class. For manufacturing applications, the BCMP networks provided limited additional modeling capability because of the restriction to a single-class network and exponential service-time distribution if FCFS nodes are present. The other complex service disciplines that are required for a more general service time distribution are not commonly found in manufacturing systems.

Although simple and easy to implement, Buzen's algorithm for CQNs suffers from some drawbacks: numerical difficulties involving overflow or underflow for extreme cases (e.g. when $M$ and $N$ are large) and difficulty in developing heuristic extensions to model more general (non-product form) cases (Suri *et al.*, 1993). The mean value analysis (MVA) approach developed by Reiser and Lavenberg (1980) addressed these difficulties. The MVA method eliminates the computation of the normalization constant and detailed joint probability distributions. It works directly with the desired performance measures, such as mean queue lengths, mean waiting times, utilizations and throughput rates. We will briefly discuss MVA for the single-class CQN with exponential, single-server, FCFS nodes.

The central result on which MVA is based is the **arrival theorem**, which says that in a product-form CQN, a customer circulating in the network 'sees' a network with one fewer customer in equilibrium. Let $X_i(n)$ represent a performance measure $X_i$ in a network with $n$ circulating customers. The following relation for the average time a customer spends at node $i$ exemplifies the arrival theorem.

$$W_i(N) = \tau_i + L_i(N-1)\tau_i, \quad i = 1, 2, \ldots, M. \tag{9.6}$$

The arriving customer at node $i$ has first to wait for the customers already present at the node ($L_i(N-1)$ on the average because of the arrival theorem) to complete their service. The first part of the RHS is the customer's own average service time and the second term is the waiting time. The above waiting time relation, together with Little's Law (Little, 1961), gives rise to a set of recursive relations. Starting with $L_i(0) = 0$, all the main network performance measures including the throughput can be calculated. It should be noted that Reiser and Lavenberg (1980) presented the MVA for a multi-class (or multi-chain) CQN. The multi-class MVA algorithm is also based on the arrival theorem; an arriving customer at a node observes the equilibrium solution with one fewer customer in the arriving customer's class. Reiser and Lavenberg (1980) also present different forms of MVA for calculating more detailed measures (e.g. queue length probabilities) and for handling more network features (e.g. load-dependent servers).

The MVA relations lend themselves naturally to heuristic extensions for handling non-product form features. For example, to handle CQNs with non-exponential service times, Reiser (1979) presented an approximation by introducing the mean residual service time (which introduces service time variance term into the equations) of the customer in service at the instant a new customer arrives. Reiser (1979) still used the arrival theorem which does not hold unless the network has product form. Computational complexity is still a limiting factor of the original MVA when used for large networks with many chains (classes) and large customer populations. Again, to address this issue, the nature of the MVA relations facilitated the development of a heuristic extension. The Schweitzer and Bard (S-B) algorithm (Suri and Hildebrant, 1984) which is iterative rather than recursive is computationally attractive for large multi-class networks. The recursive relation shown in (9.6) was replaced by,

$$W_i(N) = \tau_i + \frac{N-1}{N} L_i(N)\tau_i, \quad i = 1, 2, \ldots, M. \qquad (9.7)$$

which makes the model approximate, but enables easy implementation with low computational requirement. The FMS modeling tool, $MVAQ$, developed by Suri and Hildebrant (1984), is based on the S-B algorithm.

The late seventies and early eighties gave rise to several attempts by researchers working in the computer performance modeling and manufacturing systems modeling areas, to model non-exponential service times in queueing networks. The inability of the product-form networks to accurately model service-time variability present in the real system was seen as a major impediment to the successful application of queueing networks in practice. For a review of the several approaches developed during this period, mostly based on the convolution and MVA algorithms, the reader is referred to Agrawal (1985), Chandy and Sauer (1978), Lavenberg (1983), and Suri *et al.* (1993).

*General networks: parametric decomposition approach*

During the early to mid eighties, a fundamental shift occurred when many researchers started to focus more on the application side than on the exactness of the solution methodology. Whitt (1983) expressed the change needed in one simple, terse statement in his paper: 'A natural alternative to an exact analysis of an approximate model is an approximate analysis of a more exact model.' A comprehensive methodology for analyzing open models that included the variability present in both arrival and service processes emerged. Starting with the early work of Kuehn (1979) a large volume of publications appeared, culminating in a paper by Whitt (1983) which may be viewed as the main archival reference for details of what is now known as the **parametric decomposition** (PD) approach. An early application to manufacturing job shops is presented in Shanthikumar and Buzacott (1981). Recent research has addressed several features (in addition to general service times) that are prevalent in manufacturing systems – deterministic routing, machine failures and repair, batch service, overtime, etc.

In his 1983 paper, Whitt presented the details of the PD approach in the context of a software package called the queueing network analyzer (QNA). Although Whitt's development was for open multi-class networks, we will first describe it for the open single-class network with single-server FCFS nodes. Some additional notation is defined first. The squared coefficient of variation (SCV) of a random variable (rv) is defined as the variance of the rv divided by the square of its mean.

{Input or given quantities}
$c_{0i}^2$: interarrival-time SCV of external arrivals to node $i$, if the network is open.
$c_{si}^2$ service-time SCV at node $i$.

{Intermediate quantities – all steady-state values}
$c_{ai}^2$: interarrival-time SCV at node $i$.
$c_{di}^2$: interdeparture-time SCV at node $i$.

The PD approach involves two main steps. The first step in the decomposition approach is the analysis of the interaction between nodes to approximately determine the mean and the variance of the interarrival time at each node. The next step computes the performance measures based on $GI/G/m$ approximations that are based on the first two moments of the interarrival and service times. Each step is elaborated next.

**Computing parameters of arrival processes at the nodes.** A single node is related to other nodes in the network model by its input and output processes. The internal flow parameters (rates and variability parameters of arrival processes) approximately capture the interdependency among the nodes. First, the mean total arrival/departure rate of customers at node $i$ are obtained via the traffic rate equations representing the conservation of flow (Equation (9.2)). If utilization $\rho_i \geq 1$, then the $i$th node is unstable and the algorithm stops. Note that the calculations involving the arrival rates and utilizations are **exact**. Key approximations enter the picture while setting up the **traffic variability equations** which yield the variability parameters for the internal flows, $c_{ai}^2$. The equations are linear, and are obtained by combining renewal approximations for the basic operations – merging of flow, splitting of flow and departure or flow through a node. For further details regarding these approximations the readers are referred to Bitran and Dasu (1992), Tirupati (1992) and Whitt (1983). To summarize, this step involves the solution of two sets of linear equations – the traffic rate equations yield the total arrival rate at each node and the traffic variability equations yield the SCV of interarrival times at each node.

**Calculation of node performance measures.** This step relies heavily on the approximations developed in the queueing literature for $GI/G/1$ or $GI/G/m$ queues (Shanthikumar and Buzacott, 1980). The queues or nodes are treated (approximately) as being stochastically independent and the expected (equilibrium) waiting time at each queue is computed using approximate formulae that are based on the first two moments of the interarrival and service times. For example, if the $i$th node is a single-server node, then using the two-moment approximation for the waiting time in a $GI/G/1$ queue developed by Kraemer and Langenbach-Belz (1976) we have

$$W_{qi} = g\left(\frac{c_{ai}^2 + c_{si}^2}{2}\right)\left(\frac{\tau_i \rho_i}{1 - \rho_i}\right) \qquad (9.8)$$

where $g \equiv g(\rho, c_{ai}^2, c_{si}^2)$ is defined as

$$g(\rho, c_{ai}^2, c_{si}^2) = \begin{cases} \exp\left(\dfrac{-2(1 - \rho_i)}{3\rho_i}\dfrac{(1 - c_{ai}^2)^2}{(c_{ai}^2 + c_{si}^2)}\right), & c_{ai}^2 < 1 \\[2ex] \exp\left(-(1 - \rho_i)\dfrac{(c_{ai}^2 - 1)}{(c_{ai}^2 + 4c_{si}^2)}\right), & c_{ai}^2 > 1. \end{cases} \qquad (9.9)$$

The other network performance measures can be easily obtained using Little's Law (Little, 1961) as follows:

$$L_{qi} = \lambda_i W_{qi}, \quad W_i = W_{qi} + \tau_i, \quad \text{and} \quad L_i = \lambda_i W_i. \qquad (9.10)$$

Recent articles by Bitran and Dasu (1992) and Tirupati (1992) present excellent overviews of the PD approach with mathematical details of the flow approximations involved.

For manufacturing systems, the routing and service-time information for different part types or classes is usually available in the form of process plans. Each part type has its own route that specifies the sequence of workstations visited. Also, each part type may have its own processing time distribution at each workstation (on its route) characterized by rate and variability parameters. Also, the processing times can be different for different visits to the same workstation by the same part type. Bitran and Tirupati (1988) and Whitt (1983) explain how the information available in the form of process plans can be easily transformed into a form required for the queueing network model, that is, into a routing matrix and node service-time parameters for a **single aggregate customer class**. In fact, this is the first step in the queueing network analyzer software (AT&T, 1992; Whitt, 1983) which accepts input for different customer classes. The analysis proceeds for the aggregated model using the two-step PD process described above. The performance measures computed are for the aggregated model, and measures for particular classes can also be computed (Whitt, 1983).

The two-moment framework used by the PD approach, for the first time, has provided a much needed, common, cohesive framework for the performance analysis of a broad class of manufacturing networks. Within this framework several features relevant to manufacturing systems have been incorporated: multiproduct networks with deterministic routing (Bitran and Tirupati, 1988), machine breakdown and repair, changing lot sizes and inspection and testing (Segal and Whitt, 1989), batch service (Bitran and Tirupati, 1989), and overtime (Bitran and Tirupati, 1991). Features such as machine break down and repair and overtime have been modeled by modifying the service times. We illustrate such a procedure with an example taken from an automated assembly system context (Kamath and Sanders, 1991).

In the case of an automatic assembly station, the service time is typically fixed (deterministic) except when a station 'jam' occurs, when there is a repair distribution. Let $T$ be the overall service or processing time at the workstation. $Z$ is an indicator rv that takes the value 1 if a jam occurs during service at the workstation and 0 otherwise. $E(Z)$ is the probability of occurrence of a jam at the workstation, which is usually specified. If we let $S$ ($E(S)$ = deterministic assembly time and $var(S) = 0$) represent the deterministic portion of the service time, we have,

$$T = S + ZR$$

where $R$ is the rv representing the repair time. The rv $Z$ is assumed to be stochastically independent of rv $R$. The mean and variance of $T$ can be easily derived and are

$$E(T) = E(S) + E(Z)E(R) \qquad (9.11)$$

and

$$var(T) = var(Z)var(R) + var(Z)[E(R)]^2 + [E(Z)]^2var(R)$$
$$= E(Z)var(R) + var(Z)[E(R)]^2 \qquad \text{(as } Z \text{ is an indicator rv)}$$
$$\tau = E(T) \quad \text{and} \quad c_s^2 = var(T)/\tau^2. \qquad (9.12)$$

We have presented above the parametric-decomposition approach for analyzing general OQNs and a summary of several recent extensions to model common features of manufacturing systems. Although the PD approach involves several approximations for calculating flow parameters and performance measures, it provides remarkably accurate estimates (Bitran and Tirupati, 1988, 1989, and 1991; Shanthikumar and Buzacott, 1981; Kamath and Sanders, 1987), and is consistent with the Jackson network theory (Whitt, 1983). After the arrival process parameters are calculated, the nodes are treated (approximately) as being stochastically independent. This independence can be viewed as a generalization of the product-form solution for the Jackson network. In the PD approach, the dependence between the nodes is captured approximately through the rate and variability parameters of the node arrival processes. For a Jackson network, the PD approach is **exact**, and for general networks it becomes approximate.

**Parametric decomposition approach for general CQNs.** Over the last few years, the PD approach described above has been extended to certain classes of CQN models. In particular, approximations were developed for closed tandem queueing networks (Kamath and Sanders, 1987; Kamath *et al.*, 1988). In a closed network, the number of customers in the network is fixed and the network throughput (i.e. the output rate of a node that is identified as the output node) is the unknown quantity. In the first step of the PD approach, the arrival parameters can be computed if a value is assumed for this throughput. In the second step, the waiting times are first calculated using the approximate formula for the open network case. A correction factor (Kamath *et al.*, 1988) is applied to account for the fixed network population. Finally, using Little's Law, a relation is established between the total number of customers in the system and the mean response times at the individual nodes through the unknown network throughput. For the closed tandem case, writing this relationship in an equation form gives us a nonlinear (transcendental) equation in a single unknown. The solution to this equation can always be obtained to a reasonable degree of accuracy in a few iterations (Kamath *et al.*, 1988). Some preliminary work has been done in extending this work to general closed networks (Kamath and Basnet, 1990).

In manufacturing systems, simultaneous requirement of multiple resources to service a job is common: a part needs a pallet (a passive resource) before entering a manufacturing/assembly system and competing for other active resources (i.e. machines, operators, etc.), a part requires the machine and the services of a robot (or an operator) for loading and unloading operations, and an assembly needs the services of an operator when it experiences a jam while occupying a station. If a job requires a pallet to enter a manufacturing system, this can be easily handled using a CQN model by noting that the number of pallets imposes a population constraint on the system. However, queueing models often ignore most other instances of simultaneous resource possession (SRP) that usually involve active resources (e.g. machine and robot, workstation and operator, etc.) because of the non-standard form of the model and resulting violation of product-form assumptions (de Souza e Silva and Muntz, 1987; Jacobson and Lazowska, 1982). SRP can have a significant effect on system performance. **This is because a part may keep an active resource (e.g. machine) busy and nonproductive, while waiting for another active resource (e.g. operator, AGV).** Approximate solution techniques have been developed for analyzing specific instances of SRP. Examples of these instances are limited memory partitions and shared peripherals in computer systems (de Souza e Silva and Muntz, 1987; Jacobson and Lazowska, 1982) and simultaneous use of multiple instrument types or operator types in an automated electronic test facility (Pattipati and Kastner, 1988).

Kamath and Sanders (1991) consider a closed tandem queueing network model with general service times and SRP in the context of an automatic assembly system, where the focus is the effect of waiting time for an operator resource on the system production rate. Results from a general machine-repairman model are integrated into the two-moment framework by modifying the machine service times. The resulting model captures the effect of delays due to operator unavailability on assembly system performance. Similar ideas have been implemented in MPX (a software package) in which multiple classes of operators tend multiple classes of machines (Suri and de Treville, 1991).

The PD approach is particularly attractive for industrial applications because its data requirements are minimal. It only requires mean and variance (not distributions) information about arrival and service processes. It is computationally simple; only two systems of linear equations have to be solved. MPX (MPX, 1992), previously called MANUPLAN II (Suri *et al.*, 1986) and QNA (AT&T, 1992; Snowdown and Ammons, 1988; Whitt, 1983) are two commercial software packages that use analysis techniques based on the PD approach. MPX is designed for manufacturing operations while QNA is for any queueing situation. Because of the clever use of manufacturing terminology instead of queueing terminology and the many manufacturing features it provides, MPX is beginning to gain industrial acceptance. Although the PD solution is approximate, the net result is a tool that industry can use to obtain fast, reasonably accurate estimates of system performance to guide rapid decision making.

### 9.2.2   Other models

Petri net models are powerful, formal graph models that can be used to study the qualitative (absence/presence of deadlock, reinitializability and buffer overflows) as well as quantitative performance of manufacturing systems. See Chapter 8 for more on this approach. Balbo *et al.* (1988) integrated queueing network and Petri net models to evaluate computer system configurations with complexities such as job priorities, blocking, dynamic job routing and simultaneous resource possession. Viswanadham and Narahari (1992) present examples illustrating the application of this integrated approach in the manufacturing system context. A recent survey by Leung and Suri (1990) contains references on other analytical and hybrid approaches.

## 9.3 SIMULATION MODELS

Simulation modeling is the only viable approach to the **detailed** performance evaluation of complex manufacturing systems. Simulation models can mimic the operation of a system in as much detail as required by the user and hence can be made very accurate. Compared to other performance evaluation models, assumptions in a simulation model are closer to reality. Also, this is one of the few methodologies that can be used for the analysis of both the transient and steady-state behavior of a system. However, the simulation approach has some drawbacks:

1. Construction of new simulation models and modification of existing models are formidable tasks. Simulation model development is often a time consuming task, and verification and validation of simulation models still remain challenging research areas.
2. Analysis is computationally expensive, because several long executions are often required to obtain statistically accurate performance estimates.

Recent developments in object-oriented modeling and simulation address the first issue – development, management and reuse of simulation models – while efforts in the areas of parallel/distributed simulation and fast simulation target the execution speed issue.

The history of simulation modeling software can be broken into five periods: the era of custom programs, the emergence of simulation programming languages, the second generation of simulation programming languages, the era of extended features and the current period (Nance, 1984). In the early sixties, discrete event simulation languages such as GPSS, GASP and SIMULA were introduced (Nance, 1984; Mitrani, 1982). These languages were primarily written in general purpose languages but had built-in procedures to perform many routine simulation tasks, such as scheduling of events and statistics collection. In the late sixties a second generation of simulation languages emerged. In most cases (i.e. GPSS V, SIMULA 67 and GASP IIA), these languages were more powerful replacements of their predecessors. In the seventies, as the use of simulation modeling grew, developments in simulation languages were driven toward the extension of simulation specific languages to facilitate easier and more efficient methods of model translation and representation. Many of the languages which evolved from these developments, GPSS, SLAM, and SIMAN, are still widely used today (Law and Haider, 1989).

In addition to the developments occurring in simulation languages, changes were also occurring in the way simulation models were used within organizations. Simulation was frequently used to study smaller, short term problems and projects. This effectively increased the pressure for development of faster and more efficient modeling methodologies with higher levels of reusability and user friendly interfaces. In the eighties, rapid advancements occurred in the areas of computer hardware and software. Powerful and inexpensive personal computing environments with high resolution graphics became commonplace, and artificial intelligence (AI) techniques and expert systems gained acceptance. These changes had, and continue to have, a direct impact on simulation methodologies. Simulation modeling is now open to a much broader base of potential users through advances such as (Basnet *et al.*, 1990):

1. menu and icon driven model builders;
2. expert systems to aid in the building and debugging of models;
3. graphs and charts to display model results both during and after execution;
4. model animation to view the operation of the system as a whole or zoom in on a specific area of interest.

In the area of graphics and animation, packages such as SLAMSYSTEM, Cinema/SIMAN, and SIMFACTORY (Law and Haider, 1989) are among the leading edge competitors. The animation and graphics are typically developed and presented as an integral part of the simulation language. By contrast, software engineering, AI and expert system concepts impact simulation modeling through the use of a simulation 'front-end' or application generator. These tools interact with the user and ultimately result in code which can be passed directly to the simulation language (Endesfelder and Tempelmeier, 1987 and Thomasma and Ulgen, 1988).

### 9.3.1 Object-oriented simulation

In terms of continuing the growth of simulation modeling and expanding the use of simulation in general, construction of new simulation models and modification of existing models still provide formidable challenges to researchers. For example, modeling primitives and animation objects must be expressed in the user's language. Also, the time required to construct and validate simulation models must continue to decrease through the use of concepts such as rapid prototyping and model reusability. Object-oriented programming (OOP), a relatively new programming paradigm, is proving to be a major contributor to these areas.

*Object-oriented programming (OOP) concepts*

The principal idea associated with object-oriented programming (OOP) is that all elements in the system (program) are treated as 'objects'. An object is either a 'class' or an instance of a class. A class is that software module which provides a complete definition of the capabilities of members of the class. These capabilities are either provided by the procedures and data storage contained within the immediate class definition or inherited from other class definitions to which this class is related. An instance of a class is a realization of the class having all of the operating capabilities provided in the class definition. When a program is executing, operations are performed by sending messages back and forth between independent objects. A message is a form of procedure call in an OOP framework.

Many of OOP's characteristics can be traced to the SIMULA 1 language (Meyer, 1988). SIMULA has found a popular academic following in Europe and throughout the world, but has never gained widespread use in the commercial environment (Kreutzer, 1986, page 105). While SIMULA embodies some of the concepts of OOP, it is not a pure OOP language. Smalltalk, one of the purest OOP languages, was influenced by SIMULA's model of computation. Smalltalk added the message passing paradigm, creating a programming style which we now know as OOP (Kreutzer, 1986, page 194; Meyer, 1988, page 437).

Advantages of OOP over traditional (procedural) programming have been documented by Cox (1986) and Meyer (1988). OOP embodies four key concepts which result in making software systems more understandable, maintainable and reusable. These concepts are: encapsulation, data abstraction, dynamic binding and inheritance (Pascoe, 1986).

**Encapsulation** means that an object's data and procedures or methods are enclosed within a tight boundary, one which cannot be penetrated by other objects. The use of objects therefore improves the reliability and maintainability of system code. Message passing is a necessary result of encapsulation. In order for one object to affect the internal condition of another object, the first object must request (by sending a message) the second object to execute one of the second object's methods.

**Data abstraction** may be defined as the principle that a program should not make assumptions about implementations and internal representations. In languages supporting data abstraction, a programmer defines an abstract data type consisting of an internal representation plus a set of procedures used to access and manipulate the data (Pascoe, 1986). One may change the internal representation of an object without affecting the users of the object. The user of a data abstraction need not know the actual representation of data in storage.

Binding refers to the process in which a method or an operator and the data or operand on which it is to operate are linked. In contrast to early binding (i.e. at the time of code compilation) in conventional procedural languages, **dynamic (or late) binding** provided in OOP delays the binding process until the software is actually running. Because of this feature, the same message structure can be used for different code implementation, so that the same message can elicit a different response depending on the receiver.

**Inheritance** provides for a low-level form of software reuse. OOP classes are defined in a hierarchical tree (single inheritance) or lattice (multiple inheritance) structure. Each class inherits the methods and data storage structure of all of its superclasses. Inheritance allows the construction of new objects from existing objects by extending, reducing, or otherwise modifying their functionality. For example, forklifts and AGVs can be modeled by adding subclasses to a generic class representing a material handling device.

### Object-oriented modeling and simulation of manufacturing systems

The bulk of the work in a simulation study consists of writing, testing and debugging simulation programs. Hence, with regard to simulation programming requirements, following the object-oriented paradigm has the important advantage of preserving the bulk of developed code for general use in model building. Furthermore, the OOP paradigm provides the possibility of achieving a one-to-one mapping between the objects in the system being modeled and their abstractions in the software, thereby facilitating simulation model construction (Adiga, 1989; Thomasma and Ulgen, 1988). More importantly, the characteristics of OOP are causing us to rethink our entire approach to systems modeling using computers.

*Realistic models – explicit modeling of information and control.* Traditional simulation languages are equipped with features for modeling the flow and processing of material. They do not provide convenient structures for accurate and detailed modeling of the decision-making logic related to the management and control of manufacturing firms. Simulation models created in traditional simulation languages are difficult to use in experiments that involve investigation of different control strategies and policies. This is because the control and information constructs are frequently hard coded and dispersed into the model.

The performance of a manufacturing system is highly influenced by control policies used in its operation. In today's computer-integrated manufacturing (CIM) environment, there is even greater need for modeling tools that facilitate the easy specification and modification of decision rules, heuristics and policies that are used for operations control. Such modeling tools can bring the dream of designing and implementing an 'optimal' operations control system a step closer to reality. **The separation concept – the concept of distinctly and separately modeling physical elements, information sources, and control/decision objects** – coupled with the benefits of OOP has enabled the development of simulation environments that are far superior to any that we have known in the past. We present highlights of some long-term, on-going research efforts in this direction.

A research team at the University of California, Berkeley, based on the distinction between 'plant' and 'control' in control systems theory, imposed a strict separation between objects that represent physical devices and those that implement decision heuristics while designing a software object library for simulation of semiconductor manufacturing systems (Glassey and Adiga, 1989 and 1990). This library of objects, called BLOCS/M and written in Objective-C, has also been used to model FMSs (Adiga and Gadre, 1990). Their primary motivation for developing this library of objects was the need for a powerful simulation tool to assist the development of decision strategies for real-time resource allocation (Glassey and Adiga, 1990).

A research team at Oklahoma State University (OSU), Stillwater, proposed the separation of physical, information and control structures in order to obtain more realistic models and to facilitate the investigation of different operating policies on the system performance (Basnet *et al.*, 1990). Complete definitions (given below) of these three modeling structures were developed (Pratt *et al.*, 1991; Mize *et al.*, 1992) and the concepts were used to build a highly reusable modeling and simulation environment (using Smalltalk-80) for discrete part manufacturing systems (Bhuskute *et al.*, 1992).

A **physical object** is an object with a tangible correspondent in the real world system. An object can be classified as a physical object if the primary focus of the modeler's interest in the object is its physical extent or characteristics. Examples are workstations, parts and material handlers. An **information object** is an object which may or may not have a tangible correspondent in the real world system. In this case, the primary focus of the modeler's interest in the object is its information content. Some manufacturing examples are bills of material and routings. A **control/decision object** is a logical object which typically has no tangible correspondent in the real world system. Functions of a control/decision object include implementing decision heuristics and exercising logic algorithms. In the OSU OO modeling environment, the model representation classes have been grouped into three categories, following the separation concept:

1. Physical elements: *WorkStation, MaterialHandler, Queue, Buffer, etc.*
2. Information elements: *CustomerOrder, ShopOrder, Routing, BOM, etc.*
3. Control elements: *QueueController, WorkCenterController, etc.*

Recently, a research team at the Georgia Institute of Technology, Atlanta, has begun developing a manufacturing simulator in C++ (Narayanan *et al.*, 1992). Again, they make the distinction between physical elements or plant objects and control objects that implement decision heuristics. Their class hierarchy also contains a class database to model objects that are sources of information, for example, a factory database that contains detailed process plans. Govindaraj *et al.* (1993) present a detailed description of the architecture of controllers in the simulator. Their current implementation of controllers seems to accommodate both the hierarchical (master-slave) and the non hierarchical (client-server) architectures.

*Extensible simulation environments.* In a simulation modeling environment constructed using conventional OOP languages, the connections between objects, their behavior and their communication protocols are determined during the design of the object library and encoded in methods and instance variables. Problems arise when a modeler either needs an object that is not provided for by the environment or needs to model a new behavior for a currently available object. These are challenging problems, especially if the model construction process has to be 'programming-free' and yet have a one-to-one correspondence between software objects and real world objects. We summarize below the approaches taken by the various research groups to tackle these difficult modeling issues.

With regard to an object that is not available in the environment, in most cases a programmer has to get involved to add the required class(es) to the library. In cases where the object can be built by interconnecting some primitive elements (objects) already available in the environment, there may be a 'programming-free' solution. For example, in 'SmartSim', an icon-based simulation program generator in Smalltalk-80 developed at the University of Michigan, Dearborn, (Thomasma and Ulgen, 1988 and Ulgen and Thomasma, 1990) primitive elements and their interconnections can be grouped into a single unit called 'subsystem' represented by a single icon. The subsystem units can be stored and reused in a manner similar to the primitive elements. The subsystem concept has its roots in Zeigler's DEVS formalism (Zeigler, 1984). The separation of physical, information and control functions, and the provision of several primitive elements of the three types could also provide a capability to quickly configure higher-level objects that are not readily available in the environment (Mize *et al.*, 1992).

When a new behavior is to be modeled, an approach that is consistent with the OO paradigm is to add a subclass to an already existing class. To facilitate this, the environment should provide generic classes (e.g. machine, material handler, etc.) of modeling objects (Glassey and Adiga, 1990; Mize *et al.*, 1992). However, it should be noted that extending an existing class library involves programming. Govindaraj *et al.* (1993) have proposed an alternative approach to handling object behavior and functionality by encoding actions in templates that specify the behavior of an object rather than as methods in the object.

'SmarterSim' (Ulgen *et al.*, 1989), an extension of 'SmartSim', provides the user with a mechanism to describe certain types of behavior from scratch with little programming. This is accomplished through abstract, generic constructs such as elemental operations, requests and rules. An elemental operation is an indivisible activity that consumes some finite (non-zero) amount of time. Rules are fired at the termination of elemental operations. Rules specify actions (messages) that need to be taken upon the satisfaction of the user-specified conditions. Some examples of actions are the immediate start of new elemental operations and scheduling the start or end of elemental operations. A prototype implementation is described that provides the user capability to specify, review and modify certain types of material handling/flow logic.

In all the environments described above, the modeler is presented with an object-oriented world view, and this makes it possible for engineers with no special expertise in simulation to construct models of systems they are familiar with and are responsible for. The many advantages of OO simulation have been well documented (Thomasma *et al.*, 1990 and references therein). Current efforts are aimed at (i) exploring further the application of OO concepts in simulation modeling and (ii) expanding the role of OO simulation in the design and operations control of manufacturing systems (Adiga, 1993; Mize *et al.*, 1992).

### 9.3.2  Fast simulation

Discrete-event simulation is the only tool that can model a system in great detail as well as accurately predict its transient performance. While this should make it an ideal tool for real-time control, its execution speed prevents it from being so. Fast simulation is a promising new approach that dramatically speeds up simulation in some cases. When combined with other approaches such as parallel simulation, it could make simulation much more applicable in real-time control situations. The rest of this section provides a brief overview of the fast simulation approach, and presents some ongoing research in this area.

### Recursion models

In the event-scheduling approach (Kreutzer, 1986; Law and Kelton, 1991), the most popular for discrete event system simulation, the simulation progresses in time through the scheduling and execution of events such as customer arrival and service completion. The overhead work of a simulation program is almost entirely devoted to the creation and manipulation of a time-ordered list of future events. If this task of manipulating event lists is totally eliminated, then a dramatic reduction in run time can be expected. The **fast simulation** approach investigated by Chen and Chen (1990 and 1993) does exactly this, but only for a very special system – a tandem system with a single machine at each stage and a finite FCFS queue at each stage. For this system, they showed that fast simulation based on a **recursion approach** can save up to 80% of run time in estimating certain performance measures, compared to the traditional event-scheduling based simulation. They use recursive expressions that model fundamental relationships among variables such as customer arrival times and start and finish times of service activities. The task of generating random numbers and random variates (Bratley *et al.*, 1987) cannot be eliminated and is a part of both the traditional and fast simulation approaches. It should be noted that the fast simulation approach does not involve any approximations and hence should yield estimates identical to those obtained by the traditional simulation approach.

For simulating general manufacturing networks, one possible approach is to first develop recursive relationships for modeling the dynamics of the basic network building blocks – parallel server nodes, split and merge configurations, assembly nodes, etc. For many system configurations, a recursion model can be constructed by combining/interfacing the recursive relationships of the building blocks. The recursion models should be expressed in an algorithmic form along with the associated steps for calculating performance measures. This process will be illustrated for a widely studied manufacturing system configuration, a serial or flow line with FCFS, single-machine nodes and no capacity limits on the queues.

For a single machine with FCFS service discipline and no capacity limits on queue length (example of a simple network building block), the following (common sense) relationship is well known:

Departure time for customer $j$ = starting time for customer j's service
time + service time for customer j
= max (departure time for customer j-1,
arrival time for customer j) + service
time for customer j

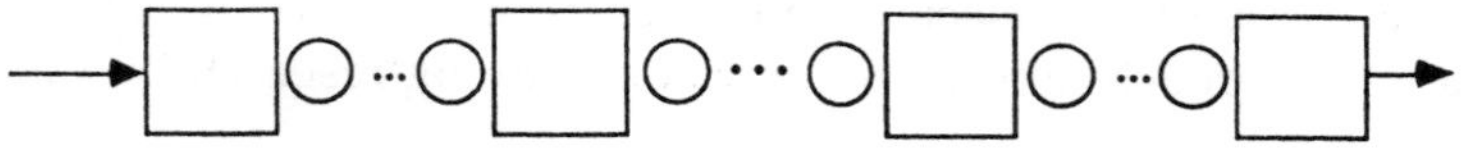

Fig. 9.3  A flow line model.

Using the above recursive relationship, a complete recursion model can be constructed for the flow line consisting of $M$ single-machine stages shown in Fig. 9.3. The recursion model is presented below in an **algorithmic form**.

**Variables:**
    Input: $M$: total number of stages
            $N$: total number of customers to be processed
            $S_{ij}$: service time for customer $j$ at stage $i$; $i = 1, 2, \ldots, M$;
            $j = 1, 2, \ldots, N$
            $A_j$: arrival time for customer $j$ at stage 1

($S_{ij}$ and $A_j$ to be generated using random number and appropriate random variate generators.)

    Output: $d_{ij}$: departure time for customer $j$ at stage $i$.

**Recursion calculations:**
Execute the following recursive relationship to calculate $d_{ij}$ for all $i$ and $j$.

$$d_{ij} = \max(d_{i,j-1}, d_{i-1,j}) + S_{ij} \qquad \{\text{Note: } d_{0,j} = A_j \text{ for all } j\}$$

**Performance measures:**

Utilization of stage $i = \left(\sum_{j=1}^{N} S_{ij}\right)\bigg/ d_{MN}$

Waiting time of customer $j$ at stage $i$, $W_{ij} = \max(d_{i,j-1} - d_{i-1,j}, 0)$

Average waiting time at stage $i$, $W_i = \left(\sum_{j=1}^{N} W_{ij}\right)\bigg/ N$

Throughput rate, $TP = N/d_{MN}$

Time in system for customer $j = d_{Mj} - A_j$

Average queue length at node $i = TP*W_i$.

Next, we present a numerical example that illustrates the computational advantage of the fast simulation approach. The example system has three flow lines with FCFS single-server nodes and infinite buffers feeding into an assembly node which needs one part from each tandem line. The assembly node is followed by a tandem line with FCFS single-server nodes and infinite buffers. A simulation program based on

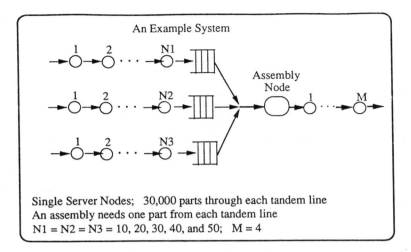

Single Server Nodes;   30,000 parts through each tandem line
An assembly needs one part from each tandem line
N1 = N2 = N3 = 10, 20, 30, 40, and 50;   M = 4

**Fast vs Traditional Simulation**

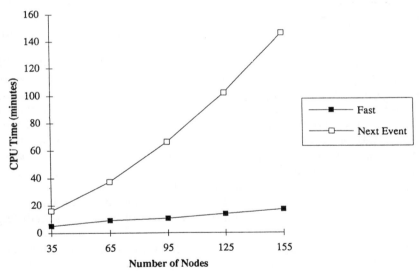

**Fig. 9.4** An example comparing fast and event-oriented simulations.

recursion models was written and its performance (execution speed) was compared to that of a model written using the traditional event-scheduling approach. Figure 9.4 shows the results of a sample experiment. Run times (CPU sec.) using both approaches are shown for five different system sizes. During this sample experiment, only one run was used to generate a data point. The simulations were run on a 386-based computer system running at 25 MHz and equipped with a numerical coprocessor. The recursion models used for the tandem lines account for a large portion of the savings in execution time.

### Hybrid (event-scheduling/fast) approach

For the simple flow line discussed earlier and the other published models, the following observation can be made. Customers cannot overtake each other, and the **original sequence of customers is preserved** throughout the system. This is not the case in many system configurations, for example, **the customer sequence can change after every FCFS parallel-server stage**, a building block that is generally found in many networks. When a parallel-server building block is used in a recursion model, the customer sequence, in almost all cases, has to be recalculated after each parallel-server stage. Deriving a computationally efficient scheme for recalculating the customer sequence is a critical step in the development of recursion models for systems containing this node (Kamath *et al.*, 1991). A recursion model of a parallel-server with FCFS discipline has been developed. This model along with the single-server case presented earlier in this section can be used for tandem lines containing any mixture of FCFS single-server and parallel-server stages.

In the ongoing research efforts at Oklahoma State University, Stillwater, fast simulation is not being limited to the execution of recursion models. Our initial hypothesis is that for certain manufacturing scenarios – for example, systems with state-dependent alternative routings and dynamic job priorities – a complete recursion-based model may be so complex that it may not be computationally superior. Hence, a **new hybrid approach** that combines the recursion based and event-scheduling based models in a single simulation model is being investigated. This hybrid approach holds a lot of promise as it combines the computational advantages of the recursion-based approach with the modeling flexibility and power of the event-scheduling approach.

For example, consider the hypothetical manufacturing system shown in Fig. 9.5. In this system, raw materials are first transformed into components by several fabrication lines and then assembled into finished goods. In any fabrication line, all parts go through the same sequence of operations. Hence, assume that each fabrication line can be modeled as a flow line. Also assume that the raw material release process and the product assembly operation are not suitable for the recursion approach. In a hybrid model of this system, the flow line operations would be expressed using recursive relationships. Let us consider the execution of an event-oriented simulation of the reduced model. When an arrival event is encountered for any flow line, the corresponding

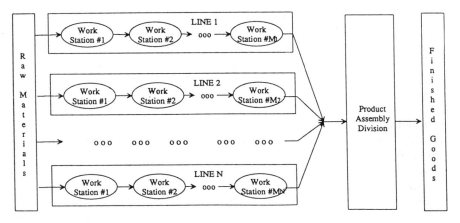

**Fig. 9.5**   A hypothetical manufacturing system.

recursion model will be invoked and a departure from the flow line would be scheduled. As far as the event scheduling model is concerned, the flow line behaves like a single stage. In a pure event-scheduling approach, each operation or workstation in a flow line would be modeled as a stage, and will generate arrival and departure events. When a new event is scheduled, it has to be inserted into the event list at the proper position which is a time consuming operation. The hybrid approach has drastically reduced the number of events that will be generated which should translate into huge savings in run time.

Finally, the philosophical differences between the fast simulation approach and other existing approaches for reducing total simulation run time should be noted. When applicable, variance reduction techniques and the regenerative simulation method (Bratley *et al.*, 1987) shorten the total run time by reducing the total run length needed to produce estimates with a desired statistical accuracy. They draw upon well developed concepts in probability theory, stochastic processes and statistics. Parallel/distributed simulation techniques use advances in computer hardware such as parallel architectures to reduce the total simulation run time (Fujimoto, 1989). The fast simulation approach reduces simulation run time by modeling the dynamics of the network operation using an approach that is fundamentally different from the event-oriented approach. The fast simulation approach seems to have the potential to cause a huge reduction in the computer time for a **single** run. When coupled with other existing approaches, this could lead to dramatic reductions in the total simulation run time.

## 9.4  CONCLUSION

The past few years have seen major advancements in the development and use of performance analysis tools for manufacturing systems. Rapid analysis software based on queueing models is continuing to gain acceptance in industry. Object-oriented concepts have caused revolutionary changes in simulation modeling and the next generation simulation environments are beginning to emerge. Because of the significant improvements in computing power and parallel processing technology, simulation and optimization approaches are likely to play a bigger role in on-line decision making and real-time control.

Traditional systems modeling has always focused on a particular problem or purpose. Often this has led to single-use, throw-away models. This mentality of modeling is obviously expensive, time consuming and wasteful. It is a barrier to the wider usage of models. There is a great need for industry to regard modeling as an integral part of its decisions/operations system (Mize *et al.*, 1992). To make this paradigm shift a reality, researchers in systems modeling are faced with the challenge of creating powerful environments that integrate a battery of performance analysis and optimization tools. Such environments would contain model representation schemes and methodologies which support all phases of a system's life-cycle, from preliminary design to final phase-out. Using the same common base representation of the system, the modeling environment would be able to accommodate both simulation and analytical methodologies. Finally, it is conceivable that embedded expert systems would be provided to guide the user through tool selection and model construction, and several iterations of model modification and execution, as well as output analysis and interpretation.

## ACKNOWLEDGMENT

The author gratefully acknowledges the support provided by the Oklahoma Center for the Advancement of Science and Technology (OCAST). During the final stages of preparation of this chapter, the author was also supported by the National Science Foundation under grant DDM-9311357. The author thanks Professor Joe Mize for providing valuable advice and encouragement.

## REFERENCES

Adiga, S. (1989) Software modeling of manufacturing systems: a case for an object oriented programming approach, in *Analysis, Modelling and Design of*

*Modern Production Systems* (eds A. Kusiak and W.E. Wilhelm), J.C. Baltzer AG, Basel, Switzerland.

Adiga, S. (ed.) (1993) *Object-oriented Software for Manufacturing Systems*, Chapman & Hall, London.

Adiga, S. and Gadre, M. (1990) Object-oriented software modeling of a flexible manufacturing system. *Jnl. of Intell and Rob Sys*, **3**, pp. 147–65.

Agrawal S.C. (1985) *Metamodeling: A Study of Approximations in Queueing Models.* Research Reports and Notes, Computer Systems Series, MIT Press, Cambridge, MA.

AT&T QNA Software. (1992) AT&T Software Solutions Group, NJ.

Balbo, G., Bruell, S.C. and Ghanta, S. (1988) Combining queueing networks and generalized stochastic Petri nets for the solution of complex models of system behaviour. *IEEE Trans on Comp*, **C-17**(10), pp. 1251–68.

Baskett, F., Chandy, K.M., Muntz, R.R. and Palacios, F.G. (1975) Open, closed, and mixed networks of queues with different classes of customers. *Jnl. of ACM*, **22**(2), pp. 248–60.

Basnet, C., Farrington, P.A., Pratt, D.B. *et al.* (1990) *Experiences in developing an object-oriented modeling environment for manufacturing systems*, in Proc. of the 1990 Winter Simulation Conference (eds O. Balci, R.P. Sadowski, R.E. Nance), IEEE, Piscataway, NJ, pp. 477–81.

Basnet, C. and Kamath, M. (1991) *A two-moment queueing network model for flexible manufacturing systems with transport vehicles*, in Proc. of the First Joint Conf. of the OR Society of New Zealand and the New Zealand Production & Inventory Control Soc., pp. 7–12.

Bhuskute, H.C., Duse, M.N., Gharpure, J.T. *et al.* (1992) *Design and implementation of a highly reusable modeling and simulation framework for discrete part manufacturing systems*, in Proceedings of the 1992 Winter Simulation Conference (eds J.J. Swain, D. Goldsman, R.C. Crain *et al.*), IEEE, Piscataway, NJ, pp. 680–88.

Bitran, G.R. and Dasu, S. (1992) A review of open queueing network models of manufacturing systems. *Queueing Systems: Theory and Applications*, **12**, pp. 95–134.

Bitran, G.R. and Tirupati, D. (1988) Multiproduct queueing networks with deterministic routing: Decomposition approach and the notion of interference. *Management Science*, **34**(1), pp. 75–100.

Bitran, G.R. and Tirupati, D. (1989) Approximations for product departures from a single-server station with batch processing in multi-product queues, *Mgmt. Sci.*, **35**(7), pp. 851–78.

Bitran, G.R. and Tirupati, D. (1991) Approximations of networks of queues with overtime. *Mgmt. Sci*, **37**(3), pp. 282–300.

Bratley, P., Fox, B.L., Schrage, L.E. (1987) *A Guide to Simulation*, Springer Verlag, NY.

Buzacott, J.A. and Shanthikumar, J.G. (1980) Models for understanding flexible manufacturing systems. *AIIE Transactions*, **12**, pp. 339–50.

Buzacott, J.A. and Shanthikumar, J.G. (1985) On approximate queueing models of dynamic job shops. *Mgmt. Sci*, **31**(7), pp. 870–87.

Buzacott, J.A. and Shanthikumar, J.G. (1993) *Stochastic Models of Manufacturing Systems*. Prentice Hall, Englewood Cliffs, New Jersey.

Buzacott, J.A. and Yao, D.D. (1986) Flexible manufacturing systems: A review of analytical models. *Mgmt. Sci*, **32**(7), pp. 890–905.

Buzen, J.P. (1973) Computational algorithms for closed queueing networks with exponential servers. *Communications of ACM*, **16**(9), pp. 527–31.

Chandy, K.M. and Sauer, C.H. (1978) Approximate methods for analyzing queueing network models of computing systems. *Computing Surveys*, **10**(3), pp. 283–317.

Chen, L. and Chen, C. (1990) *A fast simulation approach for tandem queueing systems*, in Proceedings of the 1990 Winter Simulation Conference, (eds O. Balci, R.P. Sadowski, R.E. Nance), IEEE, Piscataway, NJ, pp. 539–46.

Chen, L. and Chen, C. (1993) A fast simulator for tandem queueing systems. *Comp and Ind Eng*, **24**(2), pp. 267–80.

Conway, A.E. and Georganas, N.D. (1989) *Queueing Networks – Exact Computational Algorithms: a Unified Theory based on Decomposition and Aggregation*, MIT Press, Cambridge, MA.

Cox, B. (1986) *Object Oriented Programming: An Evolutionary Approach*, Addison-Wesley, Reading, MA.

Dallery, Y. and Gershwin, S.B. (1992) Manufacturing flow line systems: a review of models and analytical results. *Queueing Systems: Theory and Applications*, **12**, pp. 423–94.

de Souza e Silva, E. and Muntz, R.R. (1987) Approximate solutions for a class of nonproduct form queueing networks. *Performance Evaluation*, **7**, pp. 221–42.

Endesfelder, T. and Tempelmeier, H. (1987) The SIMAN module processor – a flexible software tool for the generation of SIMAN simulation models, in *Simulation in CIM and AI Techniques* (ed. J. Retti), SCS, pp. 38–43.

Fujimoto, R.M. (1989) Performance measurements of distributed simulation strategies. *Trans of The Soc for Comp Sim*, **6**(2), pp. 89–132.

Glassey, C.R. and Adiga, S. (1989) Conceptual design of a software object library for simulation of semiconductor manufacturing systems. *JOOP*, Nov./Dec., pp. 39–43.

Glassey, C.R. and Adiga, S. (1990) Berkeley library of objects for control and simulation of manufacturing (BLOCS/M), in *Applications of Object-Oriented Programming* (eds L.J. Pinson and R.S. Wiener), Addison-Wesley Publishing Company, Reading, MA, pp. 1–27.

Glassey, C.R. and Petrakian, R.G. (1989) *The use of bottleneck starvation avoidance with queue predictions in shop floor control* in Proceedings of the 1989 Winter Simulation Conference (eds E.A. MacNair, K.J. Musselman, P. Heidelberger), IEEE, Piscataway, NJ, pp. 908–17.

Glynn, P.W. (1989) *Optimization of stochastic systems via simulation*, in Proceedings of the 1989 Winter Simulation Conference (eds E.A. MacNair, K.J. Musselman, P. Heidelberger), IEEE, Piscataway, NJ, pp. 90–105.

Gordon, W.J. and Newell, G.F. (1967) Closed queueing systems with exponential servers. *Op Res*, **15**, pp. 254–65.

Gross, D. and Harris, C.M. (1985) *Fundamentals of Queueing Theory*, John Wiley & Sons, NY.

Govindaraj, T., McGinnis, L.F., Mitchell, C.M. et al. (1993) *OOSIM: a tool for simulating modern manufacturing systems*, in Proceedings of the 1993 NSF Design and Manufacturing Systems Conference, SME, pp. 1055–62.

Jackson, J.R. (1957) Networks of waiting lines. *Op Res*, **5**, pp. 518–21.

Jackson, J.R. (1963) Jobshop-like queueing systems. *Mgmt. Sci.*, **10**, 131–42.

Jacobson, P.A. and Lazowska, E.D. (1982) Analyzing queueing networks with simultaneous resource possession. *Comm of the ACM*, **25**(2), pp. 142–51.

Kamath, M. (1990) *Analytical performance modeling of asynchronous automatic assembly systems*, in Proceedings of the 1990 International Industrial Engineering Conference, IIE, Norcross, GA, pp. 557–62.

Kamath, M. and Basnet, C. (1990) A two-moment approximation for solving general closed queueing networks, Working Paper CIM-WPS-90-MK2,

Center for Computer Integrated Manufacturing, Oklahoma State University, Stillwater, Oklahoma.

Kamath, M., Bhuskute, H., and Duse, M. (1991) Fast simulation techniques for queueing networks. Working Paper CIM-WPS-91-MK1, Center for Computer Integrated Manufacturing, Oklahoma State University, Stillwater, Oklahoma.

Kamath, M. and Sanders, J.L. (1987) Analytical methods for performance evaluation of large asynchronous automatic assembly systems. *Large Scale Systems*, **12**(2), pp. 143–54.

Kamath, M. and Sanders, J.L. (1991) Modeling operator/workstation interference in asynchronous automatic assembly systems. *Discrete Event Dynamic Systems: Theory and Applications*, **1**, pp. 93–124.

Kamath, M., Suri, R. and Sanders, J.L. (1988) Analytical performance models for closed-loop flexible assembly systems. *The Int Jnl. of FMSs*, **1**, pp. 51–84.

Kraemer, W. and Langenbach-Belz, M. (1976) *Approximate formulae for the delay in the queueing system GI/G/1*, in Proceedings of the Eighth International Teletraffic Congress, Melbourne, p. 235.

Kreutzer, W. (1986) *System Simulation Programming Styles and Languages*, Addison-Wesley, Reading, MA.

Kuehn, P.J. (1979) Approximate analysis of general queueing networks by decomposition. *IEEE Trans. Comm.*, **27**, pp. 113–26.

Lavenberg, S.S. (ed.) (1983) *Computer Performance Modeling Handbook.* Academic Press, New York.

Law, A.M. and Haider, H.W. (1989) Selecting simulation software for manufacturing applications: practical guidelines and software survey. *Ind Eng*, **31**(5), pp. 33–46.

Law, A.M. and Kelton, W.D. (1991) *Simulation Modeling and Analysis*, McGraw-Hill, Inc., New York, NY.

Leung, Y.T. and Suri, R. (1990) Performance evaluation of discrete manufacturing systems. *IEEE Cont Sys Mag*, June, pp. 77–86.

Little, J.D.C. (1961) A proof for the queueing formula $L = \lambda W$. *OR*, **9**, pp. 383–87.

Liu, C.M. and Sanders, J.L. (1988) Stochastic design optimization of asynchronous flexible assembly systems. *Annals of Op Res*, **15**, pp. 131–54.

Meyer, B. (1988) *Object-Oriented Software Construction*, Prentice Hall International (UK) Ltd., Hertfordshire, UK.

Mitrani, I. (1982) *Simulation Techniques for Discrete Event Systems*, Cambridge University Press, Cambridge, UK.

Mize, J.H., Bhuskute, H.C., Pratt D.B. *et al.* (1992) Modeling of integrated manufacturing systems using an object-oriented approach. *IIE Trans*, **24**(3), pp. 14–26.

MPX Instructional Version. (1992) Network Dynamics, Inc., Burlington, MA.

Nance, R.E. (1984) *Model development revisited*, in Proceedings of the 1984 Winter Simulation Conference (eds S. Sheppard, U.W. Pooch, and C.D. Pegden), IEEE, Piscataway, NJ, pp. 75–80.

Nandkeolyar, U. and Christy, D.P. (1989) *Using computer simulation to optimize flexible manufacturing system design*, in Proceedings of the 1989 Winter Simulation Conference, (eds E.A. MacNair, K.J. Musselman, P. Heidelberger), IEEE, Piscataway, NJ, pp. 396–405.

Narayanan, S., Bodner, D.A., Mitchell, C.M. (1992) *Object-oriented simulation to support modeling and control of automated manufacturing systems*, in Proc. of the 1992 Object-Oriented Simulation Conference, SCS, pp. 59–63.

Pascoe, G.A. (1986) Elements of object-oriented programming. *Byte*, August, pp. 139–44.

Pattipati, K.R. and Kastner, M.P. (1988) A hierarchical queueing network model of a large electronics test facility. *Info and Dec Tech*, **14**(1), pp. 45–64.

Pratt, D., Farrington, P., Basnet, C., Bhuskute, H. *et al.* (1991) *A framework for highly reusable simulation modeling: separating physical, information, and control elements*, in Proc. of the 24th Annual Simulation Symposium (ed. A. Rutan), IEEE Computer Society Press, pp. 254–61.

Ramasesh, R. (1990) Dynamic job shop scheduling: a survey of simulation research. *OMEGA: Int Jnl. of Mgmt. Sci*, **18**(1), pp. 43–57.

Reiser, M. (1979) A queueing network analysis of computer communication networks with window flow control. *IEEE Trans on Comm.*, **COM-27**(8), pp. 1199–1209.

Reiser, M. and Lavenberg, S.S. (1980) Mean-value analysis of closed multichain queueing networks. *JACM*, **27**, pp. 313–22.

Segal, M. and Whitt, W. (1989) *A queueing network analyzer for manufacturing*, in Proceedings of the 12th International Teletraffic Congress, pp. 1146–52.

Shanthikumar, J.G. and Buzacott, J.A. (1980) On the approximations to the single server queue. *Int Jnl. of Prod Res*, **18**, pp. 761–73.

Shanthikumar, J.G. and Buzacott, J.A. (1981) Open queueing network models of dynamic job shops. *Int Jnl. of Prod Res*, **19**, pp. 255–66.

Shanthikumar, J.G. and Sargent, R.G. (1983) A unifying view of hybrid simulation/analytic models and modeling. *Op Res*, **31**(6), pp. 1030–52.

Snowdown, J.L. and Ammons, J.C. (1988) A survey of queueing network packages for the analysis of manufacturing systems. *Manuf Rev*, **1**(1), pp. 14–25.

Solberg, J.J. (1977) *A mathematical model of computerized manufacturing systems*, in Proc. 4th Int. Conf. Production Research, Tokyo, Japan.

Suri, R. (1985) An overview of evaluative models for flexible manufacturing systems. *Annals of Op Res*, **3**, pp. 13–21.

Suri, R. (1988) A new perspective on manufacturing systems analysis, in *Design and Analysis of Integrated Manufacturing Systems* (ed. W.D. Compton), National Academy Press, Washington DC, pp. 118–33.

Suri, R. and de Treville, S. (1991) Full speed ahead: A timely look at rapid modeling technology in operations management. *OR/MS Today*, June, pp. 34–42.

Suri, R. and Diehl, G.W. (1987) Rough-cut modeling: an alternative to simulation. *CIM Review*, **3**, pp. 25–32.

Suri, R., Diehl, G.W. and Dean, R. (1986) *Quick and easy manufacturing systems analysis using MANUPLAN*, in Proc. Spring IIE Conference, IIE, Norcross, GA, pp. 195–205.

Suri, R. and Hildebrant, R.R. (1984) Modeling flexible manufacturing systems using mean-value analysis. *Jnl. Manuf Sys*, **3**(1), pp. 27–38.

Suri, R., Sanders, J.L. and Kamath, M. (1993) Performance evaluation of production networks, in *Handbooks in Operations Research and Management Science, Vol. 4: Logistics of Production and Inventory* (eds S.C. Graves, A. Rinnooy Kan, and P. Zipkin), Elsevier Science Publishers B.V., pp. 199–286.

Thomasma, T., Mao, Y. and Ulgen, O.M. (1990) Manufacturing Simulation in Smalltalk, in *Object-Oriented Simulation* (ed. A. Guasch), SCS, pp. 93–7.

Thomasma, T. and Ulgen, O.M. (1988) *Hierarchical, modular simulation modeling in icon-based simulation program generators for manufacturing*, in Proceedings of the 1988 Winter Simulation Conference (eds M. Abrams, P. Haigh, and J. Comfort), IEEE, Piscataway, NJ, pp. 254–62.

Tirupati, D. (1992) *Application of the parametric decomposition approach in the analysis of manufacturing queueing networks*: An overview, in Proceedings of the First Industrial Engineering Research Conference (eds G. Klutke, D.A. Mitta, B.O. Nnaji, and L.M. Seiford), IIE, Norcross, GA, pp. 307–11.

Ulgen, O.M. and Thomasma, T. (1990) SmartSim: an object oriented simulation program generator for manufacturing systems. *Int Jnl. of Prod Res*, **28**(9), pp. 1713–30.

Ulgen, O.M., Thomasma, T. and Mao, Y. (1989) *Object oriented toolkits for simulation program generators*, in Proceedings of the 1989 Winter Simulation Conference, (eds E.A. MacNair, K.J. Musselman, P. Heidelberger), IEEE, Piscataway, NJ, pp. 593–600.

Viswanadham, N. and Narahari, Y. (1992) *Performance Modeling of Automated Manufacturing Systems*, Prentice-Hall, Inc., Englewood Cliffs, New Jersey.

Whitt, W. (1983) The queueing network analyzer. *Bell Systems Tech Jnl.* **62**, pp. 2779–815.

Wolff, R.W. (1989) *Stochastic Modeling and the Theory of Queues*, Prentice-Hall, Inc., Englewood Cliffs, New Jersey.

Zeigler, B.P. (1984) *Multifaceted Modeling and Discrete Event Simulation*, Academic Press, London.

# CHAPTER 10

# Qualitative intelligent modeling of manufacturing systems

*David Ben-Arieh and Eric D. Carley*

## 10.1 INTRODUCTION

This chapter presents a new intelligent modeling methodology for manufacturing systems. This methodology captures the structure, behavior and functionality of the system as a whole and of its components, emphasizing the cause-effect relations between the components.

The building blocks of the modeling approach are activities, information items that flow among the activities and rules governing the behavior of the activities. The methodology supports a hierarchical structure in which each module can be further detailed.

In addition, the methodology supports various types of analysis that are of major importance to manufacturing systems such as reachability analysis. This analysis can show the activities affected by a change upstream (for example a design change can be analyzed for its system-wide ramifications). Backward connectivity detects all possible remedies to a downstream activity (such as a bottleneck machine that can be relieved by a design change). Analysis of rules consistency ensures global consistency and completeness in addition to lower level consistency.

### 10.1.1 The modeling approach

Industrial systems can be viewed as a collection of discrete, reactive, information intensive processes. In addition, industrial systems are typically hierarchical and asynchronous. In order to improve design, control, and evaluation of the various aspects of such processes, a rigorous modeling framework is required. This modeling approach is required to capture the intricate nature of the manufacturing system and provide meaningful analysis of such a system. There are numerous contemporary modeling tools from a wide variety of disciplines, but they all give only partial answers to modeling difficulties of such systems, and their analysis capabilities do not address the special needs of the manufacturing domain.

The modeling methodology presented herein represents all aspects of the industrial systems including its structure (that can be hierarchical or nested), behavior and functionality (the distinction between the behavior and the functionality of a system is discussed in section 10.2). This modeling concept captures the activities being performed in the system, the information flow among the activities and the rules that govern the behavior of the activities. The activities and the information that flow in the system can be compound, thus representing a hierarchical structure containing a detailed set of activities or information items.

The modeling approach is targeted for the process of design, process planning and manufacturing. When modeled in sufficient detail, the features used in the design process (feature-based design) can be represented. Thus design changes can be simulated and the corresponding cause-effect relations with the process planning and manufacturing process can be evaluated. Similarly, a change in the manufacturing end can instigate a design (or any other activity) change in order to accommodate that variation. This type of modeling and analysis can lead to a better understanding of feature-based design and better integration of the manufacturing activities.

In order to model the system, a new modeling paradigm 'object oriented functional networks' is used. In this modeling approach, the entire process is represented graphically as a network of activities connected via information channels. Information items that flow in the channels trigger the activities. Each activity consumes the input information, evaluates it and generates information items to its output channels according to rules that define each activity.

By building appropriate rules, this modeling methodology can be used to implement quantitative analysis such as discrete event simulation, but more importantly, it shows a new approach for qualitative intelligent modeling. This approach can be used to assess design changes or possible solutions to process planning problems, and even be used to directly control the manufacturing system.

Due to the hierarchical modeling approach, each activity can represent an entire set of lower level activities. Each such module is referred to as a 'world' containing an entire subsystem.

The graphical representation of the model depicts activities as square nodes and information channels as directed arcs with the channel name shown as a circular node (these nodes are also termed 'information nodes'.) A double lined node represents a hierarchical structure. An example of such a network is presented in Fig. 10.1.

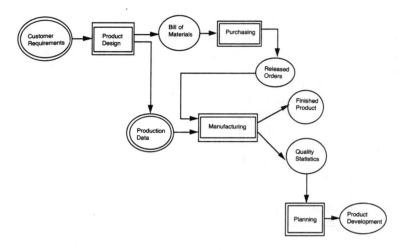

**Fig. 10.1** An example of a model network.

The modeling approach presented in this chapter supports the current effort to extend the available qualitative modeling methodologies (e.g. IDEF$_3$ and IDEF$_4$), and supplies solutions to many of the problems partially addressed by these methodologies.

The modeling approach presented in this chapter has been implemented using Common LISP on a SUN workstation. The implementation uses an object oriented programming environment (KEE) that supports frames as knowledge representation tools.

## 10.2   BACKGROUND

### 10.2.1   System modeling

Peterson (Peterson, 1981) defines modeling of a system as a logical representation of the important parts of the system. The advantages of a model are threefold:

1. a model is easier and cheaper to operate than the real system;
2. it ignores irrelevant details;
3. it is easy to analyze and base decisions upon.

A modeling tool should be expressive, mapping all the properties of interest in the system into the model. It should be easy to understand and use, and should allow analysis and descision making concerning the real world system.

There is a multitude of ways to represent real world systems, reflecting numerous aspects of the systems. However, there are three orthogonal views of a system that need to be captured in the model (Harel and Pnueli, 1985):

1. Structural view represents the structure of the system, its components, and links among the components.
2. Functional view reflects the functionality of the system as a whole. This view reflects the 'transfer function' of the system regardless of its individual states.
3. Behavioral view represents the various states the system goes through, and the flow of the process.

A good modeling tool should capture all three views of the system.

### 10.2.2 Reactive systems

Discrete event systems can be further dichotomized in many ways: sequential systems vs. concurrent ones, deterministic vs. non-deterministic systems, synchronous vs. asynchronous systems, etc. However, an important classification is **reactive** vs. **transformational** systems (Harel, 1987).

A transformational system is characterized by having a defined transformation that is applied to its input to generate the output. Such a system can be viewed as a black box, in which the input is transformed into an output, when such an action is initiated. An example of such a system can be a batch processing of a database query. In this case, all the input information is gathered, and once it is processed no further input is accepted. Reactive systems on the other hand are systems that constantly react to input signals. The system components are constantly active and every input change causes a state change in the system. Most systems in practice are reactive; however, due to modeling difficulties, some are regarded as transformational.

Modeling reactive systems is inherently difficult for the following reasons. There is no simple transformation that can represent the system. The reactive system is usually hierarchical, and its components are concurrently and asynchronously active. The communication among the components is hard to represent, and the variety of states that the system can have is typically too large to enumerate.

Unfortunately, manufacturing systems belong to this category, and therefore the need for a more advanced modeling approach is evident.

### 10.2.3 Modeling of reactive systems: related work

System modeling methods contain a wide interdisciplinary variety of methods including mathematical methods, graphical techniques, formal languages, rule based approaches, etc. However, none of these approaches is expressive enough to model all three aspects of reactive manufacturing systems.

One of the most attractive methods for reactive systems modeling is

Petri net (Peterson, 1981; Reisig, 1982). Petri net is a graphical and analytical methodology for system modeling and analysis, and a large body of modeling work has been conducted using this approach (Dubois, 1983; Ben-Arieh, 1991a).

The structure of a Petri net $N$ is defined as a four tuple $(P, T, I, O)$. $P$ is the set of places, $T$ the set of transitions, $I$ – the input functions and $O$ – the output functions that define the network.

Typically, places describe conditions or facts such as 'a job is waiting' or 'a machine is idle.' Transitions establish the actions or processes such as 'allocate next job' or 'load waiting pallet'. The input and output functions relate places and transitions within a model.

Input places $I(t_j)$ are mappings of the transitions $t_j$ to a collection of places based on the input function $I$. The output places $O(t_j)$ are mappings to a collection of places based on the output function $O$.

The inputs and outputs of a transition in a Petri model are defined as bags of places. Using bag theory instead of set theory allows multiple occurrences of a particular place as an input or output to a transition.

A place $p_i$ is an input place of a transition $t_j$ if $p_i \in I(t_j)$, and an output place of a transition $t_j$ if $p_i \in O(t_j)$. A transition $t_j$ defines an input of a place $p_i$ if $p_i$ belongs to the output bag of $t_j$. Similarly, $t_j$ is an output of a place $p_i$ if $p_i$ belongs to the input bag of $t_j$. An example using this relationship is shown in Fig. 10.2.

While a Petri net can be described using bag theory and the relationships between places and transitions, most Petri nets are also described using a graphical display of the model. In this case, places are displayed as circles and transitions are shown as boxes (or bars). The input and output functions are illustrated by directional arcs connecting places and transitions.

If an arc points from node $i$ to node $j$ (either from a place to a

$$N = (P, T, I, O)$$
$$P = (p_1, p_2, p_3, p_4, p_5)$$
$$T = (t_1, t_2, t_3, t_4)$$

$$I(t_1) = \{p_1, p_2\} \qquad O(t_1) = \{p_2, p_3, p_3\}$$
$$I(t_2) = \{p_1, p_5\} \qquad O(t_2) = \{p_3\}$$
$$I(t_3) = \{p_3\} \qquad O(t_3) = \{p_4\}$$
$$I(t_4) = \{p_4, p_4, p_4\} \qquad O(t_4) = \{p_3, p_5\}$$

**Fig. 10.2** An example of a Petri net structure.

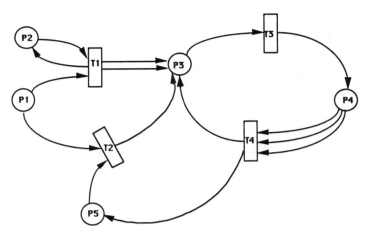

**Fig. 10.3** Graphical display of a Petri net.

transition or from a transition to a place), then $i$ is an input to $j$, and $j$ is an output of $i$. The graphical display of the network presented in Fig. 10.2 is demonstrated in Fig. 10.3.

A Petri net models the dynamics of a system as a flow of activities using tokens. Tokens are represented graphically as black dots and allow transitions to 'fire' representing the execution of activities. A transition can fire when all of its input places have tokens. The firing process removes all the tokens from the input places and places new tokens in all the output places of the transition that fires. Figure 10.4 shows a marked Petri net (a net with tokens) that results from firing transitions $t_1$ in the net presented in Fig. 10.3. The initial state of the net is one token in $p_1$ and one in $p_2$ which allows $t_1$ to fire.

The Petri net modeling approach has been expanded to include quantitative variables (timed Petri nets) (Morasca *et al.*, 1991) as well as qualitative variables using colored Petri nets (Heuser and Richter, 1992, Jensen, 1987). However, the ability of Petri nets to adequately describe a reactive environment is still severely restricted. Petri nets cannot accommodate rule activation, or even a combination of OR conditions on tokens firing.

A different approach towards reactive system modeling is shown in the **statecharts** modeling methodology (Harel, 1987). Statecharts are a continuation of work developed on a more general concept called higraphs (Harel, 1988). Statecharts separate reactive systems into states and transitions using boxes and arrows. The separation of these entities is based upon relationships that form sets of states within the model.

The relationships within the various sets of states can be represented graphically using boxes as states and arrows as transitions. The arrows

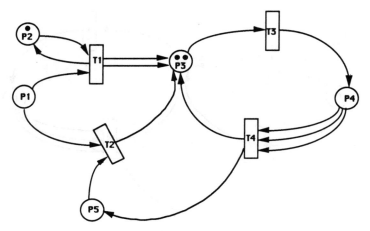

**Fig. 10.4** The net after firing $t_1$.

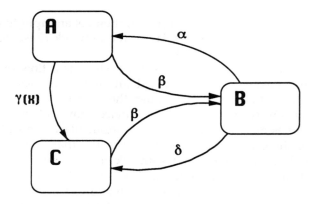

**Fig. 10.5** Statechart example.

represent the events that cause the system to change from one state to another. These events can be conditional or unconditional. An example shown in Fig. 10.5 demonstrates a system that consists of three states: $A$, $B$ and $C$. The event $\beta$ for example transfers the system to state $B$. Event $\gamma(x)$ occurring in state $A$ transfers the system from state $A$ to state $C$ only if the condition $X$ holds at the time of the transition.

This methodology is hierarchical and supports the functional, behavioral and structural views of the system. However, it does not represent the information flow in the system.

Other modeling approaches include graphical specification languages which are mainly used for information system design and verification.

There are many such tools including SAS (Lissander, 1985), SADT (Ross, 1985), SARA (Estrin, 1986), and SREM (Alford, 1977).

A different approach is taken by rule-based modeling systems. In this approach the system is described in terms of physical or operational rules (Murray, 1988; Ben-Arieh, 1988). This approach is useful when the system's details are not known, and therefore it emphasizes the functional view of the system but still lacks clear structural and behavioral description.

Computer programming languages are widely used for system modeling and simulation. In addition to the traditional discrete event simulation languages (which are not reactive by nature), one can find object oriented languages (Antonelli, 1986; Bezivin, 1987), concurrent simulation languages (Ben-Arieh, 1991b) and distributed simulation approach (Misra, 1986). These approaches do not support knowledgeable modeling of the complex interactions among the systems' components.

Distributed simulation or control languages are based on a network of asynchronous processes, communicating using time stamped messages. Some examples are COSL (Pathak, 1989) and DISS (Melman, 1984). A different approach for modeling such systems was demonstrated using concurrent logic programming (Dotan and Ben-Arieh, 1991) and a graphical-based modeling language (Ben-Arieh, 1991c).

It is important to note that analytical models such as queuing networks are not reactive by nature and therefore are not considered appropriate for modeling as described here. Similarly, discrete event simulation techniques do not capture the reactive nature of the systems modeled, the information flow and internal rules.

## 10.3 OBJECT ORIENTED FUNCTIONAL NETWORK

The **object-oriented functional network** methodology has been developed in order to encapsulate all orthogonal views of a system: structure, functionality and behavior. It was also designed to allow the user a clear inspection of a complex system at any level without loss of any of the desired views. This modeling approach also supports graphical representation by visualizing the model as a network. This modeling system remedies the shortcomings of the $IDEF_0$ modeling approach (ICAM, 1981). The limited power of the $IDEF_0$ methodology is the basis for the effort in extending its modeling power through the $IDEF_3$ and $IDEF_4$ (IDEF3, IDEF4) methods. The methodology presented herein goes beyond $IDEF_3$ and $IDEF_4$ by suggesting a unified modeling approach with extensive expressive power and inherent analysis methodology (which was never a part of the IDEF family of modeling methods).

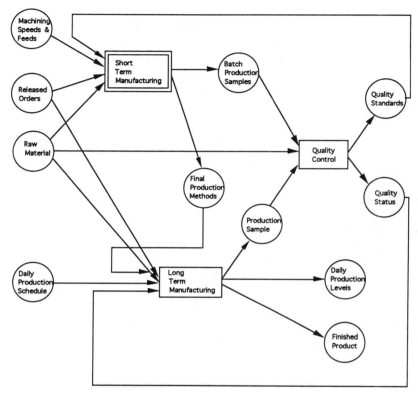

**Fig. 10.6** An example of an object-oriented functional network.

The structure of the model relies on the concept that every real-world system can be described as an information part or an action part. To maintain simplicity, the model contains only two structures: action nodes and information nodes. Action nodes describe the structure of the system while also containing the functionality of each step. Information nodes describe the communication between the action nodes and have a behavior of their own.

Each action and information node is recognized as a separate and concurrently active entity. Information nodes can only attach to an action node and similarly, action nodes can only attach to information nodes.

Any combination of inputs and outputs to a single action node makes up a **world**. A world defines a complete interaction of inputs and outputs and is capable of being expanded into lower levels. Within a model containing lower levels, a world can be exploded showing the contained action, information, input and output nodes.

The visual representation of action nodes is as squares while information nodes are represented as circles. Since each such node can have a more detailed view (depth), a node that is not atomic is double lined. For example, if an action node contains sub-modules it will have double lines outlining the square.

An example of such a simple network is presented in Fig. 10.6. In this figure the highest level network is the same as in Fig. 10.1. However, Fig. 10.6 represents an extension to the activity *manufacturing*. In the lower level presented in Fig. 10.6, the *manufacturing* activity consists of three activities: *short term manufacturing, long term manufacturing* and *quality control*. The activity *short term manufacturing* is itself an aggregate activity which is not explored further in this level. Also, the aggregate information item *production data* in Fig. 10.1 is expanded in Fig. 10.6 to *machining speeds & feeds, raw material* and *daily production schedule* detailed information items.

The object oriented functional network fires a node using tokens (information items). Unlike other modeling techniques, a token that is passed to a node (activity or information) may or may not fire that node. However, when a token is passed to a particular node it becomes active. An active node may require more than one token before it can fire to the next directed node(s), and it fires once all necessary tokens from the connected information nodes are available.

In order to account for the number of tokens waiting to pass on, it is necessary to maintain a list of tokens received and compare it with the number of tokens desired by the output information nodes. Tokens passing from action nodes depend solely on the behavioral rules of the output nodes.

The object oriented functional network presents the model graphically to the user in two distinct formats: a high-level view of the system structure termed universal view and a detailed view termed world display.

The universal display is a high-level view of the system that shows the activities and their hierarchical structure. This view is useful to explore the structure of complex systems, as demonstrated in Fig. 10.7. This figure, which corresponds to Fig. 10.1, shows the activity *manufacturing* which is detailed in Fig. 10.6 at the top of the display. The universal display allows for easy reachability searches as it is the 'road map' of activities. Since each world display describes an action occurring on the previous level, it would be difficult for the modeler to accurately track all connected activities and their lower-level details. The universal display allows the user to trace the action desired. The activities tree can then be checked along the horizontal pathways or levels of its structure to determine all possible actions connected through the information nodes.

The second type of display, the world view, shows a detailed view of

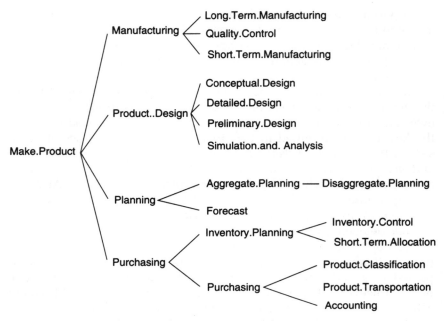

**Fig. 10.7**   Universal display of the world 'Make Product'.

the action and information nodes, as demonstrated in Figs. 10.1 and 10.6. The world display is the description of options and restrictions that occur if a certain action is taken. The information nodes represent the possible outcomes of activating a particular activity in the system. The world display supplies the necessary connections that control both functionality and behavior. This display gives the user a detailed view of the system and its components. Each sub-system is unique and therefore has its own identity with respect to the rest of the system. Deeper levels of the system can be explored by the user until the system being investigated is complete. However, a user may be interested only in the behavior of a particular sub-system and a few of its sub-levels and not in the higher levels or the most basic elements.

Both types of presentation complement each other, since the universal view emphasizes the structure of the system while the world view presents its behavior and functionality.

## 10.4   MODELING THE BEHAVIOR OF THE SYSTEM

The behavior of the system is encoded with causal rules located within the output information nodes. These rules define the required number of tokens and the identity of each token. The rules within a world are

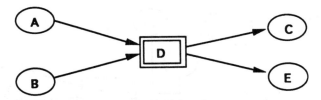

**Fig. 10.8**  An example of a simple world.

created using the causal relationships between input nodes and output nodes.

The development of causal relationships within a model supports modeling methodology that can be applied to physical systems, biological systems, economic systems or mathematical systems (White and Frederiksen, 1990).

As an example, Fig. 10.8 describes an action node D that has two input nodes A and B, and two output nodes C and E. Now assume that activity D behaves as follows:

**If** A has a token and B has a token **then** E is active.
**If** A has a token **then** C is active.

This means that information item A is required for both E and C, while input B is required only for item E. From a different point of view: E is composed of both A and B, while C is composed only of an A.

The rules that capture this behavior are represented as follows:

$$A \Rightarrow E, C$$
$$B \Rightarrow E$$

The information items (tokens) that feed the action node can be of two types: **public** or **private**. A private information item is one that cannot be duplicated to more than one node. This type corresponds to an original and unique data item, or a resource that cannot be shared. A public information item on the other hand, can be copied to all the information nodes that require it. It is important to remember that information nodes may represent people, tools, database fields, etc.

In the example, when a token is passed from A, it is split and sent to both rules that require a token from A. This implies that information item A is public.

In the case that A is private, the rules are presented in the form:

$$A \Rightarrow E$$
$$A \Rightarrow C$$
$$B \Rightarrow E$$

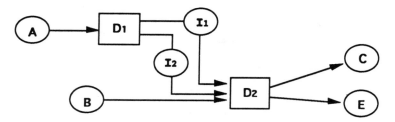

**Fig. 10.9**   A detailed view of Activity D.

This implies that A is supplied only to item E and cannot be split. Only if the first rule is already satisfied (there is an A already in E), then the second rule will be fired and an A will be applied towards a C.

### 10.4.1   Rules firing and consistency

When an activity has more detailed levels, the rules in the higher level description have to maintain consistency with the ones in the lower levels. In order to demonstrate inconsistency, assume that activity D in Fig. 10.8 is expanded as shown in Fig. 10.9.

Also assume that the internal rules are as follows:

Rule set for $D_1$:                                 $A \Rightarrow I_1, I_2$

Rule set for $D_2$:                                 $B \Rightarrow E$
$$I_1 \Rightarrow C$$
$$I_2 \Rightarrow C$$

In this case, this set of rules is inconsistent with the higher-level view of the system. In the lower-level view, A drives node C only, while in the higher-level view A is required by node E too. This inconsistency is detected by the system and can be corrected in one of the following ways:

- change the upper-level rules;
- change the lower-level rules;
- change the higher-level system structure;
- change the lower-level system structure.

If the last option is chosen, one way to correct the problem is presented in Fig. 10.10.

In this case, the rules are also modified as follows:

Rule set for $D_1$:                                 $A \Rightarrow I_1, I_2\ C$

Rule set for $D_2$:                                 $B \Rightarrow E$
$$I_1 \Rightarrow C$$
$$I_2 \Rightarrow C$$

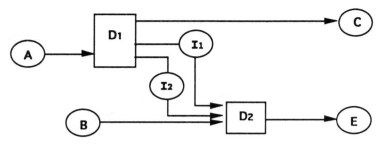

**Fig. 10.10** A modified lower level view.

### 10.4.2 Rules representation

The rules reside within the information output nodes that need to receive the tokens. Each such node has a set of rules that govern the behavior of the activity that supports that output node. When tokens arrive at a particular activity the activity scans all its output nodes and fires the rules that are feasible at that time.

Each rule consists of a listing of tokens required to fire that rule with the designated token types (public or private). The tokens within a world are passed according to these relationships which are defined by the modeler. These relationships can be described using AND/OR links and are implemented as either **sequential** rules or **default** rules within the output information nodes.

Sequential rules define those rules that must be satisfied first before the respective output node may receive a token. Within sequential rules, both AND/OR operators can be found. This allows the designer to create multiple branches within a single firing through a model. Sequential rules also allow the designer to develop strict sequences of rules that must be followed by the model. However, sequential rules are not required to exist within a world as default rules are.

Default rules define a set of rules that may or may not be satisfied by the model. At least one default rule must exist within the default set in order for the system to operate. Individual default rules consist of AND links between the causal inputs. If sequential rules are present within a world then all sequential rules must be satisfied first before any default rules can be reviewed for activation.

Figure 10.11 demonstrates a 3-input-2-output world containing both sequential and default rules.

Assume that the sequential and default rules for output node E are defined as follows:

**Sequential rules:**
Rule 1: [1] $\{3A \wedge 2B\} \vee \{6C \wedge 1B\}$
Rule 2: [3] $\{2A \wedge 3B\}$

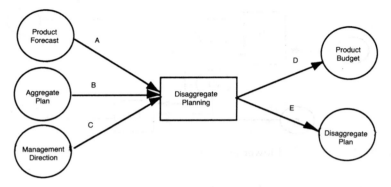

**Fig. 10.11**   An example of a rule consistency problem.

**Default rules:**
{3A ∧ 2B}
{6C ∧ 3B}
{2C ∧ 4A}

The first sequential rule is performed only once before being re-moved from the list of fireable rules. This rule has two branches that can be executed by the node depending on the available tokens. The second sequential rule has only one branch that must be satisfied, but this branch has to be executed three times. The second sequential rule reads as E and requires two tokens from A and three tokens from B in order to receive a token. Once all rules are removed from the sequential list of fireable rules, the behavior of the node is described by the default rules alone.

The default rules may be performed any number of times and thus are not removed from the node. Default rules can be thought of as a single sequential rule that is combined from a list of AND and OR operations. The default rules define the nodes' behavior in the most flexible way. Due to their ability to fire whenever a desired combination of inputs is satisfied, the modeler is forced to ensure that rules are not duplicated in any form (the issue of rules redundancy is handled by the system as is the issue of rules consistency, and is not discussed herein).

## 10.5   ANALYSIS OF THE MODEL

The main purpose of this elaborate modeling approach is to support meaningful analysis of the actual system. The more information is captured by the model, the more meaningful the analysis can become.

Due to the special modeling approach, the analysis of the models is different and unique. Currently, four types of analysis are provided:

- Analysis of events reachability (effect analysis). In this analysis mode, an event is triggered by the user, or by the system, and all the succeeding events are activated until the entire tree of affected events (or a selected subtree) is executed. This mode is useful for investigating possible outcomes of a selected event. An example of such an analysis is the investigation of all the functions affected by changing the feature **width** of a part during the design stage.
- Backwards analysis (cause analysis). In this analysis a selected node (activity or information) is triggered, and then all the possible activities affecting this node are traced. This analysis is performed up to a specified level upstream or until the entire tree of causes is found. An example of this analysis is the exploration of all possible activities that may need to be reactivated when a particular machine fails on the shop floor. Such activities can include design (change a particular design feature to fit a different machine), purchasing and process planning. This mode of analysis also highlights the information items required to counteract the downstream exception.
- Event tracing mode. Every node in the model (activity or information) has a special slot dedicated to representing conditions specified by the user. Once a node is activated all the emanating events with conditions that are fulfilled can be activated. This is a selective effect analysis.
- Quantitative analysis of selected properties. Each rule can contain either a set of conditions expressed as Boolean expressions or a computation that has to take place when the node is activated. Thus, by activating a chain of activities, some desired properties can be computed, such as cost or time factors.

The analysis of a model is planned to be performed in one of two modes: unsupervised run or a stepwise run. Using the stepwise run, the user can place information items in any portion of the model and trace them through the model. The stepwise mode requires the user to actively move all information items from node to node. Within this run mode all the analysis options are available to the user, and the results are visible after every single event. The advantage of this mode is the possibility of analyzing a particular section of the system that is of special interest, for example, the possible effects of a delayed delivery of parts.

In the unsupervised mode, the model will process the information items without user interaction. The pace of items moving between nodes will be decided by the user, but once the pace is determined, the analysis can be performed unattended. This mode is useful for system-wide investigation as well as to actually control and drive the external system represented by the model.

In order to allow flexible analysis, an analysis engine has been developed. This mechanism allows users to build their own sets of rules, analysis methods or patterns within the model structure.

### 10.5.1  Example: modeling a manufacturing system

The modeling methodology presented so far is quite general and suitable for a broad set of applications. However, one of the most intriguing domains is the domain of manufacturing systems. This section demonstrates the modeling approach using an example of a manufacturing system.

This example uses the world view of the system presented in Fig. 10.6. The rules that correspond to that view are shown in Table 10.1.

Only a single default rule is given for each output node in order to simplify the firing example of the model. In order to determine an allowable firing sequence of the activities, tokens are placed at the initial nodes *machining speeds & feeds, release orders, raw material and daily production schedule*. These tokens represent the availability of these items.

When the system is activated, only the output node *batch production samples* can be activated, and therefore it receives a token. Next, the *long term manufacturing* activity is activated. In this case, the output node *daily production levels* receives a token. Note that all other output nodes require tokens from the node *final production method* which is not active yet. Next, the activity *quality control* is enabled. This activity passes a token to the information node *quality standards*. Now that *quality standards* has been active, the activity *short term production* is evaluated again, resulting in the creation of *final production methods*.

**Table 10.1**  Behavioral rules for world manufacturing

| output node | output node rules – default only |
|---|---|
| Batch production samples | 1 machining speeds & feeds *and* 1 raw material |
| Final production methods | 1 release orders *and* 1 quality standards |
| Production sample | 1 final production methods *and* 1 raw material |
| Daily production levels | 1 release orders *and* 1 daily production schedule |
| Finished product | 2 raw material *and* 1 final production methods |
| Quality standards | 1 batch production samples *and* 1 raw material *and* 1 production sample |
| Quality status | 1 raw material *and* 1 production sample |

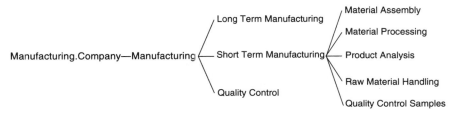

**Fig. 10.12** Cause analysis of the world 'raw material handling'.

This example demonstrates modeling the behavior of a part of a manufacturing system. It shows that even iterative behavior or feedback loops can be represented, thus allowing different behavior patterns of the same system upon different operation cycles.

The analysis of the system is timeless. Execution of activities consumes no time at all, emphasizing the feasibility of an operation sequence and cause-effect relationships among the various activities.

An example of cause analysis is shown in Fig. 10.12. In this figure the universal display of the manufacturing system is shown. The user investigates activities that have an effect on the low level activity *raw material handling*. The motivation for such an analysis can be the need to change the way raw material handling is being performed, and therefore investigation of the activities that may have to be changed. In this example, the activities highlighted are the ones that affect *raw material handling*. It is important to note that this analysis fires the rules 'backwards' to see what activities can affect the activity in question. A more detailed view of this analysis can be presented in the world view. In this case the activities and information items that are linked to the activity investigated are shown in Fig. 10.13.

## 10.6  CONCLUSION

This article presents an intelligent object-oriented modeling approach suitable for complex manufacturing processes. This modeling approach provides an improved modeling and analysis methodology for manufacturing systems. It emphasizes qualitative (cause-effect) relationships between activities with manufacturing systems, an aspect neglected by current modeling methods.

The importance of this approach lies in its ability to represent such complex industrial processes with sufficient expressiveness and detail. The model captures all aspects of the real system including structure, functionality and behavior. The significance of this type of modeling is in its contribution towards understanding the structure, controls and

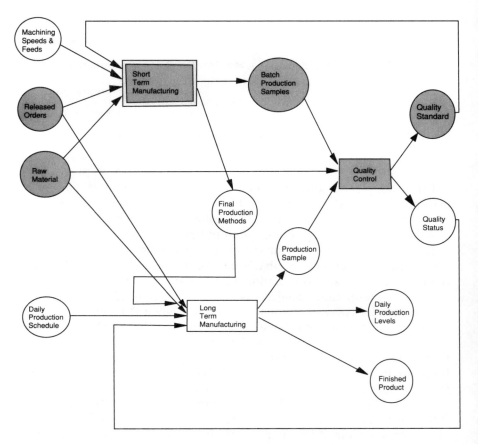

**Fig. 10.13**  A world display of affected activities and information items.

impacts of manufacturing systems. This modeling approach allows for an interactive modeling and evaluation mode in which the user can modify the model and ask questions in order to determine the appropriateness of a manufacturing system.

In the age of concurrent engineering and feature-based design concepts, this methodology can greatly enhance and support these concepts. Modeling the cause-effect relationships among the various activities in manufacturing systems, and capturing the effects of detailed information values on the activities, lead to a powerful modeling system.

The presented methodology supports new analysis methods such as evaluation of strategic decisions (e.g. which product to develop, how to structure the design process), and tactical decisions (e.g. which alternate part to purchase, what process planning changes will result from a design modification). It is even conceivable that this model could be expanded to be integrated with an entire manufacturing system and control the system instead of just modeling it.

The presented modeling approach has been implemented in a KEE environment, running on a SUN 3/60 workstation.

## REFERENCES

Alford, M.W. (1977) A Requirements Engineering Methodology for Real Time Processing Requirements, *IEEE Trans. on Software Eng.*, **SE-3**, pp. 66–8.

Antonelli, C.J., Volts, R.A. and Mudge, T.N. (1986) Hierarchical Decomposition and Simulation of Manufacturing Cells Using Ada, *Simulation*, **46**(4), pp. 141–52.

Ben-Arieh, D. (1988) A Knowledge Based Simulation and Control System, in *Artificial Intelligence: Implications for CIM* (ed. A. Kusiak), Springer Verlag, pp. 461–72.

Ben-Arieh, D. (1991a) Concurrent Modeling and Simulation of Reactive Manufacturing Systems Using Petri Nets, *Computers and Industrial Engineering*, **20**(1), pp. 45–58.

Ben-Arieh, D. (1991b) Concurrent modeling and Simulation of Multi Robot Systems, *Robotics and Comp Integ Manuf*, **8**(1), pp. 67–73.

Ben-Arieh, D. (1991c) Modeling and Control of Discrete Manufacturing Systems Using Graphical Concurrent Modeling Language (GCML), in *Cont and Dyn Sys*, **48**, pp. 47–74. Academic Press, Inc.

Berry, G. and Gonthier, G. (1988) The Esterel Synchronous Programming Language: Design Semantics, Implementation, INRIA *Research Report, no. 542*.

Bezivin, J. (1987) *Some Experiments in Object Oriented Simulation*, Proc. OOPSLA, Oct. 1987.

Dotan, Y. and Ben-Arieh, D. (1991) Modeling Flexible Manufacturing Systems: the Concurrent Logic Programming Approach, *IEEE Trans. on Robotics and Automation*, **7**(1), pp. 135–48.

Dubois, D. and Stecke, K.E. (1983) *Using Petri Nets To Represent Production Processes*, Proc. of IEEE Conf. on Decision and Control, San Antonio Texas, Dec., pp. 1062–67.

Estrin, G., Fennechel, R.S., Razouk, R.R. *et al.* (1986) SARA: Modeling Analysis and Simulation Support for Design of Concurrent Systems, *IEEE Trans. on Software Eng.*, **12**(2), pp. 293–310.

Harel, D. and Pnueli, A. (1985) On the Development of Reactive Systems, in: *Logics and Models of Concurrent Systems*, (ed. K.R. Apt) Springer, New York, pp. 477–98.

Harel, D. (1987) Statecharts: A Visual Formalism for Complex Systems, *Sci of Comp Prog* **8**, North-Holland, pp. 231–74.

Harel, D. (1988) On Visual Formalisms, *Communications of the ACM*, **31**(5), pp. 514–30.

Harel, D. (1992) Biting the Silver Bullet: toward a Brighter Future for System Development, *Computer*, Jan., pp. 8–20.

Heuser, C.A., and Richter G. (1992) Constructs for Modeling Information Systems with Petri Nets *Application and Theory of Petri Nets* (ed. K. Jensen). 13th International Conference, Sheffield, UK, June 1992, Springer Verlag, pp. 224–43.

# 284 *Qualitative intelligent modeling of manufacturing systems*

ICAM (1981) Integrated Computer Aided Manufacturing (ICAM) Functional Modeling Manual, *Report UM 110231100*, Air Force Systems Command, Wright-Patterson AFB, Ohio.

IDEF3 Process Description Capture Method Report, *Armstrong Laboratory Report # AL-TR-1992-0057*, Wright Patterson AFB, OH, 45433.

IDEF4 Object Oriented Design Method Manual. *Armstrong Laboratory Report # KBSI-IICE-90-STR-01-0592-01*, Wright Patterson AFB, OH, 45433.

Jensen, K. (1987) Colored Petri Nets, in *Petri Nets: Central Models and Their Properties*, (eds W. Brauer W. Reisig, and G. Rozenberg), Springer Verlag, Berlin, Feb., pp. 248–99.

Lissander, M., Lagier, P. and Skalli A. (1985) SAS: A Specification Support System, in *System Description Methodologies*, Elsevier Science Pub., IFIP pp. 221–37.

Melman, M. and Livny, M. (1984) The DISS Methodology of Distributed System Simulation, *Simulation*, **42**(4), pp. 163–76.

Misra, J. (1986) Distributed Discrete Event Simulation, *Computing Surveys*, **18**(1), pp. 39–65.

Morasca, S., Pezze and Trubian, M. (1991) Timed High-Level Nets, *Jnl. of Real-Time Sys*, **3**, pp. 165–89.

Murray, K.J. and Sallie, V.S. (1988) Knowledge Based Simulation and Specification, *Simulation*, **50**(3), pp. 112–19.

Pathak, D.K. and Krogh, B.H. (1989) Concurrent Operation Specification Language, COSL, For Low Level Manufacturing Control, *Comp in Ind*, **12**(2), pp. 107–22.

Peterson, J.L. (1981) *Petri Net Theory And The Modeling Of Systems*, Prentice Hall.

Reisig, W. (1982) *Petri Nets: An Introduction* (eds W. Brauer, G. Rozenberg and A. Salmaa), Springer Verlag.

Ross D.T. (1985) Applications and Extensions of SADT, *IEEE Computer*, Apr., pp. 25–43.

White, B.A. and Frederiksen J.R. (1990) Causal Model Progressions as a Foundation for Intelligent Learning Environments, *Artificial Intelligence*, **42**, pp. 99–157.

# CHAPTER 11

# Formal models of execution function in shop floor control

*Jeffrey S. Smith and Sanjay B. Joshi*

## 11.1 INTRODUCTION

Computer integrated manufacturing (CIM) has recently received a great deal of attention as an effective strategy to improve manufacturing responsiveness and quality. CIM seeks to integrate the entire manufacturing enterprise through the use of an integrated set of computer systems. However, the evolution to CIM has been slower than expected. This can be directly attributed to the high cost of software development and maintenance and the difficulty in achieving the required levels of integration between systems. These problems are especially evident in the development of the central part of a CIM system, the shop floor control system (SFCS). The SFCS is responsible for planning, scheduling and controlling the events on the shop floor. An SFCS (performing some or all of the planning, scheduling and control functions), together with the associated shop floor equipment, used as a stand-alone system is often called a flexible manufacturing system (FMS). This chapter presents two related approaches to modeling the execution function in SFCS. Implementation experience involving the use of these models in a full scale manufacturing system is also provided.

The sophisticated machine tools, robots and other equipment necessary for automated flexible manufacturing are generally available as stock items. Similarly, the computer hardware and networking equipment are also readily available. However, the **software control systems** necessary to implement integrated, flexible control over this equipment are not currently available and must be developed from scratch for each implementation. Merchant (1988) suggests that this problem exists because it has proven much more difficult than expected to integrate the so-called 'islands of automation' into a fully integrated system. It is much easier to optimize and automate these 'bits and pieces' of the system than to integrate them into a functional system. Bjorke (1988) attributes this difficulty to the lack of a theoretical basis by which the individual functions of the manufacturing systems can be analyzed.

The problems encountered in developing SFCS software are typical for large scale software development projects. The Computer Science and Technology Board (CSTB) presents the following observations on the state of software development from leading researchers and industry members in the software engineering field (CSTB, 1989):

- (My company's) most significant software problem is duplication of effort; we often write several times what appears to be essentially the same software.
- What we call the problem of requirements generation is actually a poverty of tools, techniques, and languages to assist business experts and computer scientists in collaborating on a process of transformation which adequately exploits the knowledge of all.
- The worst problem I must contend with is the inability of the software buyer, user, and builder to write a blueprint which quickly leads to a low-cost product.

These problems are all significant in the development of SFCS. Controllers exhibit very similar behavior, but this similarity has not been exploited. This problem exists largely because of the lack of software development tools customized for the shop floor control domain. Furthermore, without an adequate **formal description** of the shop floor control problem it is difficult to convey the software requirements to the software engineers in charge of system development. It is suggested that 'many of the problems experienced today reflect implicit assumptions that the flow from software system concept to implementation is smoother and more orderly than it is' (CSTB, 1989). The difficulty stems from the need for exact, unambiguous system specifications and the difficulty in creating these specifications.

In response to these needs, this chapter describes two approaches for modeling the execution function in shop floor control systems. These approaches are based on formal descriptions of the behavior of the underlying manufacturing systems and are used as inputs to the implementation of the software control systems. The configuration of the manufacturing system is first described using a set of **physical models**. A physical model describes the behavior of the underlying system in terms of states, physical tasks and the relationships between the states and tasks. The first approach then uses the physical models to create a set of system models which account for the proper preconditions for physical tasks. A context-free grammar is then developed to recognize the 'language' described by the system models. The second approach provides explicit descriptions of the communications protocol between the individual controllers and the physical devices which perform the physical tasks described by the physical models. These descriptions are then used to create the recognition programs necessary to implement the execution portion of control.

## 11.2  BACKGROUND

Computer integrated manufacturing (CIM) has received a great deal of attention from researchers all over the world for the past 15 years. CIM is viewed as a way to reduce the manufacturing costs and decrease the product design-to-market time in discrete parts manufacturing. Although a complete background in the topic would be beyond the scope of this work, this section presents the background material that forms a basis for the work presented in this chapter. For a broader introduction to CIM and related areas see Koenig (1990) and Ranky (1986, 1990).

Flexible manufacturing systems (FMS) attempt to bring the productivity of a dedicated flow line to small and medium batch manufacturing. The concept of flexible manufacturing was developed in response to demands for reductions in lead times without increases in production costs in low to medium volume discrete parts manufacturing (Smith, 1990). Typically, an FMS is made up of a small number of general purpose machine tools connected by an automated materials handling system. Ideally, the entire system is controlled by a computer or a system of computers programmed to exploit the inherent flexibility of the individual machines in order to process a wide variety of part types with little or no setup or changeover costs. FMS concepts and implementations have been described in many publications: Dupont-Gatelmand (1982); Kimenia and Gershwin (1983); Whitney (1985); Cohen (1985); O'Grady and Menon (1986); Gupta and Buzacott (1988) and Smith (1990).

Computer integrated manufacturing (CIM) is a much broader concept than flexible manufacturing. Where an FMS seeks to integrate a small set of shop floor machines, CIM integrates the entire manufacturing enterprise. The American Heritage Dictionary (1976) defines **integrate** as 'to make into a whole by bringing all parts together.' Bravoco and Kasper (1985) define **integration** in the context of manufacturing as two (or more) things which have common parts which are leveraged to provide economies and benefits. In a CIM system, production status, purchasing reports, marketing information and material availability information should all be accessible through a central computer system (the computer system is not necessarily **physically** centralized, only conceptually centralized) with well-defined interfaces and communications standards.

A **formal model** of a system is an unambiguous description of the system presented in an implementation-independent language with precisely-defined syntax and semantics. A formal model is an abstract view of a system which specifies the functionality and the behavior of the system without being constrained by implementation details. Formal models provide a means to validate system specifications and to check for system correctness prior to system implementation. Basnet *et al.* (1990) propose that a formal model gives a definite form to how and what can be expressed about a system to be modeled. Sharbach (1984) describes the benefits of using formal specification and design methods as follows:

• Increased understanding of a system's intended behavior.
• Reduction of the probability that errors are made in a development because correctness of the specification can be considered before detailed implementation is undertaken.
• Proving the equivalence of behavioral properties of various specifications to obtain alternative viewpoints of the same model, or in proving the refinement and decomposition of a design.
• Availability of vehicles for unambiguous communication of design decisions, also supported by the expressive power of good formal languages which improve conciseness.
• Starting point for the development of computerized tools for the development, simulation, verification and implementation of specifications, and testing of implementations.

Biemans and Blonk (1986) state that 'to ensure that CIM architectures leave no room for ambiguous interpretation, they should be defined in a formal language.' Furthermore, it is suggested that unless a clearly defined, well understood architecture of the system is initially developed, it is unlikely that the various components will interact correctly after installation. They present a generic CIM reference model, which describes the individual manufacturing system components. A workcell controller written in the specification language LOTOS (language for temporal ordering specifications (Van Eijk *et al.*, 1989) is also presented as a part of an implementation based on this reference model.

Naylor and Maletz (1986) present a modeling formalism for manufacturing control in the context of a 'manufacturing game.' This formalism was constructed with the goal of developing real-time control software that is generic in the sense that it can be used to control an arbitrary factory making an arbitrary collection of parts. Naylor and Maletz suggest that a carefully defined formalism is needed for at least two reasons. First, it must be possible to describe a factory floor so that the generic software has a model of the system to be controlled. Second, it must be possible to describe how parts are made (i.e. process plans). Naylor and Volz (1987) use this formalism in describing a conceptual framework for integrated manufacturing. An integrated manufacturing system is viewed as an assemblage of software/hardware components, where a software/hardware component is a generalization of the software component concept. The software/hardware components are both recursive and distributed. In other words, components may be hierarchically constructed from other components and need not all exist on the same computer.

Chaar (1990) further extends the model of Naylor and Volz by developing a methodology for implementing real-time control software for FMS. This methodology is based on a component-oriented, rule-based specification language. Chaar uses the modeling formalism of Naylor and Volz to model the software/hardware components of the manufacturing system and the part process plans as a set of first-order logic-based rules. These rules are then coded in the specification language which can be converted to executable code. Although some possibilities for automating this conversion are mentioned, none is described. Instead, the conversion is done manually by the software engineer.

Mettala (1989) also extends the initial work of Naylor and Volz by adding a system model, which maps the state of the system onto the internal state of the control computer. Mettala demonstrates that the various controllers in a manufacturing system hierarchy can be represented as a context-free grammar. Using known results showing the equivalence of deterministic finite automata and deterministic context-free languages, Mettala describes the use of context-free grammars (CFG) and subsequent language recognition as a means for FMS control. Based on these developments, he presents an automatic generation system for FMS control software. Software is generated based on an input description of the cell configuration and possible part movements within the cell. Mettala also describes an AND/OR graph-based formal model for part process plans which allows multiple alternative processing options to be considered by the cell controller.

Davis and Jones (1989) advocate a mathematical description of all manufacturing functions as a starting point for CIM development. Initial mathematical formulations for modeling individual processes, process control and process interaction are presented. It is stressed that these models provide descriptions of the underlying processes, but that they are independent of the specific CIM architecture. Although the clarity and predictability of these mathematical formulations are appealing, it is unlikely that they will provide sufficient power to be used directly in the control of FMS, whose behavior is inherently dynamic and stochastic.

Bourne (1991) describes a shop floor control software development system called the cell management language (CML). Bourne describes the dream for factory automation: 'roll a computer on to the factory floor, plug it into a set of machines from different manufacturers, start a program, and then, with absolutely no traditional programming, begin the task of integrating the machines into a cooperative cell.' CML was developed in a joint research project with Westinghouse and uses a table metaphor to describe communications and control protocols. The user fills in the appropriate tables and CML creates the lexical analyzers and parsers to recognize the protocol.

## 11.3   ENVIRONMENT

This section describes the shop floor environment for this work. We are primarily interested in developing control systems for automated discrete parts manufacturing of small to medium volume. In the development of a shop floor control system, a **control architecture** provides an overall structure for the individual controllers and the relationships between these controllers. A control architecture should completely and unambiguously describe the structure of the system as well as the relationships between the system inputs and the system outputs. Based on the need for global information for planning and scheduling, we are currently using a hierarchical control structure (Smith *et al.*, 1992). Use of hierarchical control architectures as a basis for shop floor control has been described by many authors (Albus *et al.*, 1981; Jones and McLean, 1986; Biemans and Vissers, 1989; Joshi *et al.*, 1990; Jones and Saleh, 1989; Senehi *et al.*, 1991).

One of the primary limitations of many existing hierarchical control architectures is the lack of sophisticated planning and scheduling at the lower levels (Jones and Saleh, 1989). As a result, there are essentially only two distinguishable layers, one to do the decision making and one to do the control. As a result of these inherent problems, Jones and Saleh (1989) propose a so-called 'multi-level/multi-layer' control architecture. The structure of this architecture is based on both spatial and temporal decompositions of the control system. An important aspect of this architecture is the decomposition of the control tasks at each level into **adaptation, optimization**, and **regulation** functions. Joshi *et al.* (1990) present a three-level hierarchical control model based loosely on the lower three levels of the original NIST AMRF control hierarchy (Jones and McLean, 1986). According to this 'scaleable architecture', the tasks for each controller are separated into **planning, scheduling**, and **control** tasks. We prefer the term **execution** to the term control, since control typically implies more than the simple physical control of equipment. These tasks are similar to the adaptation, optimization and regulation functions described by Jones and Saleh. Partitioning the controller tasks into planning, scheduling and execution tasks is critical and is discussed in more detail later in the context of controller structure and operation.

Among existing hierarchical architectures there is much debate over the required number of distinct levels. Based on the architectures presented by Jones and Saleh (1989) and Joshi *et al.* (1990), we identify three 'natural' levels: From the bottom to the top (Fig. 11.1) are the

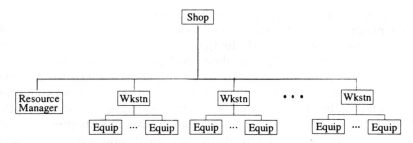

**Fig. 11.1**  Shop floor control architecture.

**equipment, workstation,** and **shop** levels. The equipment level is defined by the physical shop floor equipment and there is a one-to-one correspondence between equipment level controllers and shop floor machines. Equipment level devices are partitioned into material processors (MP), material handlers (MH), material transporters (MT), and automated storage devices (AS). This partitioning is made based on the type of the device.

The set MP is made up of machines which autonomously 'process' parts according to some set of processing instructions. This set includes NC machine tools, coordinate measuring machines, etc. The set MH is made up of machines which move parts within workstations (the term workstation as used here is defined later). This set is made up of robots and other pick and place machines. Associated with MH devices is a set of **addressable locations**. An addressable location is a physical location to which an MH device has access to pick objects up or put objects down. For example, if a robot is used to load a part on a machine tool, the machine tool fixture is defined as an addressable location. The MH device moves objects to and from addressable locations and makes no distinction between the type of device owning the addressable location. The requirement is for the MH device to be able to determine the specific control program to run based on the object description and the addressable location number. The set MT is made up of machines which move parts between workstations. This set includes automated guided vehicles (AGVs), conveyors, etc. The set AS is made up of various types of automated storage and retrieval systems. We also define a set BS of buffer storage devices. These devices are passive storage locations only and do not require explicit controllers. Additional details about this architecture are described by Smith *et al.* (1992).

The workstation level is defined by the layout of the equipment. Processing and storage machines that share the services of a material handling machine together from **workstations**. **Ports** are defined as the interface points between workstations and the shop-wide material transport system (e.g. a conveyor stop or an AGV load/unload station). Each material transport device services a set of ports and these sets may overlap if multiple MT devices can transport parts to the same physical location. Finally, the shop level acts as the top level control and interface point for the system. The shop level therefore provides the user interface to the system. The shop level receives production requirements from the master production schedule or from the MRP system, depending on the implementation. For this chapter, we concentrate on the development of the equipment and workstation level controllers.

## 11.4  PHYSICAL MODELS

The physical model describes the general behavior of the system for which the controller is being developed. The physical model was adapted from Mettala's physical graph of a manufacturing cell (Mettala, 1989). The physical model for a system defines the set of tasks performed by the system and the allowable sequence(s) of these tasks without regard for the **physical** and **logical preconditions** for tasks. Physical preconditions verify the satisfiability of physical constraints for a particular task before execution of task. Logical preconditions verify that the tasks are desirable based on the current state of the system. For example, if a robot and a machine tool belong to a workstation, a task specified in the workstation's physical model is for the robot to load a part on the machine tool. However, before performing this task in a running system, there are several preconditions which must be verified. First, the machine tool fixture must be unoccupied and the robot must be idle (**physical** preconditions). Secondly, this task only makes sense if the particular machine is part of the process plan for the part (a **logical** precondition). Physical and logical preconditions are evaluated using the physical configuration and the system status. Therefore, physical model does not specify the validity of a controller task in a **particular** configuration, only that the task is valid in **some** configuration.

A physical model is represented as a deterministic finite automaton (DFA), $P = (Q, q_0, F, \Sigma, \delta)$, where $Q$ is a finite set of **part states**, $q_0 \in Q$ is the start state, $F \subseteq Q$ is a set of final states, $\Sigma$ is a set of **system tasks**, and $\delta : Q \times \Sigma \to Q$ is a transition function. The interpretation of $\delta$ is that if a system is a state $q$ and the system task $\sigma$ is performed, then $\delta(q,\sigma)$ gives the next state of the system. The meaning of a *part state* is different at each level in the control architecture. At the equipment level the state represents the processing state of the part. At the workstation level, the state represents the equipment level entity to which the part is assigned. Finally, at the shop level, the state rep-

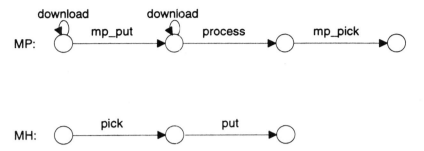

**Fig. 11.2** Physical models for MP and MH class equipment.

resents the workstation level entity to which the part is assigned. A system task is a function of the system which is specified by the controller. The system tasks are the elemental actions which the planner and scheduler are responsible for specifying.

The physical models for generic MP and MH equipment are shown in Fig. 11.2. In the MP case, the physical model includes the system tasks *mp_put*, *process*, *mp_pick*, and *download*, representing loading the part on the machine, processing the part on the machine, unloading the part from the machine and downloading processing instructions, respectively. The transition function δ is illustrated by the structure of the graph. δ specifies that the part is first loaded on the machine, processed, and finally unloaded. Prior to the start of processing, the processing instructions (e.g. the NC file) can be downloaded to the device (the download task is optional to account for the case where the processing instructions are already resident at the device). In the MH case, the physical model includes the system tasks **pick** and **put**, representing moving from the current location to the specified location and picking up a part, and moving from the current location to the specified location and putting down a part, respectively. The transition function specifies that parts are first picked up, and are then put down.

A typical workstation and the associated physical model is shown in Fig. 11.3. This workstation consists of two MP devices, one MH device, a part buffer and a workstation port. Since physical model states represent the equipment level subsystems to which parts can be assigned, the labels for the states in this model are Port, MH, MP1, and MP2, representing the workstation port, and the MH and MP equipment within the workstation. The start (and final) state is labelled as state 0 and this state represents an external entity sending parts to the workstation (i.e. the shop controller). The transition function δ specifies that parts initially arrive at the workstation port from an external source. From the port, the parts can be picked up by the MH device. Once assigned to the MH device, the parts can be put down at either of

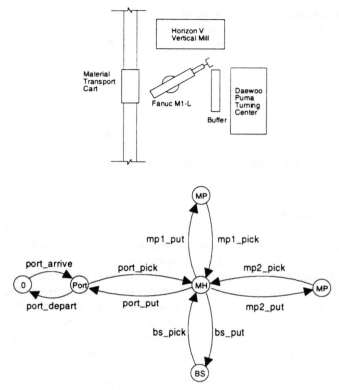

**Fig. 11.3**   Physical model for a two-machine workstation.

the MP devices or at the workstation port. Processing a part which requires processing at M1 followed by M2 would be represented by the following sequence of tasks:

arrive
port_pick
mp1_put
mp1_pick
mp2_put
mp2_pick
port_put
depart

Joshi *et al.* (1992) and Mettala (1989) describe the system tasks as *physical relations* which exist between the states and provides a formal description of how the physical models are constructed based on a textual description of the system configuration.

## 11.5  CONTEXT-FREE GRAMMAR APPROACH

This section presents a model of discrete parts manufacturing which extends the physical models described in section 11.4 to account for the proper command preconditions, such as having a robot manipulator empty prior to executing a pick operation. Figure 11.4 shows the system model for the example workstation shown in Fig. 11.3 (the part buffer is not included in the system model) (Joshi *et al.*, 1992). The system model is constructed from the physical model. Each node $k$ in the system model is labelled with $\sigma$, containing three quantities: the active entity set AE, the substructure relation set SB and the contact vector CT (this notation is similar to that used by Nayor & Volz, 1987). For our purposes, only the latter term is shown in Fig. 11.4. A system graph can consist of a maximum of $2^n$ nodes, labelled with the corresponding contact vector of $n$ bits CT = [$b^n$]. For example, each contact vector in the example case will be composed of a vector [*bbbb*], were $b$ represents a binary valued integer. In the definitions that follow, the system in Fig. 11.4 with CT = [*bbbb*] is used as an example, where machines are associated with each contact relation in the manner illustrated in Fig. 11.5.

Each arc $l$ in the system graph is labelled with some relation $R$ such that $\sigma_i R \sigma_j$ for $R \in \{pick, put, arrive, depart\}$. These relations represent the movement of parts between the equipment within the system and correspond to their physical model counterparts. Unlike the physical model however, the system model defines part movement with respect to a completely defined system state. This results in a large number of states ($O(2^n)$). Some of the possible $2^n$ nodes, can be omitted due to the fact that executing an action that could cause transition to a particular state may not result in a desired situation (hence must be avoided), or may even be redundant. For example, consider the transition from $1101 \rightarrow 1110$ caused by the pickup of part by robot from port. Now the only feasible action by the robot is a putdown, and the only available location is the port from which the part was picked. Thus even though an action was possible to cause transition from $1101 \rightarrow 1110$ there will be no 'real' part movement. Similar situations exist for transitions from $1110 \rightarrow 1111$. Hence the nodes 1110 and 1111 are not included in the system graph. Additionally, many arcs in the system graph may have the same label, denoting that the same elemental operation is required to produce a desired part movement, but that particular part movement may exist within the context of several different state changes.

The pick relation as illustrated in Fig. 11.6 specifies the event of a robot acquiring a part in the context of a running system. A system *pick* relation exists between two states $\sigma_i$ and $\sigma_j$ if and only if robot $l$ is empty in state $\sigma_i$, and some park $k$ is at either a processing machine, or a port $m$, and if robot $l$ is grasping park $k$ in state $\sigma_j$ and the relevant

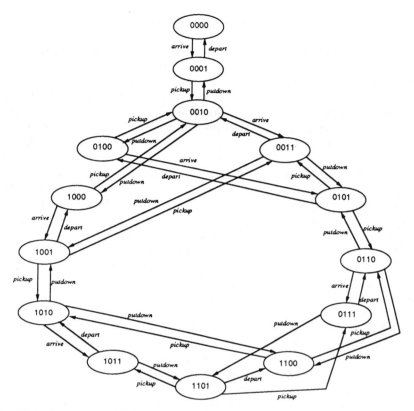

**Fig. 11.4**   System model for a two-machine workstation.

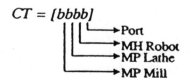

**Fig. 11.5**   Example contact vector.

processing machine or port $m$ is empty in the new state, provided that no other machine $y$ in the system changes contact state. Formally, given states $\sigma_i$ and $\sigma_j$,

$$\sigma_i \ pick_s \ \sigma_j \ iff \ \exists k,l,m \ st \ \neg CT(k,l)_i \wedge CT(k,m)_i \wedge CT(k,l)_j \wedge$$
$$\neg CT(k,m)_j \wedge V_{part(k)} \wedge V_{MH(l)} \wedge (V_{MP(m)} \vee V_{Port(m)}) \wedge \forall x,y \ (CT(x,y)_i$$
$$= CT(x,y)_j \ /x \neq k, \ y \notin (l,m), \ V_{part(x)} \wedge (V_{MH(y)} \vee V_{MP(y)} \vee V_{Port(y)}) \quad (11.1)$$

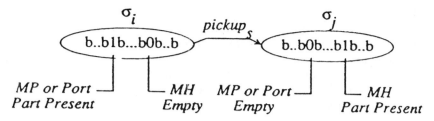

**Fig. 11.6** Pick relation.

For the example (Fig. 11.6), the state transitions formed by the (non-inclusive) contact vector pairs $[0100]_i \rightarrow [0010]_j$, $[1100]_i \rightarrow [1010]_j$, and $[1101]_i \rightarrow [1011]_j$ all represent the same $pick_s$ operation, that is, a materials handling robot picks up a part from a material processing lathe. Similar pick relations exist when a robot picks up a part from the port and milling machine. The $pick_s$, $put_s$, $arrive_s$ and $depart_s$, form the complete set of arc labels in the system graph, and are defined in an analogous manner.

Joshi *et al.* (1992) also define a set of **functional relations** associated with the equipment level devices within a workstation. These functional relations are associated with the attributes of specific machines or devices. These relations are analogous to the physical model tasks for the equipment level devices presented in section 11.3.

Having defined the state space graph models for the workstation control module and the MP, MH and AS equipment, the next step is the generation of context-free grammars that are equivalent to the graph models. A context-free grammar (CFG) is defined by $G = (V_n, V_t, \phi, S)$ where,

$V_n$ = set of non-terminal symbols
$V_t$ = set of terminal symbols
$\phi$ = set of productions
$S$ = start symbol

The productions are of the form $\alpha \rightarrow \beta$, where $\alpha$ and $\beta$ are strings from the vocabulary $V = V_n \cup V_t$, with the specific restriction that $\alpha$ must be a single symbol belonging to $V_n$.

The underlying rationale for developing control grammars for each individual controller within the manufacturing workstation stems from the fact that it is feasible to exercise control as a side effect of recognizing a context-free grammar. For each grammar two requirements must be satisfied: first, data level interfaces must be flexible, and second, mapping between the data interfaces and the dynamics of the system that must occur corresponding to each arc in the system graph must be subsumed into the grammar. These requirements are met through the development of control and interface portions of the grammars.

During the grammar generation process, the physical graph is searched to determine primitive movements that are allowed by material handling robots within the workstations. The state space graph is then searched in order to build the system relations. The relations are subsequently searched for each machine in the system, and the machine state-relation graph (the physical model) for the current machine is preserved, forming the state space for that device. The workstation grammar definition and construction process permits multiple fixtures on machines, multiple robots and multiple load/unload ports per workstation. The product of the state space search of the system graph yields a state space for each machine, and identifies the associated arcs labelled by the primitive physical relations, for each action which the particular machine under consideration is a contributor. The mapping between functional relations and system relations is made, resulting in the association of permissible functions with each state of the machine.

As an example of the context-free grammar generated let us consider the control portion of the workstation control grammar. It is responsible for ensuring collision-free entry, exit and motion of parts through the workstation by maintaining a state space model of the workstation and through an interface to the workstation scheduler. The grammatical elements consist of state **space maintenance rules, alternative rules, action rules** and **distinguished state rules**. Figure 11.7 illustrates the control rules for a workstation controller. The non-terminals symbols are denoted by lower case letters and the terminal symbols by upper case letters.

The workstation control grammar as shown in Fig. 11.7 is constructed automatically from the system state space graph (Fig. 11.5). Initially, the **start up portion** of the grammar, including the initial 'command' non-terminal is output to the target grammar file. Next, the non-terminal 'oneshot' is generated from the node of the graph, where each alternative corresponds to a node in the graph. The **distinguished state productions** correspond to the nodes in the system graph and provide the alternatives available to leave the state. The **alternate production** rules correspond to the current distinguished manufacturing state as referenced through the state maintenance non-terminal and the associated action. The semantic actions associated with these productions is the interjection of the next state into the input stream. The **action productions** correspond directly to the arc labels in the system model. The semantic actions that execute upon reduction of the action productions cause the control activities to take place. The **state maintenance productions** provide for proper maintenance of state maintenance non-

```
command :   oneshot
onehsot :       st0 |
                st1 |
                st2 |
                  .
                st13
```
⎫
⎬  Start up portion
⎭

```
st0 :   alt0
st1 :   alt1 |
        alt2
st2 :   alt3 |
        alt4 .
  .
  .
st13 : ....
```
⎫
⎬  Distinguished state
⎭  productions

```
alt0 :  st0t arrive      {Interject ("ST1");}
alt1 :  st1t pick0       {Interject ("ST4");}
alt2 :  st1t depart      {Interject ("ST0");}
alt3 :  st2t pick1       {Interject ("ST4");}
alt4 :  st2t arrive      {Interject ("ST3");}
          .
          .
          .
```
⎫
⎬  Alternative productions
⎭

```
arrive : STNID ARRIVE CARTID
depart : STNID DEPART
pick0  : PICK0 objpart
pick1  : PICK1 objpart
pick2  : PICK2 objpart
put0   : ...
put1   : ...
put2   : ...
```
⎫
⎬  Action productions
⎭

```
objpart : parttype partnum
parttype : VAR
partnum  : VAR
```

```
st0t : ST0 |
       st0t inter {Interject ("ST0");} |
       error |
       st0t error
  .
  .
  .
```
⎫
⎬  State maintenance
⎭  productions

**Fig. 11.7**  Control grammar for a workstation.

terminals, and the required interfaces to synchronization and error via the use of non-terminals 'inter' and 'error'.

The structure of the alternative non-terminals for each state begins with the state maintenance non-terminal for the current state $\sigma_i$, followed by *relat$_s$*, followed by the interjection of the next state identifier $\sigma_j$. In a running controller, the semantic action of the alternative rules forces a manufacturing state transition upon recognition of the alternative by interjecting the next state identifier.

The next step is to specify the semantic actions associated with the control, synchronization and error parsers in each module with the FMS workstation. In a manufacturing environment, a typical semantic action might be to cause a robot to move, or cause a message to be transmitted to another computer. In this section, the semantic actions associated with controlling the manufacturing system are introduced. In general, the manufacturing semantic actions are confined to action productions. Each controller in the system has its own set of semantic actions. During the code generation process the semantic actions are associated with the appropriate production. Each action production (although in different model domains) has a common generic sequence of actions that must be executed. This common generic sequence of actions associated with each class of action productions is created as a text file. During the grammar generation process the generic semantic actions are instantiated based on the model domain (workstation, MH, MP, Port, etc.), a from state, a transition terminal and a destination state corresponding to states in the model domain. The local variables to update, messages to be transmitted and message destinations are determined automatically during instantiation. This approach permits the semantic functions to have direct access to variables that are passed during the running of the system, and execution of various functional routines in the context of a grammar recognition state space, or where appropriate, within the context of distinguished manufacturing states.

As an example of semantic actions consider the following 'put' action non terminal in the workstation control grammar.

```
put0:PUTO objpart {dbmove(currparttype, currpartnum, "R1",
                   "CT1", "P1", slotid); workstationcmd =
                   sched("workstation1 put0");
                   if(workstationcmd)
                   send_command(workstationcmd, R1COM);}
```

The 'dbmove' operator updates the workstation control database, recording the movement of the current part (captured in the 'objpart' non-terminal) from robot R1 contact CT1 to port P1, at slot slotid. The slotid is set by the previous invocation of the scheduler, where the command resulting in the put0 operation is constructed. The dynamic scheduler is invoked, resulting in a workstationcmd. If the command is null, then this event has resulted in no feasible movement of a part or tool by a robot. If the command is non-null, then it is sent to the R1 robot, using the send_command function. Typical commands are generated in the form:

**move** cndl pn123 **from** port1 slot3 **to** m1 **end**;

This one example illustrates the semantic actions of database updates, scheduler interface, and communications interface.

The control grammar and semantic actions are automatically generated from the formal model of discrete part manufacturing discussed in the earlier sections. The advantages of automatic control grammar generation are many. The main advantage arises from the simplicity of specifying a system, and generation of cooperative finite state controls to exercise control in a manufacturing environment. Additional details of this approach are provided by Mettala (1989).

## 11.6 MESSAGE-BASED PART STATE GRAPH APPROACH

The physical model of a system describes the system tasks, but doesn't consider the proper physical preconditions or context for performing those tasks. The system model on the other hand, considers the context for performing system tasks but doesn't specify explicitly how these tasks are performed. In a distributed control system, control is exercised by passing **messages** between controllers and by performing **controller tasks**. The set of messages coming into and going out of a controller along with the set of controller tasks will be referred to collectively as a set of **controller events**. Furthermore, a **processing protocol** for a part is defined as a sequence of controller events required to process the part within the scope of a controller. For example, Table 11.1 shows the

**Table 11.1** Example processing protocol

| event | source | message/task | destination |
|---|---|---|---|
| 1. | workstation | assign part | machine controller |
| 2. | machine controller | at location | workstation |
| 3. | workstation | put part | robot controller |
| 4. | robot controller | put task | robot |
| 5. | robot controller | put done | workstation |
| 6. | workstation | grasp part | machine controller |
| 7. | machine controller | grasp part task | machine |
| 8. | machine controller | grasp done | workstation |
| 9. | workstation | clear | robot controller |
| 10. | robot controller | clear task | robot |
| 11. | robot controller | clear done | workstation |
| 12. | workstation | clear ok | machine controller |

sequence of controller events required for a robot to load a part on a machine tool under the direction of a workstation controller. In this situation, the workstation asks the machine tool where to place the part, and then synchronizes the actions of the machine tool and the robot during the transfer of the part.

There are three distinct processing protocols associated with the dialogue in Table 11.1. The first is a **workstation** processing protocol, which includes all messages received by and sent from the workstation controller (events 1, 2, 3, 5, 6, 8, 9, 10, 11, 12). The second is the **machine tool** processing protocol, which includes all messages received by and sent from the machine controller as well as all tasks issued to the machine by the equipment controller (events 1, 2, 6, 7, 8, 12). The last protocol is the *robot* processing protocol which includes all messages received by and sent from the robot controller and the tasks issued to the robot by the robot controller (events 3, 4, 5, 9, 10, 11). The individual controllers are each responsible for implementing their respective processing protocols. The interaction of these controllers during system operation creates the dialogue illustrated in Table 11.1. We seek to provide formal definitions of the processing protocols and then to use these formal definitions to automatically generate a software implementation of the execution module.

### 11.6.1   MPSG definition

A **message-based part state graph** (MPSG) provides a formal description of the processing protocol. As such, an MPSG describes the behavior of a controller from the parts' point of view. A MPSG extends the physical model by adding specific implementation details to the system tasks. For example, where the workstation physical model represents loading a part on a processing machine as a single activity (*mp_put*), the corresponding MPSG explicitly defines the interaction of the individual workstation, **MP** and **MH** controllers in terms of specific controller events.

An MPSG is a modified DFA similar to a Mealy machine. There are two primary modifications of the DFA structure required to model shop floor controllers. First, the input mechanism must be modified to accept input from multiple streams. This is accomplished by using the set of controller events as the input alphabet and by creating multiple input mechanisms for reading events. Second, the transition function must be modified so that controller tasks can be performed during state transitions similar to a Mealy machine. This is accomplished by introducing a set of **controller actions** and a corresponding **controller action transition function**.

An MPSG $M$ is defined formally as the octuple, $M = (Q, q_0, F, \Sigma, A, P, \delta, \gamma)$, where $Q$ is a finite set of states, $q_0 \in Q$ is an initial or start state, $F \subseteq Q$ is a set of final or accepting states. $\Sigma$ is a finite set of controller events and serves as the input alphabet. Additionally, $\Sigma$ is partitioned into a set of input messages $(\Sigma_I)$, a set of output messages $(\Sigma_O)$, and a set of controller tasks $(\Sigma_T)$. A is a finite set of controller actions, where each $\alpha \in A$ is an executable function which performs some controller action. P is a finite set of **physical preconditions** for controller actions. P is partitioned so that for each $\alpha \in A$ there is a corresponding $\rho_\alpha \in P$, where $\rho_\alpha$ is a function which returns either true or false. $\delta : Q \times \Sigma \to Q$ is a state transition function. Similarly, $\gamma : Q \times (\Sigma_O \cup \Sigma_T) \to A$ is a controller action transition function. These individual components are described in more detail in the following paragraphs.

In a standard DFA implementation the input symbols are read from a single input stream (e.g. a compiler reads a source code file). In an MPSG implementation however, the input symbols are read from two different streams: the controller communications module and the task list. The communications module receives incoming messages from the supervisory and subordinate controllers and the task list is the communications medium between the execution module and the planning and scheduling modules within the controller. The input alphabet $(\Sigma)$ is therefore partitioned into three sets of controller events based on the event type. $\Sigma_I$ is the set of valid input messages which are received from other controllers. $\Sigma_O$ is the set of valid output messages which are sent to other controllers in the shop. $\Sigma_T$ is the set of controller tasks which identify controller actions to be executed.

$\Sigma_O$ and $\Sigma_T$ events are specified by the scheduler by placing them on the task list. $\Sigma_O$ and $\Sigma_T$ are scheduled events because they can initiate physical action. For example, moving a robot into a machine tool is a task from the robot's point of view. From the machine tool's point of view, granting permission to enter the machine's work volume is an output message. Both of these can initiate the physical action of moving the robot. By contrast, incoming messages can only change the state of the part. As such, an output message is essentially a special case of a task, where the task is to send a message to an external system. The distinction is made because output messages from one controller become input messages to another controller, whereas tasks are specific to a single controller.

The set of controller actions A, is a set of executable functions $\alpha$, which perform the controller tasks specified by $\Sigma_O$ and $\Sigma_T$. For example, the $\alpha \in A$ which corresponds to the *grasp_part* task is an executable function which closes the machine tool's part fixture. The implementation of these functions is device specific. The controller task transition function $\gamma$, maps state-task pairs $(Q \times (\Sigma_O \cup \Sigma_T))$ onto A. The interpretation of the controller action transition function, $\gamma(q, b) = \alpha$, where $q \in Q$, $b \in \Sigma_O \cup \Sigma_T$, and $\alpha \in A$, is that the MPSG $M$, in state $q$ and scanning the event *b*, executes function α. Upon successful completion of α, *M* undergoes a state transition specified by $\delta(q, b)$.

The feasibility of many controller tasks depends on the current **system** state rather than the **part** state. For example, putting a part on a fixture is only feasible if there is no other part currently on the fixture. The physical preconditions are used to verify the system state prior to executing controller tasks. Many of these conditions involve other parts within the system. Since the MPSG models part states rather than system states, a mechanism for identifying external constraints is required. P is a set of physical preconditions which is partitioned according to the set A such that, associated with each controller action α there is a physical precondition $\rho_\alpha$ which must evaluate to true in order for the corresponding controller action α to be called. For example, if the controller task is to start (*cycle_start*) the machine tool, the following physical precondition must hold:

$$\rho_{cycle\_start} = (\rho_1 \wedge \rho_2 \wedge \rho_3)$$

where:

$\rho_1$ – the part must be in the chuck
$\rho_2$ – the chuck must be closed
$\rho_3$ – the machine tool work volume must be clear.

If this condition does not evaluate to true, a cycle start cannot be sent to the machine tool, the specified task is not consumed by *M* and no state change occurs. The physical preconditions are used to prevent **catastrophic errors** in the system. A catastrophic error involves a collision between machines or parts within the system. For example, if a robot is holding a part and is instructed to pick up another part, a collision will occur between the part in the gripper and the part to be picked up unless the robot knows to queue the pickup task until the gripper is unoccupied. The physical precondition for the robot pick task provides this information and will prevent this collision from happening.

The state transition function δ of an MPSG is represented by a graph where each node represents a state and each arc represents a transition. An arc label represents the controller event ($\sigma \in \Sigma$) which causes that transition to occur. Incoming and outgoing message events contain a suffix (_*sd*) which defines the message path. The first character *s* in this suffix represents the source of the message and the second character *d* represents the destination for the message. The valid characters are *s*, *w*, *e*, *r* and *d* representing a shop, workstation, equipment, robot and device, respectively. Controller tasks contain the prefix *t_*. For example, the event *t_grasp* represents the *grasp* task.

A set of MPSG for generic controllers in the proposed shop floor architecture has been developed (Smith, 1992). This set includes MPSG for generic **MP**, **MH**, **AS** and **MT** equipment, processing, storage and transport workstations and shops. These MPSGs can be used as a basis for specific implementations of controllers for these devices/control levels. Complete presentation of these MPSGs would be beyond the scope of this paper. However, the following sections present two of the example MPSGs: one for generic **MP** equipment and the other for generic processing workstations.

### 11.6.2  MP class equipment MPSG

Figure 11.8 shows an example MPSG for a generic **MP** class equipment level controller. Initially, the workstation controller assigns the part to the equipment controller (*assign_we*). In response to the assign message, the equipment controller performs the assign task (*t_assign*), which determines the specific location to which the part should be delivered. This action is performed through consultation with the planner/ scheduler. The equipment controller then responds to the workstation request with the addressable location at which the part is to be loaded. There are two types of part load tasks specified by the MPSG: synchronized loads and unsynchronized loads. The synchronized load (initiated by the *@loc_ew* message) is specified when coordination between the workstation **MH** device delivering the part and the **MP** device is required. The unsynchronized load (initiated by the *@loc_ns_ ew* message) is used when there is no coordination required.

Regardless of whether a synchronized or unsynchronized load is used, receipt of the *clear_ok* message from the workstation signifies that the workstation **MH** device has cleared the work volume of the processing device. At this point, the part is completely under the control of the equipment controller. From this point, the processing can be started (*start*) or the processing instructions can be sent to the device (*dnld*). Once the device has been started (state 8), the controller waits for the signal from the device signifying that processing has been completed (*finish*). (If necessary, the controller can poll the device status and, when the status changes, simulate receiving the *finish* message.) After receiving the *finish* message the part can undergo additional processing by downloading additional instructions and performing another *start* task. Once all processing has been completed, the equipment controller informs the workstation controller that the part is finished (*done*). When the workstation is ready to unload the part, it responds with a *remove* message which initiates the part unload sequence. The implementation of the physical model *mp_pick* task is similar to the implementation of the *mp_put* task. Formally, the generic **MP** class equipment MPSG $M_{MP}$, (Fig. 11.9) is defined as follows:

$$Q = \{0, 1, 2, \ldots, 17\}$$
$$q_0 = 0$$
$$F = \{17\}$$
$$\Sigma = \{\Sigma_I \cup \Sigma_O \cup \Sigma_T\}$$

where

$\Sigma_I$ = {*assign, grasp, clear_ok, finish, remove, release*}
$\Sigma_O$ = {*@loc, @loc_ns, grasp_ok, done, release_ok*}
$\Sigma_T$ = {*assign, grasp, start, stop, remove, release, delete, dnld*}

$A$ = {o_@loc( ), o_@loc_ns( ), o_grasp_ok( ), o_done( ),
   o_release_ok( ), t_assign( ), t_grasp( ), t_move( ), t_start( ),
   t_stop( ), t_remove( ), t_release( ), t_delete( ), t_dnld( )}

For the description of the MP physical preconditions, two additional predicates are defined. First, the predicate *running( )* returns true if the device is processing a part. Formally, *running( )* iff $\exists$ $x$ st $V_{part}(x) \wedge state(x) = 8$. Second, the predicate *blocked(loc)* returns true if access to the location *loc* is blocked while the machine is running. *blocked(loc)* is defined by the physical configuration of the device.

$$P = \{\rho_{@loc}, \rho_{@loc-ns}, \rho_{start}\}$$

where:

$\rho_{@loc}$ = $\neg OC(destination) \wedge (\neg running( ) \vee \neg blocked(destination))$
$\rho_{@loc-ns}$ = $\neg OC(destination) \wedge (\neg running( ) \vee \neg blocked(destination))$
$\rho_{start}$ = $CT(MP\_device, processing\_instructions) \wedge CWV (MP\_device)$.

The physical preconditions $\rho_{@loc}$ and $\rho_{@loc\_ns}$ prevent the **MP** controller from providing an occupied delivery location to the workstation controller. Without these preconditions, the equipment controller could specify an occupied delivery location to the workstation, creating the potential for a collision. Therefore, if the scheduler specifies that a part should be delivered to an occupied location, $\rho_{@loc}$ and $\rho_{@loc\_ns}$ will force the execution module to wait until the current part is removed before providing the delivery location to the workstation controller. These preconditions also verify that either the machine is not running, or the specified location is not **blocked**. This prevents the controller from giving an external device permission to enter the work volume of a running machine. The physical precondition $\rho_{start}$ insures that the processing instructions for the part are resident in the machine controller and that the work volume is clear (e.g. the MH device used for loading the part has cleared the **MP** device). For specific devices, other physical preconditions could be specified. For example, when grasping a part on a machine tool with a pneumatic chuck, checking the air pressure might be a physical precondition for the *grasp* task.

MP Class Equipment

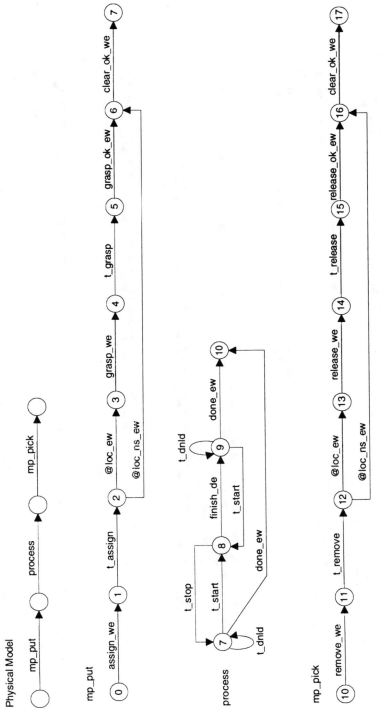

Fig. 11.8 MPSG for MP equipment.

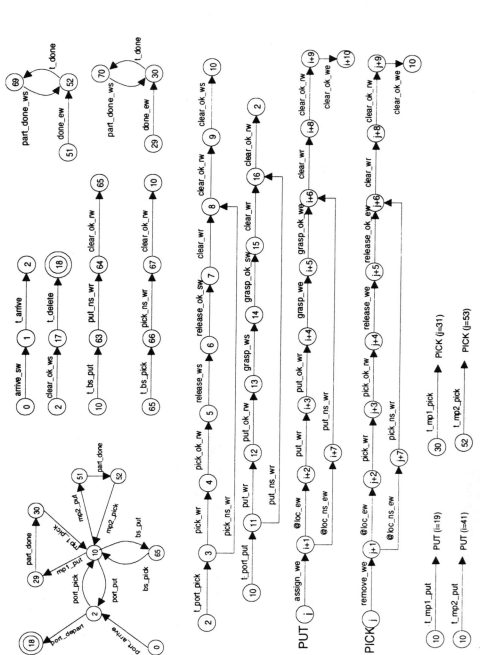

**Fig. 11.9** MPSG for a two-machine workstation.

### 11.6.3 Workstation MPSG

Figure 11.9 shows an MPSG for the processing workstation shown in Fig. 11.3. The graph in this figure has been partitioned so that it will be more readable. The upper left corner of the figure contains a modified version of the physical model where the node labels have been replaced with the corresponding state numbers from the MPSG. Each of the expanded system tasks are shown separately. These expanded arcs can be related to the physical model through the node labels on the ends of the expanded arcs. Furthermore, since the *pick* and *put* tasks for the two **MP** devices are identical, the corresponding expanded tasks are shown using the variables $i$ and $j$ in the node labels. Formally, the processing workstation MPSG $M_{WP}$, is defined as follows:

$$Q = \{0, 1, 2, \ldots, 17\}$$
$$q_0 = 0$$
$$F = \{18\}$$
$$\Sigma = \{\Sigma_I \cup \Sigma_O \cup \Sigma_T\}$$

where

$\Sigma_I = \{arrive, depart, pick\_ok, release\_ok, clear\_ok, put\_ok, grasp\_ok,$
$@loc, @loc\_ns, done, remove\}$

$\Sigma_O = \{pick, pick\_ns, release, clear, put, put\_ns, grasp, assign, clear\_ok,$
$remove\}$

$\Sigma_T = \{arrive, depart, port\_pick, port\_put, mp\_put, mp\_pick, bs\_put,$
$bs\_pick, delete, remove\}$

$A = \{o\_pick(\ ), o\_pick\_ns(\ ), o\_release(\ ), o\_clear(\ ), o\_put(\ ),$
$o\_put\_ns(\ ),$
$o\_grasp(\ ), o\_assign(\ ), o\_clear\_ok(\ ), o\_remove(\ ),$
$t\_arrive(\ ), t\_depart(\ ), t\_port\_pick(\ ), t\_port\_put(\ ),$
$t\_mp\_put(\ ),$
$t\_mp\_pick(\ ), t\_bs\_put(\ ), t\_bs\_pick(\ ), t\_delete(\ ),$
$t\_remove(\ )\}$

Note that, since the workstation controller does not deal directly with the physical equipment, the majority of these controller action functions are used to compile and send messages to their equipment level subordinates. Similarly, since the individual equipment level controllers are responsible for the distribution of parts within their local storage, the workstation physical preconditions do not typically verify specific location properties. Instead, they typically check the remaining capacity and operational status of the equipment within the workstation. For example, the physical configuration of the workstation specifies the capacity of each device and the workstation locations associated with each device. However, in a running system, the workstation has no knowledge about the specific location of individual parts assigned to each device. This information is stored in the equipment level system status. Instead, the workstation status includes only the number of parts assigned to the equipment and the equipment capacity. Therefore when evaluating the feasibility of an *mp_put* task, there is no workstation precondition specifying that the delivery location not be occupied ($\neg OC(dest)$). Instead, the workstation precondition is that there be remaining capacity on the device. The equipment controller preconditions verify that the delivery location is unoccupied and not blocked prior to sending the *@loc* or *@loc_ns* messages. Formally, the processing workstation physical preconditions are defined as follows:

$$P = \{\rho_{port-put}, \ \rho_{port-pick}, \ \rho_{mp\_put}, \ \rho_{mp\_pick}, \ \rho_{bs\_put}, \ \rho_{bs\_pick}\}$$

where:

$$\begin{aligned}
\rho_{port\_put} &= V_{pon}(destination) \wedge Capacity(destination) > 0\\
\rho_{port\_pick} &= V_{MH}(destination) \wedge Capacity(destination) > 0\\
\rho_{mp\_put} &= V_{MP}(destination) \wedge Capacity(destination) > 0\\
\rho_{mp\_pick} &= V_{MH}(destination) \wedge Capacity(destination) > 0\\
\rho_{bs\_put} &= V_{BS}(destination) \wedge \neg OC(destination)\\
\rho_{bs\_pick} &= V_{MH}(destination) \wedge Capacity(destination) > 0
\end{aligned}$$

Part removal requests from the shop controller are not shown in the graph in Fig. 9. In the normal case, the shop would signal that a part is to be removed once processing has been completed on the part. However, in the case where the shop wants to preempt the parts, the removal request could come at any time during processing. From an execution point of view, a removal request does not have any effect on the allowable controller tasks. It is up to the planning/scheduling module to either react to the request or to ignore it. When a *remove* message is received from the shop, the execution module sets a flag specifying that a part remove has been requested by the shop controller. Receipt of the *remove* message would cause a state transition from state *i* to state *i+j*, where *j* is some fixed number which is used to avoid confusion in the graph representation. The *t_remove* task simply sets the remove request flag in the part record and causes a transition back to state *i*. The planner/scheduler has access to this flag and can respond appropriately.

## 11.7　IMPLEMENTATION EXPERIENCE

Both of the shop floor control models presented in this chapter have been implemented at the Pennsylvania State University's Computer Integrated Manufacturing Laboratory (PSU CIM Lab). This laboratory includes a full-scale automated flexible manufacturing system (shown in Fig. 11.10). The equipment included in this laboratory is described in

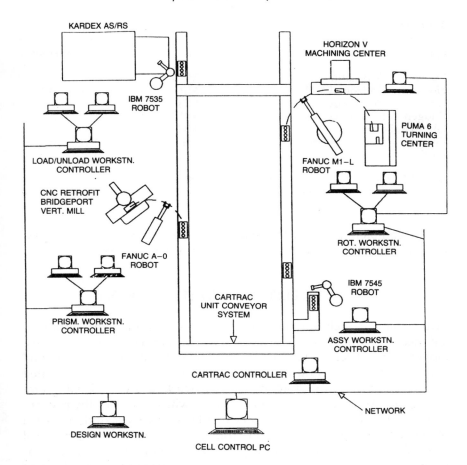

**Fig. 11.10** PSU CIM lab.

Table 11.2. The controllers for this system were developed to run on personal computers running the DOS operating system and were developed using the C++ language. Communication between controllers was performed using serial communications.

The first generation control system was developed using the context-free grammar approach outlined in section 11.5. Under this approach, a 'controller compiler' was developed so that the control software for implementing the system graph logic could be generated automatically from the input description of the equipment. However, the communication protocols required to implement the physical model tasks were 'hard coded' into the software generation environment. The message-based part state graph model was then developed to represent these processing protocols as described in section 11.6. The MPSG allows the

**Table 11.2** Equipment in the PSU CIM Lab

| equipment name | equipment class | workstation |
|---|---|---|
| Daewoo PUMA Lathe | MP | W1 |
| Horizon V Horizontal Mill | MP | |
| GE FANUC M1-L Robot | MH | |
| Bridgeport Retrofit NC Mill | MP | W2 |
| GE A0 Robot | MH | |
| Cartrac Conveyor System | MT | W3 |
| Kardex ASRS | AS | W4 |
| IBM 7535 Robot | MH | |
| IBM 7545 Robot | MP | W5 |

user to completely specify the sequence of events for each system task. A recognition table is then generated from these event sequences.

## 11.8  CONCLUSIONS AND FUTURE WORK

Two related models for the execution portion shop floor control in automated manufacturing systems have been presented. These two models have been developed to facilitate the implementation and maintenance of the control software necessary to run fully automated flexible manufacturing systems. Current work is involved with combining the two models into a single model which provides the primary benefits of both. Ideally, this model should accept a description of each of the processing protocols and from that, generate an operational controller completely automatically. Further work currently in progress is towards integration of execution modules with the other modules of the shop floor controller (scheduling, planning, error recovery, etc.)

## ACKNOWLEDGMENT

This work was partially supported through National Science Foundation (NSF) awards DDM-9009270 and DDM-9158042, both awarded to Dr. Sanjay Joshi, and by DARPA grant ARPA #8881 awarded to Dr. Richard Wysk, Dr. Sanjay Joshi and Dr. Dennis Pegden.

## REFERENCES

Albus, J., Barbera, A. and Nagel, N. (1981) *Theory and Practice of Hierarchical Control*, Proceedings of the 23rd IEEE Computer Society International Conference, Washington, DC, pp. 18–39.

*American Heritage Dictionary* (1976) Houghton Mifflin Company, Boston.

Basnet, C., Farrington, P.A., Pratt, D.B. *et al.* (1990) *Experiences in Developing an Object Oriented Modeling Environment for Manufanturing Systems*, Proceedings of the 1990 Winter Simulation Conference, pp. 94–105.

Biemans, F. and Blonk, P. (1986) On the Formal Specification and Verification of CIM Architectures using LOTOS, *Comp in Ind*, **7**, pp. 491–504.

Biemans, F. and Vissers, C.A. (1989) Reference Model for Manufacturing Planning and Control Systems, *Jnl. of Manuf Sys*, **8**(1), pp. 35–46.

Bjorke, Oyvind (1988) Towards a Manufacturing Systems Theory, *Rob & Comp Integ Manuf*, **4**(314), pp. 625–32.

Bravoco, R.R. and Kasper, G.M. (1985) *An Overview of How Information Flow and Architectures Support Computer-Integrated Manufacturing*, Proceedings of CIMCON '85, Anaheim, CA, April 15–18.

Bourne, D.A. (1991) The Cell Management Language, *Manufacturing Cells: Control, Programming, and Integration* (eds D.J. Williams and P. Rogers), Butterworth-Heinemann, Oxford.

Chaar, J.K., Teichroew, D. and Volz, R.A. (1990) Developing Manufacturing Control Software: A Survey and Critique, *Int Jnl. of Flex Manuf Sys*, **5**, pp. 53–88.

Cohen, P.A. (1985) *Trends in Flexible Manufacturing Systems*, Proceedings of CIMCOM'85, Anaheim, CA, Apr.

CSTB (Computer Science Technology Board) (1989) National Research Council, *Scaling Up: A Research Agenda for Software Engineering*, National Academy Press, Washington, DC.

Davis, W.J. and Jones, A.T. (1989) A Functional Approach to Designing Architectures for CIM, *IEEE Trans on Sys, Man, and Cybernetics*, **19**(2), Mar./Apr., pp. 164–89.

Dupont-Gatelmand, C. (1982) A Survey of Flexible Manufacturing Systems, *Jnl. of Manuf Sys*, **1**(1), pp. 1–16.

Gupta, D. and Buzacott, J.A. (1988) A Framework for Understanding Flexibility of Manufacturing Systems, *Jnl. of Manuf Sys*, **8**(2), pp. 89–97.

Jones, A.T. and McLean, C.R. (1986) A Proposed Hierarchical Control Architecture for Automated Manufacturing Systems, *Jnl. of Manuf Sys*, **5**(1), pp. 15–25.

Jones, A. and Saleh, A. (1989) *A Decentralized Control Architecture for Computer Integrated Manufacturing Systems*, IEEE Symposium on Intelligent Control, pp. 44–9.

Joshi, S.B., Wysk, R.A. and Jones, A. (1990) *A Scaleable Architecture for CIM Shop Floor Control*, Proceedings of CIMCON '90 (ed. A. Jones), National Institute of Standards and Technology, May, pp. 21–33.

Joshi, S.B., Mettala, E.G. and Wysk R.A. (1992) CIMGEN – Computer Aided Software Engineering Tool for Development of FMS Control Software, *IIE Transactions*, **24**(3), July, pp. 84–97.

Kimenia, J.G. and Gershwin, S.B. (1983) An Algorithm for the Computer Control of a Flexible Manufacturing System, *IIE Transactions*, **15**(4).

Koenig, D.T. (1990) *Computer Integrated Manufacturing: Theory and Practice*, Hemisphere Publishing Corporation, New York, NY.

Merchant, M.E. (1988) The Percepts and Sciences of Manufacturing, *Rob & Comp Integ Manuf*, **4**(1/2), pp. 1–6.

Mettala, E.G. (1989) *Automatic Generation of Control Software in Computer Integrated Manufacturing*, PhD thesis, Pennsylvania State University.

Naylor, A.W. and Maletz, M.C. (1986) The Manufacturing Game: A Formal Approach to Manufacturing Software, *IEEE Trans on Sys, Man, and Cybernetics*, **SMC-16**(3), May/June, pp. 321–34.

Naylor, A.W. and Volz, R.A. (1987) Design of Integrated Manufacturing Control Software, *IEEE Trans on Sys, Man, and Cybernetics*, **SMC-17**(6), Nov./ Dec., pp. 881–97.

O'Grady, P.J. and Menon, U. (1986) A Concise Review of Flexible Manufacturing Systems and FMS Literature, *Comp in Ind*, **7**, pp. 155–67.

Ranky, P.G. (1986) *Computer Integrated Manufacturing*, Prentice Hall International, Englewood Cliffs, NJ.

Ranky, P.G. (1990) *Flexible Manufacturing Cells and Systems in CIM*, CIMware Limited, Guilford, Surrey England.

Scharbach, P. (1984) *Formal Methods and the Specification and Design of Computer Integrated Manufacturing Systems*, Proceedings of the International Conference on the Development of Flexible Automation Systems, July.

Senehi, M.K., Barkmeyer, E., Luce, M. *et al.* (1991) Manufacturing Systems Integration Initial Architecture Document, National Institute of Standards and Technology, *NIST Interagency Report NISTIR 4682*, Gaithersburg, MD.

Smith, J.S. (1990) Development of a Hierarchical Control Model for A Flexible Manufacturing System, Master's thesis, The Pennsylvania State University.

Smith, J.S., Hoberecht, W.C. and Joshi, S.B. (1992) A Shop Floor Control Architecture for Computer Integrated Manufacturing, IMSE Working Paper Series, Pennsylvania State University, University Park, PA.

Van Eijk, P.H.J., Vissers, C.A. and Diaz, M. (eds) (1989) *The Formal Description Technique LOTOS, Results of the ESPRIT/SEDOS Project*, North-Holland, New York.

Whitney, C.K. (1985) Control Principles in Flexible Manufacturing, *Jnl. of Manuf Sys*, **4**(2), pp. 157–66.

# CHAPTER 12

# Object-oriented design of flexible manufacturing systems

*Gabriele Elia and Giuseppe Menga*

## 12.1 INTRODUCTION

Modeling the problem space, promoting reuse and shortening the development timecycle are particularly significant aspects in the design of CIM systems, where heterogeneous elements have to be integrated in a complex control architecture.

In the field of software engineering the OO approach has proved to be a winner with respect to those crucial aspects. Originally these goals were pursued by focusing the attention on system components, and component reuse was achieved through libraries of reusable classes organized in hierarchies of inheritance: what is called **a frame-work of classes**. The Smalltalk language (Goldberg and Robson, 1983) offers a classical example of basic data structures from which other, more application-oriented libraries can easily be envisioned (e.g. referring to manufacturing, libraries with classes such as shop, cell, workstation, or order, product, lot, piece and so on).

However, more recently the emphasis in this field has been shifting from component to architecture reuse through the realization that any design is deeply rooted in the patterns of relationships between its elements. The noted architect Christopher Alexander (1979) observes that what characterizes a design more than its elements are the patterns of relationships linking them. These 'design' patterns are the natural consequence of well defined 'need' patterns, and they can be found repeatedly in designer implementations, indicating the style of that designer. By linking 'need' and 'design' patterns, Alexander introduces the concept of '**pattern**'.

In an OO context, elements are objects and 'patterns' indicate clusters of cooperating objects linked by certain fixed relationships.

A pattern can be expressed by the triplet:

- statement of one aspect of the problem (the 'need' pattern);
- assessment of the related design alternatives;
- rules leading to the solution in terms of a 'design' pattern.

The complete structured set of patterns needed to approach a class of problems is indicated as a **pattern language**. Each pattern describes a specific design aspect of the problem; however, on the whole, the resulting pattern language becomes a design method which covers the development process from analysis to implementation.

The idea of pattern adopted here is the one postulated by Alexander (1977, 1979) and transferred to software engineering by Peter Coad (1992) and Ralph Johnson (1992).

The problem addressed by this paper is the **design** of the control architecture of flexible manufacturing systems made up of layers of **concurrent** modules. These have to be implemented on a **distributed** computer architecture exploiting an evolutionary development process that transforms a prototype of the **logical design** into the **physical design** (Booch and Vilot, 1991).

A certain number of design 'patterns' structured in a pattern language have been studied with a view to approaching CIM applications and are adopted here to solve the problem. This has been conceived also as a contribution to the long-standing project being conducted by international standardization bodies such as the National Institute of Standards and Technology (NIST), ISO and CEI, in their attempt to define a standard **reference model** (McLean *et al.*, 1983) and standard design rules for CIM.

The paper is organized as follows: section 12.2 describes a typical FMS, section 12.3 describes the patterns and contains the design of the FMS. An introduction to OO programming, explanation of our graphical notation and description of the generally accepted OO software lifecycle are contained in the appendix.

## 12.2   THE FLEXIBLE MANUFACTURING SYSTEM (FMS)

We consider a typical FMS for machining mechanical cylindrical pieces. This is one cell belonging to a larger production shop. The cell is composed of a set of numerically controlled lathes, an automatic inventory system for raw and finished pieces and an automatic guided vehicle (AGV) system. Integration is achieved through a computer network which encompasses shop and cell computers, along with CNC and PLC for the machine, inventory and AGV controls.

### 12.2.1   The structure of the FMS

The production is represented by small lots of different pieces. Each piece undergoes a sequence of turning operations on the same or on different lathes, according to the type of operation.

The lathes are classified according to the subset of operations they can execute. An admissible operation is assigned to the lathe by re-tooling the machine, and it is not dismounted until the completion of the lot of pieces. The operations which are simultaneously present on the same machine are limited by the size of the tool buffer.

An automatic transportation system handles pieces inside the cell.

### 12.2.2   The logic of control

The shop defines a plan of production of lots of different piece types. According to the plan and with a real time scheduler, it prepares and assigns to the cell loading conditions which are constrained by the availability of materials and number of operations on the lathes. A loading condition is defined by the configuration of operations mounted on the lathes and by the set of lots simultaneously active inside the cell.

The cell, with a real time dispatcher, assigns pieces to the lathes as soon as they become idle, and issues requests of missions to the transportation system, monitoring the shop floor.

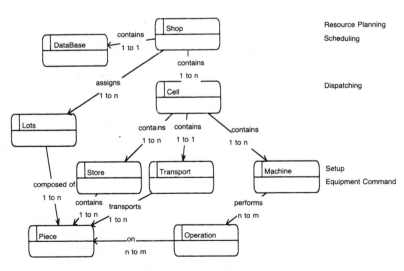

**Fig. 12.1**   Analysis of the FMS.

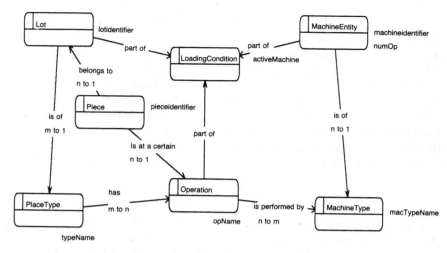

**Fig. 12.2**   Entity-relationship diagram of the FMS.

### 12.2.3   The conceptual model of the FMS

An entity-relationship (E-R) (Chen, 1976) model is adopted for the system analysis, with a formalism described in appendix 12.A.3.

Figure 12.1 describes the structure of the FMS. The organization of the elements expresses the hierarchy of controls, which are formalized in section 12.3.1.

Figure 12.2 details the database of the FMS. A piece belongs to a unique lot; a lot is composed of many pieces. Every lot is composed of one type of piece, but more than one lot may have the same piece type. A piece type defines a sequence of operations on the pieces of a lot; different types may share the same operations. The machine has a type which defines the set of operations which the machine may perform. Machines of different types may perform the same operation. At one point in its production, a piece is at a certain operation. A ternary relationship between lot, operation and machine states a loading condition (i.e. assignment of operations of lots to the machines).

## 12.3   CIM PATTERNS

It is a common observation that designers are always driven to making choices through patterns. They synthesize the designer experience gained in the solution of a certain class of problems. However, patterns can be shared to become collective knowledge in a certain environment.

To formalize the intuitive notion of patterns and make them transferable, they have to be structured internally and in their relationships with each other.

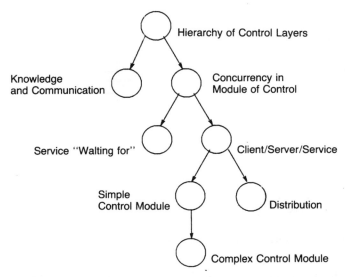

**Fig. 12.3** Pattern language for the design of the FMS.

The template we use to describe a pattern basically follows the proposal of Johnson (1992), and has the following fields:

- the pattern name;
- the statement of the problem which the pattern addresses;
- the technical description of the pattern in terms of the framework of classes, supporting the architecture;
- the exemplification of the pattern in the case study;
- the definition and design rules of the pattern.

Each pattern describes some of the elements of the framework of classes, which offers the building block of the architecture. However, as Alexander suggests, patterns from general to specific can be organized in a graph where each pattern leads to a series of others (Fig. 12.3). Then the whole set of patterns, together with their structuring principles, becomes a language and a design method which accompanies the design from the conception to the final implementation.

This section describes the patterns which we have identified for designing an FMS and shows them in the example given.

### 12.3.1  Pattern 1: hierarchy of control layers

An FMS organizes its functionalities within an architecture of hierarchical layers of controls. Each controller performs services, requests services from other controllers or signals events so as to inform the system of its state.

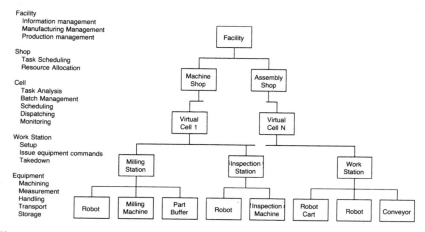

**Fig. 12.4**  The hierarchical reference model.

This pattern explains how the hierarchical structure of FMS is translated, in OO design, exploiting the relationships of **use** and **inclusion**.

### The CIM reference model

Control modules in CIM are arranged in a hierarchy of controls. Each controller takes commands from only one higher-level system, but it may direct several others to the next lower level. Long-range tasks enter the system at the highest level and are broken down into subtasks, which are to be executed at that level or transferred as commands to the next lower level (McLean *et al.*, 1983). Following a study made by the US National Bureau of Standards, a reference model for CIM is now widely accepted, where the following layers are identified (Fig. 12.4):

- **Facility** – the complete factory with the strategic function of satisfying order requests for final products.
- **Shop** – a shop-floor module to schedule production and manage resources.
- **Cell** – a production unit for real-time dispatching and routing of lots of pieces.
- **Workstation** – an elemental machine, transport or buffer for operations on a single piece of the lot.
- **Equipment** – a component with an independent control unit.

### OO design of manufacturing systems

Each control layer represents control modules offering a similar family of functionalities.

A control module is characterized by the presence of:

- autonomous decision-making capabilities;
- a series of available services (planning orders, scheduling lots, operating on pieces (McLean *et al.*, 1993));
- a pool of controlled resources.

In an OO approach modules are objects, and the program results from the dynamic exchange of messages between them. A message is sent to and object to request a service. This is called USE relationship. In fact, the **'separation of concerns'** principle (Parnas *et al.*, 1985) suggests that one object may delegate part of its functionality by requesting of other objects, in turn, the execution of specific operations.

Objects can be shared between several others. However the **'information hiding'** principle (Parnas *et al.*, 1985) postulates that, when possible, resources should be hidden. This is the case of one object being used by only one other, then it is conveniently encapsulated as a private resource of that object class. This is called relationship of inclusion (INC).

Use and inclusion are the two structuring principles of an OO architecture, giving rise to hierarchies of seniority of use and hierarchies of parenthood (parent/child) of inclusion (CRI-CSI, 1987).

From the previous observations, the hierarchy between modules of control resulting from encapsulation bears a particular significance as it mirrors the logical structure of levels of the control. Strictly speaking in fact, in a hierarchical architecture a control module at a higher layer should always encapsulate all lower-layer controls of its concern. However there may be exceptions such as a centralized database server shared among several modules.

Figure 12.5 translates in use/inclusion design decisions the shop analysis of Fig. 12.1.

The FMS cell, fms, is included (encapsulated) by the upper-level module in the reference model, the shop; the database, db, whose structure is depicted in Fig. 12.2, is included by the shop but used also by the cell for dispatching pieces.

**All entities of a FMS systems are objects structured according to the relationships of inclusion and use.**

### 12.3.2   Pattern 2: knowledge and communication between control modules

A FMS by its very nature follows a bottom-up development process in which preexisting control modules must be integrated. As an integrator of lower-level controllers, it must have full knowledge of them. Lower-level controllers such as machines, transports and buffers, taken from different vendors, have been designed without any prior knowledge of the environment in which they will operate.

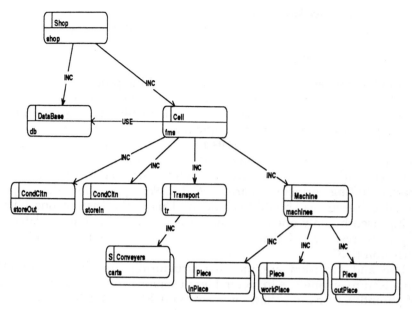

**Fig. 12.5** OO design of the shop-cell-database modules.

Moreover modules communicate, and knowledge of each other has important implications on the way the communication between modules is established.

This pattern gives guidelines for dealing with the communication between FMS entities.

Control modules in communicating with each other must be imperative, as in the case of a command issued by one module to the other, or must be reactive, as in the case of events monitoring (Fig. 12.6).

These two situations correspond to the following two mechanisms:

- direct communication, i.e. client/server (C/S), where the caller asks for services from the receiver;
- indirect communication, in the style of dependencies (Goldberg and Robson, 1983) or callbacks (Young, 1989), here called broadcaster/ listeners (B/L), where the **broadcaster** signals events tagged by a symbolic name, and joined to data. One or more **listeners** can monitor the same object while they wait for a certain event to occur and, when they receive it, they execute in turn one of their operations.

The C/S mechanism is involved when an object invokes another object method and is deeply rooted in OO programming. The B/L

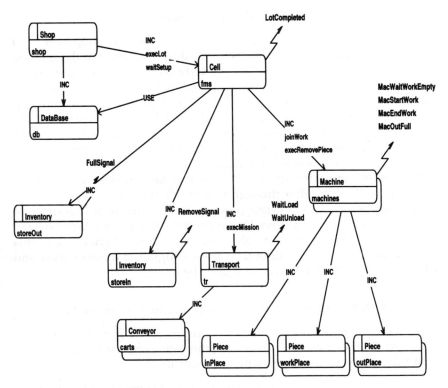

**Fig. 12.6**   The communication mechanisms.

mechanism is created by giving all objects of the framework the capability of broadcasting and listening to events. In particular, listening to an event is a state into which an object enters, or from which it exits, at a certain instant of time by a request issued by itself, by the broadcaster or by a third party.

There is no redundancy in these two mechanisms as their choice has important consequences on the reusability of the design.

The need to integrate lower-level controllers designed without any prior knowledge of the environment in which they will operate, suggests organizing the controls in such a way that:

• two controls at the same level should not need to communicate directly;
• when a higher-level control addresses a command to a lower level, this is done explicitly;
• when a lower-level control answers a higher-level one, e.g. by sending monitoring information, this is done indirectly.

*The FMS example: communication*

The example proposed has three levels of control: shop, cell and workstation. The shop asks the cell to produce lots through the method *execLot(Lot\* aLot)*. The cell notifies the termination of a lot through the event *LotCompleted*. The cell (Fig. 12.6) offers its services direct to machines (*joinWork( )*, *execRemovePiece( )* and so on) and to transport (*execMission( )*), but the evolution of the cell has to be driven by the events broadcast by its components, e.g. *MacOutFull* when a piece enters the output buffer of a machine. For a complete description of the behavior of the cell, see section 12.3.7.

**All control modules of a CIM system are entities which communicate with each other through C/S or B/L mechanisms.**

**In a bottom-up CIM design process, reusability in a hierarchy of layers of control will improve if two controls at the same layer are not allowed to communicate directly. Communication between controls takes place using a C/S mechanism from a higher to a lower layer and a B/L mechanism from lower to higher layer.**

### 12.3.3   Pattern 3: concurrency in modules of controls

The essence of a control module is represented by: the services offered to the outside, carried on through concurrent activities, and the series of resources which it controls.

This pattern describes the intrinsic properties of concurrency of modules of control. This leads to the distinction between active and passive objects.

A control module offers services in concurrency controlling a pool of shared resources. Concurrency is defined as the situation where multiple activities, composed of sequential operations, cooperate to achieve a common goal (Briot, 1992).

It is necessary to distinguish between concurrency and parallelism (Briot, 1992). Concurrency is the logical simultaneity of actions which are potentially or virtually contemporary. When this is achieved allowing processes to share one (or more) processor, it is called multiprogramming (Andrews and Schneider, 1983).

Parallelism is the physical simultaneity of actions, on computers with *ad hoc* hardware, e.g. a multi-CPU system, or a group of workstations connected through a local area network. In this case, the term distributed programming is used.

Along with physical entities, logical entities, like a concurrent activity, can be seen as objects. The independent threads of execution are the abstraction of concurrency, i.e. separate execution environments with their own data for the object operations. The operations of one object in fact, can be executed by the thread offered by the client which

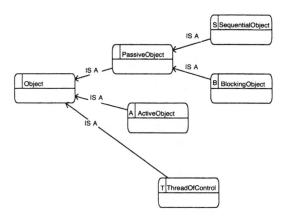

**Fig. 12.7** Objects and threads.

issued the request such as the subfunction call in sequential programming, or by an internal thread owned by the object itself, independent from the client thread.

The two possible ownership relations between objects and threads of execution lead to a classification becoming classical, where those entities which rely on the client thread for their activities are called **passive objects** (sequential or blocking) and the ones which possess their own private threads, **active objects** (Fig. 12.7).

**Passive objects**
Passive objects, which depend on a client thread for the execution of their operations, are further divided in two groups:

1. *Sequential objects* whose semantics are defined in the presence of a unique client thread, and for this reason they have to be bounded to a unique thread of control, or protected by a blocking object, as seen below.
2. *Blocking objects* whose semantics are preserved in the presence of multiple client threads. Their function is to block an incoming client until some (external) condition is satisfied. They must be used as shared resources so as to allow synchronization and the exchange of data between threads.

**Active objects**
They are the backbone of any concurrent application; in fact control modules are in themselves active objects in that they possess their (usually multiple) threads of control.

Two specifications in particular characterize any active object (CRI-CISI, 1987):

1. *The interface* – the synchronization between client and server during the exchange of a request of service is specified by the interface of the object.
2. *The implementation* – the dynamic behavior of the services offered is expressed in the body of the object.

### Threads of control

*ThreadOfControl* is the class whose instances are independent threads. This is neither a passive nor an active object; its instances simply lend their structure for the concurrent execution of the services of the objects which possess or borrow them.

### The FMS example: modules of control

A simple analysis shows that in order to carry on its activities, shared resources of a control module can be:

- entities where service operations store data; e.g. the shop contains a database with the list of lots which have to be produced;
- entities where services enter in conflict to share limited resources, like buffers of pieces in the machines or a pool of carts for transportation in the transport system;
- other control modules which perform (sub)services developed over time; e.g. the cell manages (and uses) machines, transport system, stores.

In the FMS example, we identify the following modules (Fig. 12.8):

- the shop;
- the cell;
- the machine;
- the transport systems.

They will be active objects because they have autonomous decision-making capabilities and have to manage different activities simultaneously. In fact the FMS manages different lots at the same time, (e.g. the machine works a piece while waiting for some other pieces to be removed from its output buffer; the transport handles as many different missions as the number of available carts.

On the contrary, stores for semi-worked and finished pieces (named *storeIn* and *storeOut*) are modeled as blocking objects, with the behavior of a bounded buffer. Blocking objects in the cell are described in section 12.3.4.

The database is an example of a sequential object; as several control modules interact with it, it must be protected by encapsulating it with a blocking object.

**Properties of modules of control of CIM applications, which involve synchronization of resources and concurrency management, suggest a**

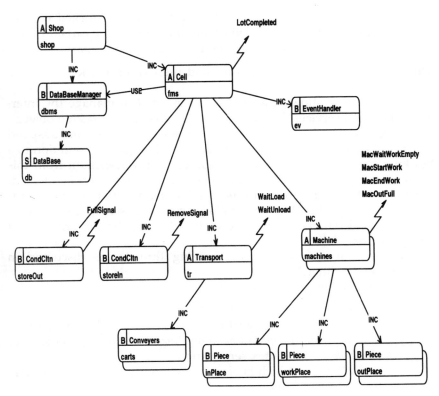

**Fig. 12.8** FMS high level design.

classification of objects, based on their relationship with threads of control, in passive and active. Passive objects are then divided in sequential and blocking.

Active objects have to be chosen for entities which have autonomous decision-making capabilities and have to manage different services at the same time. Blocking objects are used for synchronization and concurrency control of activities. Sequential objects offer the data structures of the application.

### 12.3.4 Pattern 4: Services 'Waiting for'

In their evolution over time, services, offered by a control module, may need to wait for a condition to occur or for data being transferred from/to a shared resource, before performing one of their operations in concurrency with others.

The former case is exemplified by the service of producing a piece by a cell controller while waiting for the event 'WaitOutFull' from a machine; the latter is exemplified by the mission of transport provided by the conveyors on the shop-floor being blocked while waiting for the first cart available from the pool of free carts, or for the first slot available in a machine buffer to unload a part.

This pattern explains the properties and the role of 'blocking' passive objects.

Such situations show how entities are needed to manage the inter-action between services. These entities are, in fact, the blocking objects.

In an OO world the only way a service can enter a state and suspend its flow of execution is to issue a request of waiting for a condition to a blocking object. The flow will be resumed with a transition from this waiting state whenever the condition on the blocking object is satisfied.

The *Condition* is the abstract class used to define any blocking object, from which specific implementations such as semaphores, event handlers and shared queues can be derived by inheritance or encapsulation.

Conditions as a way of synchronizing threads are discussed in Andrews and Schneider, 1983.

Two typical cases are of interest here.

1. *A queue of services waiting for a condition*
   The state of waiting for a synchronization condition is modeled:
   (i) inheriting from the class *Condition*. Representative examples are:
       - *Semaphore*, the mutex semaphore;
       - *EventHandler*, which is blocking while waiting for one of a set of possible events.
   (ii) encapsulating a *Condition* object in a *ThreadOfControl* to interlock two services and allow a message between them to be ex-changed directly without the use of external shared resources e.g. in a synchronous request of service.
2. *A queue of messages waiting for a service*
   A state of waiting to store or retrieve data is modeled by the class *CondCltn* (Condition Collection), i.e. a collection which encapsulates a *Condition*, and redefines the collection *add*( ) and *remove*( ) methods. The *CondCltn* has the property of a bounded buffer (Andrews and Schneider, 1983): a service which tries to add a piece is suspended if the buffer is full and restarts when a place is available; meanwhile, a service is suspended when it tries a remove operation and the buffer is empty; it will be restarted as soon as an object is available in the buffer.

*The FMS example: blocking objects*

In the FMS example, we find different blocking objects:

- the *storeIn* and *storeOut* buffers are *CondCltn*, owned by the *Cell* class. They act as a bounded buffer: the cell service automatically stops if it tries to remove a piece when the *storeIn* is empty or to add a new piece when the *storeOut* is full.

- *inPlace*, *workPlace* and *outPlace* are, similarly, *CondCltn* acting as blocking objects of the *Machine* services.
- *Cell* has an *EventHandler* used to block its services waiting for the events from transport, machines and stores (section 12.3.7). Here the cell controller uses an eventhandler to monitor its resources (e.g. *MacOutFull* or *WaitNotFull*) and to take the proper consequent actions.

**Blocking objects have to be introduced to model shared resources of the control module any time interaction between its concurrent services is needed. They represent queues of services waiting for a message or queues of messages waiting for a service and they are modeled by specializing or encapsulating a *Condition*.**

### 12.3.5 Pattern 5: the Client/Server/Service

A common representation of the control modules at the different levels of the hierarchy is strongly needed for standardization purposes, to enforce reusability and facilitate the task of distributing the application.

This pattern abstracts the properties of active objects and states the correspondence between modules of control and active objects.

Such a pattern plays a pivotal role for the whole architecture. In particular, the model of a control module which is needed in a CIM application must have the capability of encapsulating a resource definition and the services that manipulate them, as well as of representing a system as a structure of hierarchically-layered virtual machines (Parnas *et al.*, 1985). The Client/Server/Service model proposed in this pattern has been inspired by SR (Andrews *et al.*, 1988).

Two classes are needed to support it: the *Service* and the *Server* (Fig. 12.9).

**The *Service***
The *Service* extends the *ThreadOfControl* to become an extended finite state machine which describes the dynamics of the application; it always belongs to a unique server owner, maintains an internal symbolic state value and a series of sequential objects for data, broadcasts events with the incoming state name any time a state transition occurs. It rarely needs to be redefined as it delegates the execution of its dynamic behavior to a private method of the server owner.

**The *Server***
The *Server* is the abstract class which defines active objects and hence control modules; it has to be redefined in order to model concrete

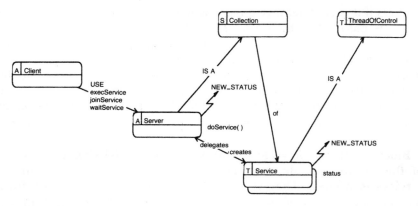

**Fig. 12.9**    Client/server/service.

classes. It is a *Collection* of service objects: each service which it contains executes the server operations in concurrency, while sharing the same set of resources. A server service can, in turn, use another (sub)server and in this case becomes a client for that server.

The *Server* offers for all active objects a standard interface which is represented by three public methods: *execService( )*, *joinService( )*, and *waitService( )*, respectively, asynchronous, future (asynchronous with the client being able to accept results later) and synchronous requests. Asynchronous request is used when the client is not interested in recovering any data from the service asked of the server. Future request is needed when the client may have some interaction in the future with the service. Synchronous requests represent the case when the client block itself is waiting for the response.

When invoked, each one of these methods starts a new service in the server.

The server has a private behavior method *doService( )* that contains the service operations. Such a method is deferred, as its implementation will depend on the concrete objects derived from the server. Moreover the server relays its service events externally so that its state can be monitored by other objects.

The services possessed by the server can be generated at the server's instantiation with ancillary roles, and then they have the same life span as the server, or they can be dynamically created at each service request from a client and will exist up to the conclusion of the specific service.

When a new Service is created, it is added to the list of services of the *Server* and it invokes the execution of the behavior method, *doService( )*, of its server.

Any control module of the hierarchy is represented as a server which renders concurrent services (i.e. sequences of operations which it develops over time) to its clients. Each new service in a server is an independent thread executing in concurrency with the other previously requested.

### 12.3.6 Pattern 6: simple control module

In a CIM system there is a variety of control modules managing different pools of shared resources and offering their specific service.

This pattern explains how concrete modules of control can be programmed. It also shows how simple control modules can be obtained as a subtype of a server.

When a control module offering a unique kind of service has to be designed, the following problems must be considered.

1. The resources needed by the control must be identified; the decision must be taken as to whether they must be encapsulated, or whether they should be referred to as external use depending on information hiding concepts.
2. The service behavior must be specified in terms of a finite state machine extended by sequential objects which may eventually be needed bounded to the service.

   A graphical notation to specify the service behavior will add expressiveness to the formalism and by animating the prototype, debugging will be easier. Figure 12.11 shows an example of the specification of a service behavior.

Because of the concurrent environment, the resources shared by the services of an active object can only be blocking objects (*Condition* or *CondCltn*) or other active objects (*Server*). The absence of sequential objects inside the server should be noted. In fact, they should be encapsulated into blocking objects or bounded to the services.

**Control modules are implemented as a subtype of the server, encapsulating or referring to shared resources, which must be blocking objects or other servers, and specifying and redefining the *doService( )* method of the server in terms of an extended finite state machine (Fig. 12.11).**

### 12.3.7 Pattern 7: complex modules of control

Certain control modules offer more than one kind of service to their clients.

This pattern explains how modules of control capable of rendering services of different kinds can be programmed.

Though the basic server supports a unique kind of service, derived classes can easily extend this model to offer multiple kinds of services as well (see the class cell of Fig. 12.10).

When a new active object is conceived the different kinds of services it offers must be identified. Then for each kind of service:

- the new class is specified inheriting from the server. The derived class is not a subtype because a different interface has to be defined for it;
- at most, three interface methods for each behavior method are defined. These methods delegate on a one-to-one basis the implementation of the synchronization specification to the standard server interface;
- a new behavior method for each kind of service is added, to which the interface refers through a proper *doService( )* method, for each one of the different dynamic behaviors.

*The FMS example: the classes* Cell *and* Machine

The class *Cell* inherits from *Server*; its member data (Figs 12.8 and 12.10), are the transport system *tr*, instance of class *Transport*, the

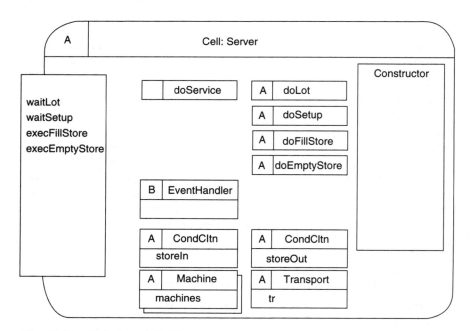

**Fig. 12.10**  Cell class definition.

collection of machines *machines*, stores of semi-worked and finished pieces (*storeIn* and *storeOut*, of class *CondCltn*). *Transport* and *Machine* classes inherit from *Server* as well.

The service of routing of pieces between machines and stores is expressed by the method *doLot( )*. Its behavior is depicted in Fig. 12.11. The main points of the service are:

- the first piece, the lot, is removed from the *storeIn*;
- a transport request is issued;
- the service enters into the state *WaitEvent*;
- the service accepts the events.
  - *WaitLoad*, issued by the transport when it has reached the source from which to remove a piece; the piece is removed from the source (*machine* or *storeIn*) and added to the transport which starts its mission;
  - *WaitUnLoad*, issued by the transport when it reaches a 'destination' (machine input place or *storeOut*); in the former case, the *joinWork( )* method of the machine is called;
  - *MacOutFull*, issued by the machine when a finished piece is stored in the output buffer; the cell answers by issueing a new transport service;
  - *removeSignal*, issued by the *storeIn*; this is used for starting the work of a new piece of the lot (until pieces are in the *storeIn*);
  - *FullSignal*, issued by the *storeOut* when the lot is completed and the *doLot( )* service terminates;
- the *doLot( )* service terminates.

Let us consider now the definition of the class *Machine*. Two services are identified: working of a piece (which involves also the loading of it) and removing a piece from the machine.

We first define the data structure of the class *Machine*: from the requirement specification, it comes that a machine owns an input buffer, a working place and an output buffer. The dimensions of these buffers are parameters that depend on the machine type. They are instances of class *CondCltn*, named *inPlace*, *workPlace* and *outPlace*. The semantic of *CondCltn* is the following: they represent bounded buffers where objects (in this cases, pieces) are stored; an *add(aPiece)* operation blocks the executing thread if there isn't any slot available, and the thread will restart automatically when a slot becomes empty; a *remove( )* operation is dual with *add( )*, that is, it is blocking when the collection is empty.

Two different private behavior methods, *doWork( )* (Fig. 12.12) and *doRemovePiece( )*, are activated by two public methods, *joinWork( )* and *waitRemovePiece( )*, defined for *Machine*. Removing a piece is requested with a synchronous interaction, working with a deferred synchronous

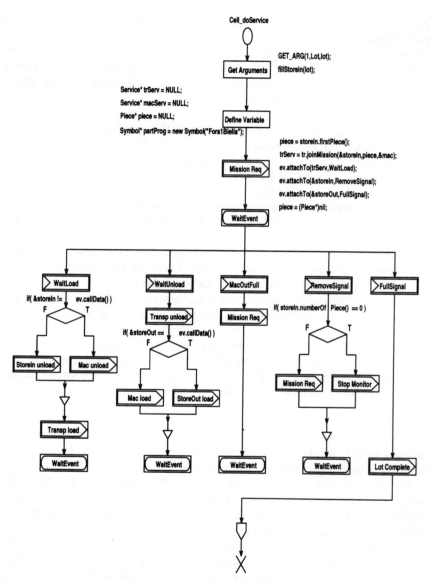

**Fig. 12.11**   Cell: doLot( ) method.

one (the cell doesn't stop but receives a reference to the new service started in the machine, and can re-synchronize in the future).

We briefly describe the *doWork( )* method depicted in Fig. 12.12: the piece, parameter of the service, is first put in the *inPlace* buffer, then in

Machine_do Work

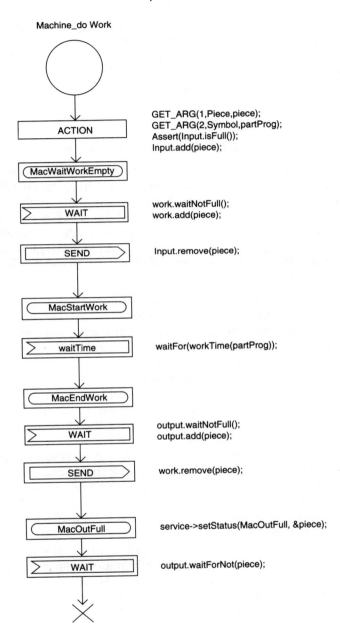

**Fig. 12.12** Machine: doWork( ) behavior description.

the *workPlace*, and finally in the *outPlace*. The logic is that pieces flow as soon as places are available in the buffers.

**Multi service control modules inherit from a *Server* without being its subtype and are designed by defining a private behavior method and the corresponding interface required for each kind of service offered.**

### 12.3.8    Pattern 8: distribution

A complex CIM application requires the prototyping and simulation of the different physical elements which have to be integrated before deriving the physical implementation. Active objects, representing control modules and used for simulation on a single computer, have to be transformed in or reside on remote computers or peripheral devices interconnected through a common communication network, and define the physical distributed architecture.

This pattern explains how the Client/Server/Service pattern is well suited for distributed implementations.

The pattern which follows considers the transition from logical to physical design and implementation in a distributed environment. The units of distribution are active objects.

A prototype of the logical design and the physical implementation are two distinct aspects of the same problem, often simultaneously present during the development, as when prototypes are used to emulate parts of the physical process during the phase of installation of certain critical software systems.

With regards to distribution, a message to a remote (active) object can be sent directly through the communication network or, to maintain continuity with the logical design and the problem description network, can be hidden in standard library classes. We adopted this second choice, introducing the classes classes *Stub* (Gibbson, 1987) and *RemoteContext* (Fig. 12.13).

The *Stub* is the polymorphic view of a server; it encapsulates the communication protocol of a physically remote object in the network and acts as the communication interface to the remote reality. In a program, a *Stub* substitutes the model (the prototype) of a physically remote active object with a new implementation (e.g. *ShopStub*, *CellStub*, *WorkstationStub*).

The *RemoteContext* is a large-grain object (Chin and Chanson, 1991) which offers the context environment for the remote active objects of the real implementation.

*Stub* and *RemoteContext* are a matched pair belonging to the framework, each encapsulating an instance of class *Network*. The two former

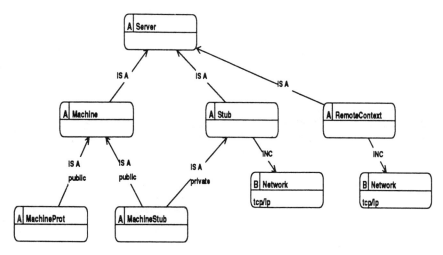

**Fig. 12.13** Stub and RemoteContext.

classes are rarely modified, while the *Network* must have different versions for each different communication network.

**Evolution to the physical design and implementation is achieved by maintaining two representations of the same entities (the prototype and the reality) and transforming the prototype into the physical reality by exploiting polymorphism.**

**The evolution from the prototype to a distributed implementation is a process which makes use of the *Stub* to create polymorphic representations of remote objects to substitute to the model, and of *RemoteContext* which offers the context environment to physical processes and encapsulates the remote objects. These two classes are offered by the environment, and they delegate to a *Network* object, which wraps the communication network, and should be provided for each different installation, the implementation of the low-level protocols.**

## 12.4  CONCLUSION

We have presented a novel OO approach for the design of FMS. The originality of our approach is the proposal of a high-level language based on patterns, called pattern language, which drives the design from the analysis to the physical implementation, and stresses reusability at the highest level of the architecture. This approach has been motivated by the need to offer simple rules and add clarity to the design of complex highly-layered control structures, as in the case of

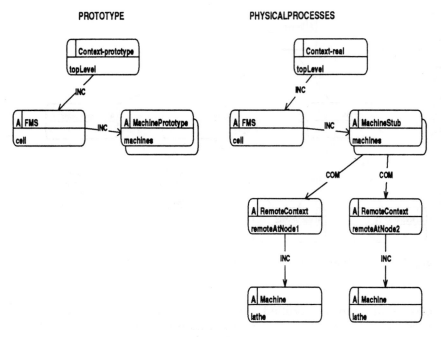

**Fig. 12.14** Prototype and physical distribution.

computer integrated manufacturing. At the same time it has eased the burden of the designers when they have to make difficult architectural choices regarding modularity and reusability.

Finally, one contribution which this paper makes – the embedding of the design into a well-defined framework – is to allow dynamic behavior to be expressed as a property of the whole architecture, instead of being considered as an attribute in isolation of each single object.

The framework of classes, the design method induced by the patterns and the notation described here are the basis of a CASE environment for the C++ programming language called G++, described in Menga and Lo Russo (1990); Menga *et al.* (1991); Menga *et al.* (1993) and Gaved *et al.* (1993).

APPENDIX A.12: OO PROGRAMMING AND THE OO
SOFTWARE LIFE-CYCLE

### 12.A.1  OO programming

The basic concepts of an object orientation are class, instance, method, message and inheritance.

A **class** contains the definition of the properties (data attributes and functions to manipulate them) common to one or more objects, which are called **instances** of the class; the instances store only their own copy of the data and refer to the class for the properties which are common to the class. The class is the extension of the concept of type in procedural programming languages to OO approaches.

A **message** is the external request for a service offered by an object to the outside world; a **method** is the implementation of a service, i.e. a function which manipulates data local to the object or which contains calls for messages to other objects. The activation of a method in an object is obtained by sending a message to it: as the message response is local to an object, the same name can generate in objects of different classes different behaviors; this is at the basis of a functionality called **polymorphism**. **Information hiding** is the encapsulation of the implementation aspect of an object inside its class capsule. In certain cases, in order to avoid modifying a class just to change the implementation of a method, an object can ask an included object to execute one of its functionalities. This mechanism is called **delegation**. The execution of an OO program is therefore obtained by the sequential or concurrent exchange of messages between objects and the resulting execution of operations.

In general, OO languages and thus C++, distinguish between **public** and **private** methods. The first are accessible from the outside and represent the interface of the object, the second are only for the class itself.

**Inheritance** is at the basis of the relationship of generalization/specialization through which classes can be ordered in a hierarchy, where a subclass inherits attributes from one (or more) superclass. The concepts of **abstract class**, **deferred method** and **overriding a method** are used to exploit inheritance. An **abstract class** defines an abstract data type and its distinctive properties, but does not yet define an implementation. This will be made by its subclasses. A **deferred method** is one whose signature has been defined but whose implementation has been deferred to derived classes. A subclass can **override**, i.e. redefine, a superclass method with another one with the same signature. Overriding of attributes is not allowed in C++ language.

### 12.A.2 OO analysis and design

The development process of a software system proceeds through a sequence of steps from problem specification to delivering the final software. Traditionally they are: analysis, logical design (prototyping) and physical design. Analysis describes the problem. During the analysis, entities of the problem domain are identified and their classes are represented with their attributes, linked to each other by relationships of generalization, association and aggregation.

The logical design expresses architectural choices, and is properly the mapping of the analysis onto the patterns of a chosen framework of classes. Analysis relationships between entities are transformed in the design into precise architectural informations. An OO architecture is characterized by three structuring principles where objects and their classes are organized in graphs of inheritance:

1. a subclass derives from a superclass;
2. use – an object sends a request for a service to another object;
3. inclusion – a class encapsulates instances of other classes as its internal data.

Usually the logical design can be prototyped, resulting in an executable model of the reality (simulation).

The physical design, finally, maps the logical design into the physical hardware architecture, which in a CIM application is constituted by computers and device controllers connected through local area networks.

### 12.A.3   The notation

In the text, we have followed the convention of indicating classes and instances with an italic font; class names start with an capital letter.

The object icon is represented by the rectangle with rounded vertices in Fig. 12.15. It indicates either a class or an instance according to the presence of one or both identifiers. The character on the left hand label indicates whether the object is active, sequential, blocking or persistent. Pure data attributes, private methods and names of events issued are represented on one side of the icon.

A collection of objects of the same class is indicated by an icon with a double contour. The class name is one of the included objects.

Relationships are indicated by arcs with a label. Their semantics are defined either in the problem space (e.g. association, aggregation, generalization) as for the analysis, or in the logical design space such as inheritance (ISA), use (USE), data encapsulation (unshared use, INC), or in the physical design space, such as the interconnection through a communication network (COM) of physical modules. In the logical and physical design space USE and INC relationships can be accompanied by the names of the message requests flowing between the entities.

Figure 12.10 shows the icon used for the detailed design of a class; at the top there are the class name and the superclass; on the left side an 'export box' shows the public methods and members; inside the box, there are the private and protected members; active *behavior* methods are labeled with a capital A.

The specification of *behavior* methods is expressed through state transitions diagrams. The formalism (Figs 12.11 and 12.12) is an

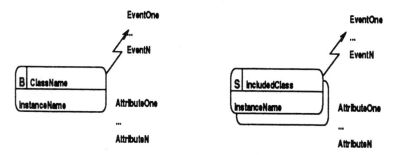

**Fig. 12.15** Object and collection icons.

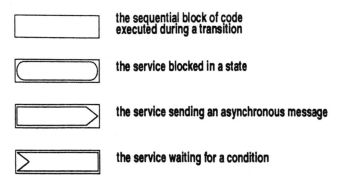

**Fig. 12.16** Symbols used for the SDL macro extension.

OO extension of SDL (the specification and description language standardized by CCITT (CCITT, 1986). Its symbols are described in Fig. 12.16. They are properly macros extending the original SDL notation.

## REFERENCES

Alexander, C. (1977) *A Pattern Language: Towns, Buildings, Construction*, Oxford University Press, New York.

Alexander, C. (1979) *The Timeless way of Building*, Oxford University Press, New York.

Andrews, G.R. and Schneider, F.B. (1983) Concepts and notations for concurrent programming, *Computing Surveys*, **15**(1), pp. 3–41, March.

Andrews, G.R., Olsson, R.A., Coffin, M. *et al.* (1988) An overview of the SR Language and Implementation, *ACM Transactions on Programming Languages and Systems*, **10**(1), pp. 51–86.

Booch, G. and Vilot, M. (1991) *Object Oriented Design with Application*, Benjamin/Cummings, Reading MA.

Briot, J.P. (1992) A Tutorial on Object Oriented Concurrent programming, ECOOP '92 Tutorial.

CCIT (1986) *Specification and Description Language – Recommendation Z. 100.*

Chen, P. (1976) The Entity-Relationship Model: towards a unified view of data, *ACM Trans. on Db Sys*, **I**, pp. 9–36.

Chin, R.S. and Chanson, S.T. (1991) Distributed Object-based Programming System, *ACM Computer Surveys*, **23**, pp. 91–124, Mar.

Coad, P. (1992) Object Oriented Patterns, *Communications of the ACM*, Sept.

CRI-CISI Ingenierie Matra (1987) 2 Rue Jules Vedrine, Toulouse, France, *HOOD Manual.*

Gamma, E., Helm, R., Johnson, R. *et al.* (1993) *Design Patterns: Abstraction and Reuse of Object Oriented Design*, in ECOOP '93 conference proceedings, Kaiserlautern, Germany, April.

Gaved, A., Elia, G. and Menga, G. (1993) Object Oriented Modeling of a Robotized Manufacturing Cell in *The International Handbook on Robot Simulation Systems* (ed. Dieter Wloka), John Wiley & Sons, London.

Gibbson, P.B. (1987) A Stub Generator for Multilanguage RPC in Heterogeneus Environments, *IEEE Trans on Sftwre Eng*, **13**(1), pp. 77–87, Jan.

Goldberg, A. and Robson, D. (1983) *Smalltalk80: the Language and its Implementation*, Addison-Wesley, Reading MA.

Johnson, R. (1992) *Documenting Frameworks using Patterns*, in OOPSLA '92 conference proceedings, Vancouver, British Columbia, Canada, Oct.

Krasner, G.E. and Pope, S.T. (1988) A Cookbook for Using the Model-View-Controller User Interface Paradigm in Smalltalk-80, *Jnl. of Obj-Oriented Prog.*, Aug./Sept.

Lippman, S. (1991) *C++ Primer*, 2nd edn. Addison-Wesley, Reading MA.

McLean, C., Mitchell, M. and Barkmeyer, E. (1983) A Computer Architecture for Small-batch Manufacturing, *IEEE Spectrum*, **20**(5), pp. 59–64.

Menga, G. and Lo Russo, G. (1990) *G++: an Environment for Object Oriented Analysis and Prototyping*, TOOLS '90, June 26–9.

Menga, G., Picchiottino, Gallo, P. *et al.* (1991) Framework for Object Oriented Design and Prototyping of Manufacturing Systems, *Jnl. of Obj. Oriented Prog – Focus on Analysis and Design*, pp. 41–53.

Menga, G., Elia, G. and Mancin, M. (1993) An Environment for Object Oriented Design and Prototyping of Manufacturing Systems in *Intelligent Manufacturing: Programming Environments for CIM* (eds W. Gruver and G. Boudreaux), Springer Verlag, London.

Parnas, D.L., Clements, P.C. and Weiss, D.M. (1985) The Modular Structure of Complex Systems, *IEEE Trans on Sftwre Eng* , **SE-11**(3), pp. 259–66, March.

Rumbaugh, J. (1991) *Object Oriented Modeling and Design*, Prentice Hall.

Young, D.A. (1989) *The X Window System Programming and Application with Xt.*, Prentice Hall, Englewood Cliffs, NJ.

# CHAPTER 13

# Efficient and dependable manufacturing – a software perspective

*J.K. Chaar, Richard A. Volz and Edward S. Davidson*

## 13.1 INTRODUCTION

Designing efficient and dependable manufacturing systems has always been a goal of modern manufacturing. To be efficient, such systems must achieve high throughput while meeting all deadlines. To be dependable, such systems must have the ability to detect and correct faults, such as hardware failures and their effects and the divergence of actual operations from planned operations.

In recent years, the goal of efficient and dependable manufacturing has generated considerable enthusiasm for computer-integrated manufacturing, and has led to significant advances in providing adequate hardware and software support (e.g. computer networks and communication protocols) for integrating the various devices of such systems. Moreover, computer-integrated manufacturing revealed the complexity of designing and implementing the **control software** of these systems.

Traditionally, control software has been assigned the task of monitoring manufacturing devices by coordinating their operations and by handling the faults that may occur while these operations are being performed. Planning the sequence of operations of such devices and scheduling the execution of this sequence have traditionally been carried out independently, prior to developing their control software.

Efficient and dependable manufacturing can be attained by integrating these planning, scheduling and monitoring activities. The integration of planning and scheduling can improve the utilization of manufacturing devices and the processing of manufacturing jobs. Hence, more efficient manufacturing is achieved. On the other hand, the integration of scheduling and monitoring can help resume the original schedule following fault detection and recovery. The integration of planning, scheduling and monitoring is essential when alternative recovery procedures are needed following fault detection. Hence, more dependable manufacturing is achieved.

In order to achieve efficient and dependable manufacturing, the steps of a methodology for developing the control software of efficient and dependable manufacturing systems and a set of software tools that can assist in applying these various steps are proposed. The methodology improves both the flexibility and the modularity of resulting software systems by structuring them as a set of modular software components. In addition to this standard software engineering concept, the methodology includes additional concepts and approaches. In particular,

1. manufacturing control software is designed and implemented as a set of components that can encapsulate the hardware devices of the system and are referred to as **software hardware components**;
2. **generic** algorithms use the **formal models** of both manufacturing systems and process plans in planning the sequence of operations of the jobs to be processed on these systems prior to scheduling and dispatching these jobs;
3. **simulation** facilitates the testing process of control software by emulating the concurrency and delays in the operation of manufacturing systems, and
4. the control software is implemented in a **general-purpose high-level distributed language** that can support the development of software components, and can interface with the variety of device-specific languages of the controllers of manufacturing devices.

The software tools are integrated in an environment to be used by software engineers to develop both a good plan and an optimal schedule for processing a batch of manufacturing jobs. The plan details the sequence of operations to be executed, and the schedule determines a feasible time for initiating the execution of such plan. Several iterations through planning, scheduling and monitoring are required to implement a good overall control strategy. The control software of the underlying manufacturing system is then used to perform and monitor the operations of the plans and schedules of such jobs.

A synopsis of some current approaches to developing the control software of manufacturing systems is presented next. The limitations of these approaches are highlighted and an attempt to improve their practices is proposed. Hence, the steps of the proposed methodology are outlined and the software tools that can aid in performing some of these steps are presented. These sections are preceded by defining the following technical terms:

**Manufacturing operation.** An operation that performs machining, assembly tasks, or inspection on the component materials.

**Transport operation.** An operation that transfers component materials between the locations of a manufacturing system.

**Process plan.** A process plan is an acyclic graph. The nodes of this graph specify the manufacturing operations to be performed by the devices of a system while the arcs of this graph denote the precedence relationships between these operations.

**Process plan step.** A process plan step is a node of the process plan graph that details the complete sequence of tasks that must be executed by a manufacturing device when performing an assigned operation of the process plan.

**Plan.** A plan expands the acyclic graph of an associated process plan by capturing the sequence of transport operations that precede the manufacturing operations of the process plan.

**Formal model.** The formal model of a system consists of i) a set of entities, ii) a set of predicates in first-order logic, and iii) a set of rules that are composed of pre-conditions and post-conditions. The set of all the values of these predicates at any instant of time is the state of the system. The semantics of each operation of the system are captured in a rule that specifies the conditions required for executing this operation, the duration of the operation and its effect, when executed, on the state of the system.

**Job.** A job is the execution of a plan that transforms specified materials into a product.

**Cyclic schedule.** A schedule determines the time intervals between successive job initiations in a system. The schedule is cyclic whenever these time intervals form a **periodic** sequence.

## 13.2 RELATED WORK

Traditionally, the sequence of operations executed by the control software of manufacturing systems has been captured by ladder logic diagrams (Bollinger and Duffie, 1988; Elsenbrown, 1988; Gayman, 1988). These diagrams specify the input and output procedures of the programmable logic controller (PLC) that drives and cycles the operations of a manufacturing device. All possible combinations of PLC inputs must be captured by the ladder diagram. Consequently, these diagrams grow so complex that locating the cause when a problem is detected becomes extremely difficult.

The use of commercial general-purpose computers to control modern

**Table 13.1**   Current approaches to developing manufacturing control software

| Acronym | specification language |
| --- | --- |
| System 90<br>(Kompass, 1987; Vasilash, 1987; Fisher, 1989) | Zone logic |
| ISIS/OPIS<br>(Bourne and Fox, 1984; Le Pape and Smith, 1987;<br>Ow *et al.*, 1988; Smith, 1988) | rule-based |
| KBSc<br>(Shaw and Whinston, 1985a, 1985b, 1986; Shaw, 1981) | STRIPS<br>(Fikes *et al.*,<br>1971a, 1971b,<br>1972; Sacerdoti,<br>1974) |
| AISPE<br>(Bruno *et al.* 1984, 1985a, 1985b, 1986a, 1986b, 1986c,<br>1987a, 1987b) | PROT nets |
| FLEXIS DESIGNER<br>(Dove, 1988; Thomas and McLean, 1988; Wilczynski,<br>1988) | GRAFCETs<br>(AFCET, 1977) |
| o-r Petri nets<br>(Beck, 1985; Beck and Krogh, 1986; Ekberg and Krogh,<br>1987; Krogh *et al.*, 1988; Willson and Krogh, 1990) | o-r Petri nets |
| Colored Petri nets<br>(Kamath and Viswanadham, 1986; Martinez *et al.*, 1987;<br>Viswanadham and Narahari, 1987; Camurri and<br>Franchi, 1990) | Colored<br>Petri nets |
| SA Colored Petri nets<br>(Corbeel *et al.*, 1985; Gentina and Corbeel, 1987; Jafari,<br>1990) | SA Colored<br>(Petri nets) |
| Extended Petri nets<br>(Crockett *et al.*, 1987; Ahuja and Valavani, 1988;<br>Kasturia *et al.*, 1988; Zhou *et al.*, 1989, 1990) | Extended<br>Petri nets |
| SECOIA<br>(Atabakhche *et al.*, 1986; Sahraoui *et al.*, 1986, 1987) | High-level<br>Petri nets/rules |
| $G^{++}$<br>(Menga *et al.*, 1991) | $C^{++}$ Class<br>libraries |

manufacturing systems is increasing, and has led to a search for alternatives to ladder logic diagrams. The adoption of the Petri net-based function charts as a standard for describing the control logic of manufacturing devices is considered a major step in this direction (International Electrotechnical Commission 1984).

In the research domain, real-time control software has been modeled as a variant of Petri nets (Table 13.1). The choice of Petri nets for modeling the structural and behavioral properties of manufacturing control software permits the use of the theoretical properties of these nets in software verification. This choice, however, offers only limited support for planning and scheduling the operations of efficient and dependable manufacturing systems. The creation of a Petri net model for a manufacturing system is equivalent to planning the sequence of operations of a job. Repeatedly firing the transitions of this Petri net is equivalent to scheduling the execution of the operations of a batch of this job whenever performing these operations becomes feasible. Monitoring is performed by checking, prior to executing the operations of the Petri net, for the feasibility of executing these operations; a fault is detected whenever this feasibility cannot be established.

An integrated approach to planning, scheduling and operation monitoring of efficient and dependable manufacturing systems is essential to the development of the control software of these systems. Consequently, the life-cycle model of control software must include these activities, and a methodology for developing this software must provide a coherent approach to performing these activities. The planning process generates a sequence of operations, labeled a **process plan**, for processing a job in a manufacturing system (Allen, 1987), and is viewed as a special case of the general planning problem of artificial intelligence (Georgeff, 1987; McDermott, 1987). Here, capturing the physical and logical constraints of the system and the duration of the operations performed by the devices of this system is vital to the generation of feasible and/or optimal plans for this system (Descotte and Latombe, 1985; Jones *et al.*, 1987; Nau, 1987; Wolter, 1989).

Two approaches to scheduling the processing of a set of jobs in a manufacturing system have been developed. The first approach assumes the presence of a plan for each scheduled job, and generates, a priori, a complete schedule for processing these jobs by using some Operations Research techniques (Karmakar and Schrage, 1985; Maxwell and Singh, 1986; Roundy, 1988). The second approach extends the planning approach to cover the operations of all the jobs of the set (Bispo and Sentieiro 1988; Liu and Labetoulle, 1988; Subramanyam and Askin, 1986; Van Brussel *et al.*, 1990). Advocates of this approach try to perform real-time control of the factory floor by basing their scheduling decisions on the current state of the factory. Each of the two approaches has its merits and its shortcomings. The first approach advocates an off-line scheduling strategy and has a higher probability of generating optimal schedules. However, the execution of this pre-planned schedule may be affected adversely by the occurrence of a fault in the system. The on-line scheduling strategy of the second approach is more adaptive and solves the problem of invalidating the pre-planned sequence of operations of a schedule due to the occurrence of a fault. However, achieving optimality when applying this second approach is unlikely, given the complexity of the planning problem and the short interval of time for determining the set of operations that will be executed next by the manufacturing system.

The use of expert systems in performing the monitoring activity of real-time manufacturing control software is a common strategy of all current approaches to this problem (Meyer, 1986; Chintamaneni, *et al.*, 1988; Huber and Buenz, 1988; Hasegawa *et al.*, 1990; Russel, 1990). The normal course of operation of a manufacturing system is captured as a variant of Petri nets. The control sequence of this net is then followed until a fault is detected in the system. The detection of this fault triggers a search of the knowledge base of an expert system for the appropriate set of corrective measures for recovering from this fault. The efficiency of this search can be increased by classifying the set of probable faults that may occur in the manufacturing system (Chang *et al.*, 1989, 1990; Takata and Sata, 1986; Viswanadham and Johnson, 1988).

Table 13.1 presents a summary of some current methodologies for developing real-time manufacturing control software. The steps of these methodologies cover the phases of the software life-cycle, and can be implemented with the help of a set of software tools that form the software environments of these methodologies. Tools for verifying the functional and timing requirements of manufacturing control software, prototyping, automatically generating subsets of the implementation code of this software, and simulating the performance of manufacturing systems have been integrated in many of these software environments, e.g. Flexis Designer, Advanced Industrial Software Production Environment (AISPE), Specification and Emulation for Computer Integrated Automation (SECOIA), and $G^{++}$. A complete discussion of both the steps of these methodologies and the functionality of their software environments can be found in Chaar *et al.* (1990, 1993a, 1993b).

## 13.3  SOFTWARE/HARDWARE COMPONENTS

A software component is an object-oriented construct characterized by an interface and a body (internals) (Booch, 1986). The interface specifies the services provided by the object or set of objects encapsulated by the component; an object may be a manufacturing device, a database, etc. These services can be used to build other components that are called the **users** of the original component. The services, in turn, are implemented in the body of the component, and, following the standard software engineering principle of information hiding (Parnas, 1972), the implementation details **must** be inaccessible to the user(s) of the component. Furthermore, both interface and body must be compilable separately from each other and from other software components of the system.

A software/hardware component (Naylor and Volz, 1987; Ben Hadj-Alouane *et al.*, 1990, 1991) generalizes the concept of a software component by allowing the internals of the component to encapsulate hardware (Volz *et al.*, 1984a, 1984b); i.e. these internals are interfaced to and drive hardware devices. The hardware devices of a manufacturing system include computer numerically controlled (CNC) machines, robots, programmable logic controllers (PLCs) and material handling and storage/retrieval systems.

Both software and hardware interfaces are independently accessible by the users of the component, and provide distinct yet complementary views of their component. The services that are offered by its software interface can be used to control the devices of this component through its hardware interface.

The state of a software/hardware component and the semantics of the operations specified in the software interface of this component are captured, at the logical level, by the formal model of this component (Naylor and Maletz, 1986). The logical level is concerned with logical conditions and transformations of logical conditions. A typical condition might be stated as 'The robot is at the machining center'. Actions such as moving the robot from one place to the other cause a change in condition. This logical level is captured by a simple first-order logic rule-based model. The state of the system captured by the model consists of a set of predicates (relations) and values for the named variables.

Rules, analogous to those of the familiar artificial intelligence paradigm (Fikes *et al.*, 1971a, 1971b, 1972), are used to describe the semantics of the operations of a component. A rule consists of a set of pre-conditions followed by a set of post-conditions, and is associated with a logical variable and a duration (Naylor *et al.*, 1986, 1987). Whenever the pre-conditions of a rule are satisfied by the current state of the system, and the rule is selected by enabling its logical variable, the post-conditions of this rule become satisfied by the state of the system after the specified duration (time delay). Satisfying the post-conditions of a rule indicates the occurrence of an **event** of the system that affects the current state of this system.

Assembling a set of software/hardware components requires the design and implementation of a new software/hardware component; the assembled component (Fig. 13.1). The software interface of this component specifies the set of services performed, individually and/or collectively, by its constituent components. Moreover, the services collectively performed by the constituents of the component are implemented, together with their required communication software, in the internals of the assembled component. These internals encapsulate also the hardware devices of the constituent components together with any additional manufacturing and computer hardware that interconnects these devices. Finally, the hardware interface of the assembled component is composed of the union of the hardware interfaces of its

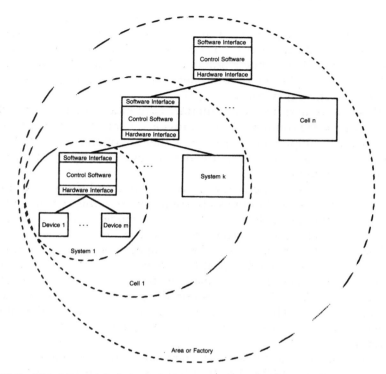

**Fig. 13.1**  The hierarchical control structure of a manufacturing system.

constituent components. A software/hardware component that is not assembled out of any other components is referred to as a **simple component**.

There are several ways of creating the formal model of an assembled software/hardware component from the formal models of its constituent components. The most useful are based on a combination of **unions** and **views** (Naylor and Maletz, 1986). The union of two or more models is obtained by taking the union of the states and the union of the sets of changes of these models. Because an assembled component has, in general, additional functions with respect to the union of the functions performed by its constituent components, or hides some of the functions performed by these constituent components, unions are augmented with views that show part of the state and parts of some changes. Hence, the union is obviously a view of the assembled component.

The process plans of a manufacturing system are segregated from its hierarchical control structure by creating a separate formal model for these process plans. The predicates of this formal model define the acyclic graph of a process plan by specifying the order in which the process plan steps are executed, and the devices that are assigned to these process plan steps. On the other hand, the rules of this formal model are used to set the goals that must be attained when deriving the sequence of operations of the plan of a job. A goal is achieved by deriving the sequence of transport operations that must precede the execution of the process plan step associated with the goal and executing the sequence of manufacturing operations associated with this step.

## 13.4 METHODOLOGY AND SOFTWARE ENVIRONMENT

This section discusses the steps of the proposed methodology and the software tools of its associated software environment. The methodology extends the traditional software lifecycle to cover planning, scheduling and operation monitoring in efficient and dependable manufacturing. Figure 13.2 presents an overview of the major activities of this methodology together with the list of steps that are performed during such activities. The sequences of such steps are detailed in Fig. 13.3.

To assist the software engineer in designing, implementing and testing the software/hardware components of a manufacturing system, software tools such as editors, database management systems, compilers and assemblers are required. Moreover, additional software tools are needed to support integrating the planning, scheduling and monitoring activities (steps 5 through 11) of the methodology. These tools are grouped in the **manufacturing software environment** (Fig. 13.4). This environment consists of a **planner system** (Fig. 13.5), a **cyclic scheduler system** (Fig. 13.6), and an **operation monitor system** (Fig. 13.7), and can be integrated with traditional software tools. Databases L1–L6, processes P1–P11, switches S1–S15, and steps 1–12 are numbered consistently throughout all figures of this chapter. Significant parts of the manufacturing software environment have been implemented and tested on a real manufacturing cell.

### 13.4.1 Requirements specification

In the methodology, the functional and performance requirements of a manufacturing system are specified first. These requirements are written by the manufacturing engineers, and specify the processing environment, the required software functions, performance constraints on the software, exception handling, implementation priorities and the acceptance criteria for the software. They are based on the selected range of products to be manufactured by the system, the acquired manufacturing devices and computers of this system and the layout of the factory floor. Natural language statements, supplemented by the ap-

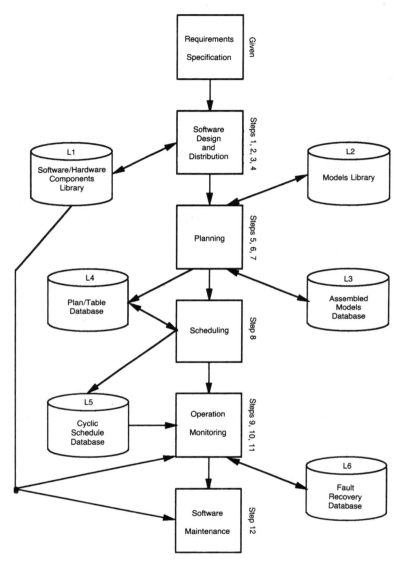

**Fig. 13.2** An overview of the methodology.

propriate set of tables, figures and charts are used to express these requirements. The steps of the requirements specification activity are labeled 'Given' in Figs 13.2 and 13.3 because it is assumed that the functional and performance requirements of the manufacturing system have already been specified and verified.

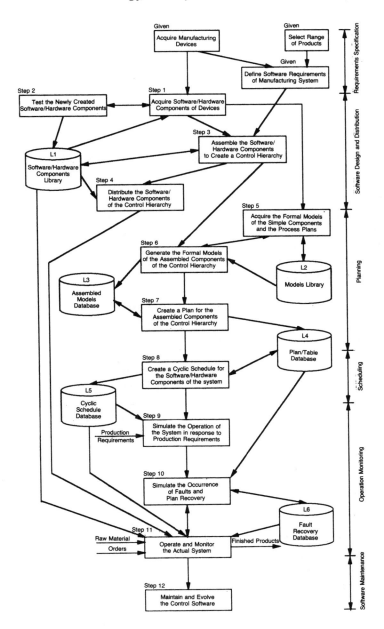

**Fig. 13.3** The steps of the methodology.

## 13.4.2  Software design and distribution

The requirements specification step of the methodology is followed by designing, implementing and distributing the software/hardware components of the manufacturing system. This software design and distribution activity is performed in steps 1, 2, 3 and 4 of the methodology (Fig. 13.3). The software/hardware components of all the individual devices of the system are acquired by the software engineer in step 1 of the methodology; some of these components may be designed and implemented by this software engineer, while others may be retrieved from the *software/hardware components library* (L1) of the manufacturing system. When a device is encapsulated in a new software/hardware component, the operations of this device are simulated and tested, online, in step 2. In step 3 of the methodology, the set of acquired software/hardware components is assembled into higher level components to form a control hierarchy for this manufacturing system. The components of this hierarchy are assigned in step 4 of the methodology to the appropriate nodes of the underlying distributed network of the system.

Starting with the functional requirements of each device of the manufacturing system, the software engineer identifies the operations that must be performed by this device together with their required parameters and their output data (step 1). Moreover, any timing constraints on the execution of these operations is deduced from the performance requirements of this device. These operations are included in the software interface of a software/hardware component. The *software/hardware components library* (L1) is queried for the component that can perform this set of operations. This query may be guided by the software engineer or be fully automated. If no component with the required specifications is found in the library, the software engineer must implement and test the operations of the new software/hardware component.

In order to implement the operations of a software/hardware component that encapsulates a manufacturing device, an interface to this device is built by the software engineer. When numerically controlled machines are encapsulated in software/hardware components, the software engineer has to interface with complex special-purpose controllers that are often programmed in APT-like languages (Koren, 1983). Likewise, enclosing robots in software/hardware components requires programming their controllers, typically implemented on general-purpose computers, in VAL (Unimation, 1982), Karel, or AML-like languages (IBM, 1981; Bonner and Shin, 1982), or, even more likely, teaching these robots. Similarly, enclosing programmable logic controllers (PLCs) in software/hardware components requires programming their special-purpose hardware, typically in ladder logic (Bollinger

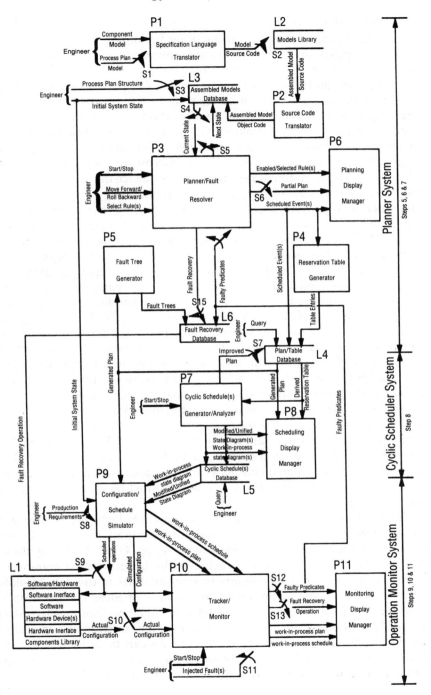

**Fig. 13.4** The manufacturing software environment dataflow.

and Duffie, 1988; Elsenbrown, 1988; Gayman, 1988). Material handling and storage/retrieval systems on the other hand, are usually controlled by general-purpose computers using languages such as assembly, Fortran, Pascal, or C (Naylor and Volz, 1987).

The operations of the software/hardware component are translated into appropriate sequences of commands to the encapsulated device. The sequence of commands of an operation may be fixed, or may depend on the operating environment of the device. When fixed, the sequence of commands of an operation is included by the software engineer in the internals of the software/hardware component. Otherwise, the software engineer generates this sequence interactively during the planning activity of the proposed methodology (Fig. 13.2) and implements a generic algorithm for interpreting the operations of the generated sequence. The above sequences of commands are then executed in order to test, on-line, the operations of the device (step 2).

In step 3 of the proposed methodology, the software engineer assembles the software/hardware components that encapsulate manufacturing devices into new software/hardware components. These assembled components encapsulate cells, areas and factories and create a hierarchical control structure for the system (Fig. 13.1). To design an assembled component, the software engineer specifies first the list of constituents of this component. The operations that can be performed by the assembled component, together with their input parameters and their output data, are then added to the software interface of this component. Some of these operations may be performed by a single device, while others may require the collaboration of many devices. When an operation of the assembled component consists of a fixed sequence of operations that should be performed by the constituent components, this operation is implemented, in procedural form, in the internals of the assembled component. In contrast, when this sequence of operations is not fixed, these operations must be planned in a later step of the methodology.

The software/hardware components of a manufacturing system are assigned next to the nodes of the computer network that control their devices or their constituent components (step 4). Distributing these components over the network is achieved by inserting their communication software, and resolving any incompatibilities in the data formats of their nodes. If the communication software of these components is inserted manually, the distributed software must be thoroughly tested to verify whether the inserted communication software is operating properly. However, communication software can be automatically inserted when the software/hardware components of the system are implemented in a distributed language (Goldsack *et al.*, 1992).

### 13.4.3 Planning

To process a job in a manufacturing system, a sequence of operations must be performed on specified materials and components by the devices of the system. This sequence of operations is captured in a plan. The output product of a job may be a single part, a pallet of parts, or a mechanical assembly of parts. Furthermore, the operations of the job can be divided into **manufacturing operations** and **transport operations**. Planning the sequence of these operations, and scheduling the execution of this sequence must precede any processing of this job. During processing, the operations must be monitored in order to detect any hardware failures that may occur during the execution of a scheduled operation of the job.

Traditionally, the sequence of manufacturing operations of a job is described a priori in the form of a **process plan**. Hence, a computer-aided process planning (CAPP) system can be used to plan the details of the manufacturing operations specified in the process plan steps of this process plan. However, process plans do not specify any transport operations that may precede the execution of their manufacturing operations. In the proposed methodology, the software engineer extends the process plan of a job into a plan by simultaneously specifying the transport operations and the details the manufacturing operations of this process plan. The plan is used subsequently to schedule the processing of batches of this job.

To derive the plan of a job, the planner system (Fig. 13.5) is presented with a **formal model** of the manufacturing system in which this job will be processed. This formal model is assembled from the formal models of the software/hardware components of the hierarchical control structure of this system, and is expressed in a component-oriented rule-based language (Chaar and Volz, 1993c). The models of these components form a rule-based system that captures, at the logical level, the state of the manufacturing system and the semantics of the operations that can be performed by this system. In steps 5 and 6 of the proposed methodology (Fig. 13.3), these models are either implemented by the software engineer or retrieved from the *models library* (L2).

Step 7 of the methodology is then used to plan the sequence of operations of a job that can be processed on this system. The planner system (Fig. 13.5) can assist the software engineer in performing these steps. This system consists of the *specification language translator* (P1), the *source code translator* (P2), the *planner/fault resolver* (P3), the *reservation table generator* (P4), the *fault tree generator* (P5), and the *planning display manager* (P6). When planning the sequence of operations of a job, the planner system operates with switches S4, S5, and S6 closed and switches S14 and S15 open. Furthermore, switches S1 and S3 are closed whenever a process plan is available, and S2 is closed when the formal model of a component is to be added to the *models library* (L2).

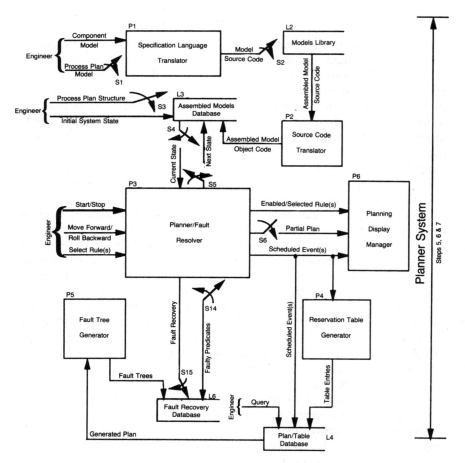

**Fig. 13.5**   The planner system dataflow.

Steps 2, 3, and 5 of the proposed methodology populate the *software/ hardware components library* (L1) and the *models library* (L2) of a manufacturing system with an equal number of components and formal models; a formal model is created for each software/hardware component of the hierarchical control structure of a manufacturing system. Furthermore, a formal model that defines the acyclic graph of a process plan (the **process plan structure**), and includes a set of rules that derive the current set of process plan steps to be executed by the manufacturing system is also created. The assembled model of the manufacturing system software/hardware components and process plans is, in turn, added to the *assembled models database* (L3).

The state of the formal model of a software/hardware component together with the preconditions and post-conditions of the rules of this model are stored in a data structure which is accessed by the *planner/fault resolver* (P3) software tool during planning. To attain this goal, the *specification language translator* (P1) is used to translate the component-oriented rule-based specification of formal models into library modules (S2 closed). These modules implement, in a general-purpose high-level programming language, the sequence of commands for creating the data structures that encapsulate these models, and are added to the *models library* (L2). The source code of these modules is then translated by the *source code translator* (P2) into object code. When executed, this object code creates and populates the data structure of the assembled model of a manufacturing system; this data structure is added to the *assembled models database* (L3).

To avoid the use of a procedural language in implementing the software/hardware components of a manufacturing system, and a declarative language in implementing the formal models of these components, a high-level procedural language can be extended and used to implement both components and models. These extensions involve adding a set of rule-based constructs to the high-level procedural language, and implementing a software tool to translate these constructs into a set of statements that can be processed by the compiler of this language. The high-level procedural language program generated by this translation can be used to build a generic simulator that can plan and execute the operations of the system. The YES/L1 (Cruise *et al.*, 1987) language that extends PL/1, and the XC (Nuutila *et al.*, 1987) language that extends $C^{++}$ are examples of such extensions. The component-oriented rule-based specification language of this methodology (Chaar and Volz, 1993c) can also be viewed as a rule-based extension to the Ada programming language.

Several approaches (Allen, 1987) can be used to plan off-line the sequence of operations of the jobs to be processed on a manufacturing system. One popular method consists of coupling the formal model of this system with an inference engine to form a rule-based system. To plan such a sequence, the inference engine is assigned the goal of executing, in their appropriate order, the manufacturing operations of the process plan steps of a job. In order to achieve the goal of executing a process plan step, this inference engine must select and execute a sequence of transport operations that can transform the current state of the manufacturing system into a state where the manufacturing operation of this process plan step can be performed. The inference engine can then plan the details of execution of the current manufacturing operation.

The use of an inference engine in planning the processing of a set of jobs offers an opportunity to schedule the operations of these jobs dynamically based on the current state of the system and the current set of feasible operations of this system (Naylor *et al.*, 1986, 1987). However, the large number of rules that may be required to fully specify the operations of a complex manufacturing system, and the large number of jobs that may be processed by the devices of this system can tremendously degrade the performance of this inference engine. Furthermore, the sequence of operations planned by the inference engine may not result in an optimal schedule for processing these jobs.

The inference engine uses a search process to derive the sequence of operations. The available search strategies (Georgeff, 1987) implicitly create a search tree of all the sequences of feasible operations that can be executed by the manufacturing system, and select a sequence of operations that leads from the initial state to the goal state of this system. In the absence of any assistance, the inference engine must select and emulate each feasible operation of the system when building the search tree. Hence, the time complexity of the planning algorithm of this inference engine is exponential in the number of rules and jobs involved.

To reduce the complexity of planning the operations of a batch of jobs, the methodology groups these jobs into classes. All jobs in a class can share the same plan; those in different classes require different processing in their manufacture. Next, a sample job is selected from each class, and a separate plan for processing each job class is derived by the software engineer. These plans are used subsequently by the software engineer to create a schedule for processing all the jobs in the batch. A single batch may contain a mix of jobs from various classes. Compared to planning for the concurrent processing of all jobs in the batch, this approach reduces both the size of the current set of feasible operations of the system and the total number of operations that must be planned prior to processing these jobs. The size of the current set of feasible operations is reduced because only a single job is present in the system during plan generation. Furthermore, the total number of operations that must be planned is greatly reduced because the number of job classes is usually very small compared to the total number of jobs.

The plan for each selected job is generated interactively by the software engineer with the help of the *planner/fault resolver* (P3) software tool. To derive this plan, the initial state of the manufacturing system and the process plan structure of the current job are specified first by the software engineer. The hierarchical control structure of the system is then traversed in a top-down manner while the software engineer refines the sequence of operations that must be performed by each level of the hierarchy when processing this job. The duration of an operation that is assigned to an assembled component of this hierarchical control structure may depend on the duration of the sequence of operations that are assigned to the constituents of this component. Consequently, the durations of some operations are assigned by revisiting, bottom-up, the assembled components of this hierarchical control structure. When completed, this tree-structured plan is added to the *plan/table database* (L4). This database stores also the time intervals, relative to the initiation of the job under consideration, during which the devices of the system perform their assigned operations. These time intervals are derived by the *reservation table generator* (P4) software tool, and are subsequently grouped in a matrix called the **reservation table** (Chaar and Davidson, 1990).

The task of planning the sequence of operations of a system can be enhanced by coupling the planner system with a user-friendly graphical interface. This interface typically would include a set of windows that are equipped with menu bars, scroll bars, pull-down menus, pop-up menus and buttons, and should be designed with the aim of displaying a sorted list of the data items that support the decision-making process of the software engineer. These useful data items would include a list of the current set of feasible operations of a software/hardware component, the semantics of some possible combinations of these operations and a list of the subset of feasible operations that have been selected by the software engineer together with their semantics. A tree-structured diagram of the partially derived plan of a job should also be presented to the software engineer. This graphical interface would be implemented in the *planning display manager* (P6) software tool. A modified version of the entity-relationship (ER) data model of database design can be used to display the semantics of the operations of software/hardware components (Chaar and Volz, 1993c).

### 13.4.4 Scheduling

Despite the fact that a large variety of jobs may be processed by a manufacturing system, it is standard practice to process in sequence a batch of identical jobs before switching to a different batch of jobs. This processing strategy is generally adopted because of the large set-up times that are usually involved in switching between the jobs of different classes. The process plan of a job class of the batch is extended by the software engineer during the planning activity of the methodology to capture, in a tree-structured plan, the complete sequence of transport and manufacturing operations that must be performed by the devices of the system when processing this job. This plan is then repeatedly executed whenever a batch of identical jobs is to be processed by the system.

When deriving a schedule, the software engineer assumes that each job has a **deterministic** flow through the manufacturing system. This flow is determined by the sequence of operations of the plan of this

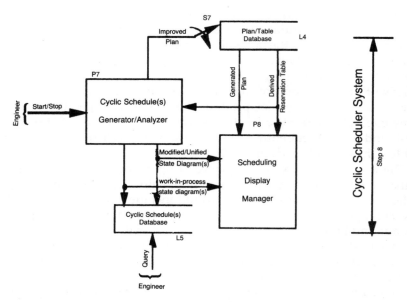

**Fig. 13.6**   The cyclic scheduler system dataflow.

job, and may require revisiting some devices of the system several times while not visiting others at all. These assumptions create a job shop model of the manufacturing system. The schedule is then known as a **job shop schedule**; it is called a **cyclic schedule** if the time intervals between successive job initiations form a **periodic** sequence.

In step 8 of the proposed methodology, the cyclic scheduler system (Fig. 13.6) can generate the set of all possible cyclic job shop schedules of a manufacturing system from the reservation tables of the job classes of this system. This system consists of the *cyclic schedule(s) generator/ analyzer* (P7) and the *scheduling display manager* (P8). When deriving a cyclic schedule, switch S7 is closed whenever the improved plan of a job is to be included in the *plan/table database* (L4). All the sequences of job initiations that lead to no collisions in the system are captured in a state diagram that displays the state of this system upon each job initiation; a **collision** occurs when two or more jobs attempt to use a manufacturing device simultaneously.

The state diagram is called a **modified state diagram** (Chaar and Davidson, 1990) when a single job class is processed or a **unified state diagram** when a mix of jobs from several job classes is processed. These diagrams are generated by the *cyclic schedule(s) generator/analyzer* (P7) software tool and are added to the *cyclic schedule database* (L5). By analyzing all the paths of a modified or a unified state diagram, particularly those that form closed loops (cycles), a cycle can be found that maximizes the throughput of the system while not violating any other constraints, such as deadlines and work-in-process constraints, on the processing of a batch of jobs. Each cycle of the modified or unified state diagram corresponds to a cyclic job shop schedule. Efficient algorithms exist for locating high throughput cycles in these state diagrams (Chaar, 1990; Davidson and Chaar, 1993).

During the processing of a job, the location of this job in the manufacturing system can be tracked based on the operation that is currently being performed on this job. The **work-in-process state diagram** captures the locations of all jobs currently being processed by this system. Hence, this work-in-process state diagram can also be used to keep track of the number of jobs that are concurrently processed by the system, and can be used to select a cycle of the modified state diagram that does not violate any work-in-process constraints that may be imposed on the cyclic schedule.

The complexity of deriving an optimal cyclic schedule is substantially less than that for deriving an optimal arbitrary schedule because these schedules assign time slots only to initiations of the jobs processed by the system as opposed to assigning time slots to the individual operations of these jobs. Once initiated, a job simply follows the plan for its job class. Moreover, the cyclic structure of the schedule imposes more regularity on the behavior of the system than an arbitrary schedule, and simplifies its control software by reducing the complexity of the data used in monitoring the operations of this system, and in detecting and correcting any faults that may occur in the system.

Deriving a plan for a job, and selecting an optimal cyclic job shop schedule for processing a batch of these jobs may or may not result in an optimal overall schedule for processing these jobs. Given a plan, cyclic job shop scheduling achieves optimum throughput asymptotically as the number of jobs in a batch increases. In contrast, conventional job shop scheduling is far more complex; although it could achieve higher throughput schedules, a heuristic approach is usually used in practice due to this complexity. The optimal cyclic schedule may actually be far from the overall optimum due to a poor plan. The proposed methodology, however, includes methods for modifying plans and rescheduling to more closely approach optimum overall throughput. These methods improve the throughput of the system to reach a productivity goal by inserting a minimal number of time delays between the operations of the plan of a job class.

The *scheduling display manager* (P8) will be used to display the outcome of the cyclic scheduling activity. In particular, separate windows should be used to display the plans, the reservation tables, the modified or unified state diagrams and the work-in-process state diagrams of the jobs and job mixes that are processed by a manufacturing system. The set of cycles of a modified or a unified state diagram together with their relevant properties should be displayed in a well-ordered tabular form by this software tool. One or more such cycles are used to schedule the processing of associated batches of jobs.

### 13.4.5  Operation monitoring

The occurrence of a fault in the manufacturing system while performing the operations of the overall schedule may invalidate the rest of the derived schedule. This problem is caused by the fact that the plans and the cyclic job shop schedules of the jobs that are to be processed by the manufacturing system are generated before the processing of the jobs is started. Hence, the execution of the scheduled operations must be monitored in order to detect, diagnose and correct any faults that may occur in the system while performing these operations. The operation monitoring activity is performed in steps 9, 10 and 11 of the proposed methodology with the help of the operation monitor system (Fig. 13.7) which consists of the *configuration/schedule simulator* (P9), the *tracker/ monitor* (P10) and the *monitoring display manager* (P11) software tools.

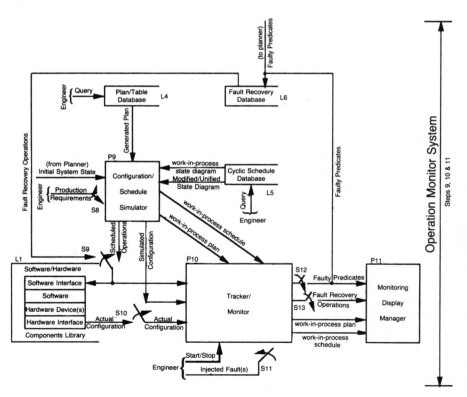

**Fig. 13.7**  The operation monitor system dataflow.

This activity includes simulating the behavior of the system in response to production requirements (S8 closed) and in the presence of faults, as well as observing the actual operation of this system. These production requirements specify the number of jobs in the batch to be processed by the system and a deadline for processing these jobs.

An added advantage that is gained from casting manufacturing hardware into software/hardware components is the ability to alternate their simulation and on-line testing. This alternation reduces considerably the idle time of the manufacturing devices that can be manually operated while the control and integration software of the system is being independently developed and tested. Furthermore, the complexity of the task of testing the control and integration software is reduced by eliminating hardware-related errors from the software testing process.

Simulation is performed by assigning simulated internals to the hardware-dependent components of a manufacturing system. This constitutes the first step in testing the control and integration software of the system. The next step consists of gradually replacing each simulated internal by an interface to the real hardware device. On-line testing is completed whenever the whole system is fully operational.

To be able to monitor in real-time the execution of the operations of a manufacturing system, an efficient monitoring scheme that can detect the presence of faults in this system is required. In the proposed methodology, the plans of the jobs in process determine the current set of operations to be performed on these jobs. The monitoring scheme is executed by the *Tracker/monitor* software tool when checking the values of the predicates specified in the conditions that must precede the execution of each operation against the actual values of these predicates as determined from the current monitored state of the actual system. When the actual value of a predicate differs from its required value, a fault is detected in the system. The monitoring scheme is efficient because the number of predicates that must be checked prior to executing an operation is very small compared to the total number of predicates of the state of this system. However, a penalty is incurred: a fault may not be detected unless it blocks the execution of the predetermined sequence of operations of the plan of a job. Hence, this monitoring scheme is a **plan-oriented** monitoring scheme.

The monitoring scheme provides a compromise between off-line scheduling and dynamic on-line scheduling. This compromise is achieved by augmenting the plans of the job classes to be processed by a manufacturing system with the sequence of operations that must be performed by this system when recovering from particular detected faults. When such a sequence of operations allows the execution of the original cyclic job shop schedule of the system to be resumed after recovery, this resumption is termed a **match-up** (Bean *et al.*, 1991).

In the operation monitor system, S9 and S10 act as three-way switches to select one of two possible inputs or remain open to disable the *tracker/monitor* (P10). S9 should present a software/hardware component with its current set of scheduled operations during normal system operation, or its fault recovery operations whenever a fault occurs. The output of S9 goes both to the software interfaces of the *software/hardware components library* (L1) and to the *tracker/monitor* (P10). S10 presents the *tracker/monitor* (P10) with the simulated configuration of a manufacturing system or the actual configuration of this system when the actual devices of the system are being operated. Thus, the system can be operated in either a simulation mode (for debugging) or an actual operation mode. Both the scheduled operations and the simulated configuration of a system are generated by the *configuration/schedule simulator* (P9) software tool.

To generate the fault recovery sequences of a plan, a set of **fault trees** is linked to the plan of each job class when this plan is derived by the software engineer. These fault trees are automatically generated by the *fault tree generator* (P5) software tool of the *planner system* (Fig. 13.5) and are stored in the *fault recovery database* (L6). The nodes of the fault trees enumerate all the combinations of faulty predicates that can block the execution of a scheduled operation of this plan; a predicate is declared faulty whenever it has non-matching simulated and actual values. When the number of predicates that can block the execution of this scheduled operation is $n$, the number of nodes of the fault tree of this operation is $2^n$. Typically, $n$ is small (e.g. $n \leq 5$) compared to the number of predicates of the state of a system. A node of this fault tree is accessed whenever a combination of faulty predicates that model the presence of a fault of the system is detected. This node encapsulates the sequence of operations that must be executed when recovering from the fault.

By augmenting the plan of a job class with a set of fault trees, a combination of one or more faulty predicates that can block the execution of the operations of this plan is captured by these fault trees. The sequence of operations to be performed when recovering from this fault may be created a priori by the software engineer with the help of the *planner/fault resolver* (P3) software tool (S4, S5, S12, S13, S14, and S15 closed). The augmented plan is then added to the *fault recovery database* (L6), and the above sequence can be automatically executed whenever the fault with which the sequence is associated is detected during the processing of a job. Some combinations of faulty predicates can sometimes be guaranteed not to occur because of some physical constraints imposed by the system. Hence, the paths of the fault tree that include these combinations can be eliminated to reduce the number of nodes of this fault tree.

For complex plans, deriving the sequence of operations of the fault trees of these plans can be complicated by the large number of possible faults in these plans. Hence, it is desirable a priori to derive only these sequences that correspond to the most probable faults of the plan. The effect of a fault on the state of the system is captured by **fault injection** (S11 closed), i.e. explicitly modifying the values of one or more predicates of the state of this system and simulating the system with such injected faults. The sequence of operations used to recover from other less probable faults can be derived when the presence of these faults is detected during the actual operation of the system. To derive such sequences, the manufacturing software environment offers the software engineer the capability to alternate between simulating the behavior of a manufacturing system and monitoring the actual operation of this system. The time interval required for recovery is used to find a different scheduling path of the modified or unified state diagram of the system. This dynamically generated path is then used to match-up with the original cyclic schedule of the system. Furthermore, the software engineer can incrementally add any missing operation that is needed for fault recovery to the appropriate formal model and software/ hardware component of the system.

By these means, a generic algorithm that interprets the data represented in the cyclic job shop schedules of a batch of jobs, the plan of each job of the batch and the sequence of operations of the fault trees of this plan can be implemented in the control software to achieve efficient and dependable manufacturing. The interpreted data is then mapped by this generic algorithm into an appropriate set of operations that are executed by the hardware devices of these manufacturing systems. Hence, a set of routines that provide detailed control for these hardware devices is required. Furthermore, this control software must include a generic algorithm that implements the plan-oriented scheme used to monitor their operations.

The hierarchical exception handling mechanism designed by Antonelli (1989) is similar to the hierarchical control structure of the assembled software/hardware components of a manufacturing system. Hence, the use of this mechanism in interpreting the sequence of operations of fault trees may be investigated. This mechanism allows an exception that occurs when executing the operation of a component to be directly handled by the appropriate component of the hierarchical control structure. After execution, this handler may either resume the old operation, signal an exception to another software/hardware component, or abandon this old operation.

A fault is handled locally by executing its fault recovery plan. However, the effect of this fault on the processing of current jobs and on the initiation of any future jobs must be handled globally by the control software of the manufacturing system. This is achieved by delaying the initiation of any new jobs until the set of operations performed on the jobs in process is identical to that specified by **scheduled operations** (Fig. 13.7). This technique is labeled matchup scheduling (Bean *et al.*, 1991) and allows the resumption of the original schedule. When the original schedule cannot be resumed, the processing of the initiated jobs is eventually completed and a new plan is introduced for processing any remaining jobs of the batch, or the operation of the manufacturing system is halted altogether in the absence of such a plan.

The actual processing of a batch of jobs by the system can be performed by executing the steps of the following algorithm:

**Algorithm 1   Process a batch of jobs**

Starting at time t = 0, follow the chosen cycles of a state diagram of the manufacturing system until all the jobs of a batch are processed.

1. Initiate the processing of a job of the batch.
2. For each operation of the tree-structured plan of this initiated job, check the simulated values of the predicates of the preconditions of this operation against their actual values.
   (a) If these values are identical, execute the procedure of the software/hardware component of the hierarchical control structure of the system that implements this operation.
   (b) Otherwise, do not initiate any new jobs in the manufacturing system. Recover from the fault by executing the appropriate sequence of operations as specified by a fault tree of the plan, or create such a sequence if it is missing.
3. Check whether the initiation of any new jobs in the system has been disabled.
   (a) If the current set of operations is identical to the set specified in **scheduled operations**, go to step 1 of the algorithm.
   (b) Otherwise, introduce a new plan for processing any remaining jobs and go to step 1 of the algorithm, or halt the operation of the manufacturing system in the absence of such a plan.

Monitoring the operation of a manufacturing system is enhanced by the presence of a graphical interface to the operation monitor system. This interface is coupled with the *tracker/monitor* (P10), and will be implemented by the *monitoring display manager* (P11). The **work-in-process plan** is created by merging the plans of all the jobs currently being processed on the system and the **work-in-process schedule** is generated by superimposing the reservation tables of these same jobs. Both should be displayed in separate windows by the monitoring display manager (P11). The current operations and the devices performing them should also be highlighted.

A different set of operations should be performed when recovering from a fault; these are also displayed when S13 is closed. In these faults, the modified entity-relationship data model is used to display the faulty predicates of the system (S12 closed). Displaying these faulty predicates can prove vital to locating the software/hardware component where the fault has occurred, and subsequently to handling this fault.

### 13.4.6 Software maintenance

The last activity of the proposed methodology maintains and evolves the manufacturing control software generated by applying the previous steps of this methodology. More specifically, this activity (step 12 of Fig. 13.3) must deal with changes in the control software of a manufacturing system that may result from processing new products by the system, or modifying the functionality of this system by adding, replacing, or removing manufacturing devices. Furthermore, the activity must specify the sequence of steps of the methodology that must be repeated when modifying this control software.

When a plan for processing a new product is to be created by the software engineer, the formal models of the software/hardware components of the manufacturing system are retrieved from the *assembled models database* (L3), and the process plan structure is updated to capture the specifics of the process plan of this new product. These formal models are then used to generate a plan for processing the product. This plan is subsequently used to create a cyclic scheduling strategy for processing the job batches that include this new product, and the fault trees of this plan are augmented by the appropriate sequences of operations. Hence, the introduction of new products, i.e. new job classes, in the manufacturing system requires a repetition of step 7 through step 12 of the methodology (Fig. 13.8).

On the other hand, the addition, replacement or removal of a manufacturing device from the system may affect the structure of many software/hardware components of the hierarchical control structure of this system. In this case, the software engineer may have to repeat all the steps of the methodology (step 1 through step 12) in order to

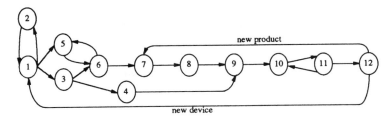

**Fig. 13.8** Sequencing through the steps of the methodology.

develop the control software of this new system (Fig. 13.8). Hence, the software/hardware components of this system, the augmented plans of the jobs of this system, the cyclic job shop schedules of these jobs and the fault trees may have to be modified.

### 13.4.7 Summary

In summary, the activities of the proposed methodology transform the requirements specification of a manufacturing system into an implementation of the control software of this system. This implementation promotes the efficiency and dependability of this manufacturing system, and consists of a set of software/hardware components, a plan for sequencing the operations of these components when processing a job and a cyclic job shop schedule for initiating the processing of batches of these jobs. Furthermore, the methodology accounts for monitoring the processing of these jobs by detecting, diagnosing and recovering from any faults that can occur during the operation of the system.

Note that the activities of the proposed methodology can be independently applied in developing the control software of a manufacturing system. This feature is enhanced by the presence of the six separate databases L1 through L6. These databases hold the results of their associated steps, and these results can be retrieved when necessary in order to independently carry out subsequent activities of the methodology. However, the steps of an activity must be performed in strict order. Sequencing through the steps of this methodology is captured in the following algorithm:

**Algorithm 2   Execute the steps of the methodology**

1. Define the software requirements of the manufacturing system, (specified by the manufacturing engineer).
2. For each device of the manufacturing system,
   (a) acquire the software/hardware component of the device (step 1);
   (b) simulate the operations of this software/hardware component (step 2);
   (c) test this software/hardware component on-line (step 2).
3. Assemble the software/hardware components of these devices into new components in order to create a hierarchical control structure of the system (step 3).
4. Distribute the software/hardware components of this hierarchical control structure (step 4).
5. For each device of the manufacturing system,
   (a) acquire the formal model of the software/hardware component that encapsulates this device (step 5).

6. Acquire the formal model of the process plan of this system (step 5).
7. For each assembled software/hardware of the hierarchical control structure,
    (a) generate the formal model of this component (step 6).
8. Using these formal models,
    (a) create a tree-structured plan for processing each job class on this manufacturing system (step 7);
    (b) based on this plan, derive a reservation table for the system for each job class (step 7).
9. Use these reservation tables to generate the unified state diagram that displays the set of all feasible cyclic job shop schedules (step 8).
10. Select a cycle set of this diagram, and simulate the operation of the system when the feasible and optimal cyclic schedule, composed of the cycles of this selected set, is used to initiate the processing of a batch of jobs (step 9).
11. Derive the sequence of operations of the fault trees of this tree-structured plan by simulating the presence of any probable faults of the system, and recovering from these faults (step 10).
12. Operate and monitor the operation of the actual manufacturing system by executing the control software of this system (step 11).
13. Interpret the augmented plan and the modified state diagram or the unified state diagram of the job batch while executing the control software of the manufacturing system (step 11).
14. Maintain and evolve the control software of this manufacturing system (step 12).

## 13.5  CONCLUSION

A methodology for developing the control software of efficient and dependable manufacturing systems, and a set of software tools that enhance the applicability of this methodology have been proposed. These tools aid in planning the operations of a manufacturing job, generating a cyclic schedule for processing a batch of jobs, and monitoring the operations of the system while this batch is being processed. They are integrated in a manufacturing software environment that can be considered a computer-aided software engineering (CASE) environment for developing software for real-time systems.

To achieve efficiency, the reservation table technique is used to create optimum cyclic job-shop schedules for processing a batch of identical jobs or a mix of jobs from several classes on a manufacturing system. A reservation table represents the plan of each job class. These tables are then used to produce a unified state diagram from which an optimum initiation strategy for the batch is derived.

To achieve dependability, a plan-oriented fault detection and correction strategy is proposed. This strategy can automatically handle any combination of faults that may be detected when monitoring the execution of the scheduled operations in the manufacturing system.

As a proof of concept, a prototype implementation of the software tools of the planner system of this environment (Fig. 13.5) has been completed. This implementation has been carried out on a Unix-based system in the Ada programming language. A prototype *specification language translator* (P1) has been developed to translate the statements of the component-oriented rule-based specification language into Ada code. This code was compiled by an Ada compiler which acted as the *source code translator* (P2) of the prototype. The planning component of an interactive *planner/fault resolver* (P3) software tool has been coupled with the prototype and used to expand the process plan of a job into a detailed plan for this job.

The *software/hardware components library* (L1) has been populated with the Ada implementation of the software/hardware components of the prismatic machining cell in Ben Hadj-Alouane *et al.* (1990, 1991). Moreover, the formal models of the software/hardware components and process plans of this cell have been coded in the component-oriented rule-based specification language and used to populate the *models library* (L2). The specification of this cell was automatically translated into Ada code by the *specification language translator* (P1). An assembled model of the prismatic machining cell was created by compiling and executing this Ada code. This assembled model was then added to the *assembled models database* (L3) and used to interactively expand into a plan the process plan of a specific job class.

## REFERENCES

AFCET (1977) Pour une Représentation Normalisée du Cahier des Charges d'un Automatisme Logique, *Automatique et Informatique Industrielles*, 6(61/62), Nov/Dec.

Ahuja, J.S. and Valavanis, K.P. (1988) *A Hierarchical Modeling Methodology for Flexible Manufacturing Systems Using Extended Petri Nets*, The Proceedings of the 1988 International Conference on Computer Integrated Manufacturing, Troy, NY, pp. 350–56, May.

Allen, D.K. (1987) An Introduction to Computer-Aided Process Planning, *CIM Review*, pp. 7–23, Fall.

Antonelli, C.J. (1989) Exception Handling in A Multi-Context Environment, PhD thesis, University of Michigan, Ann Arbor, Mich., *Technical Report RSD-TR-2-89*.

Atabakhche, H., Simonetti Barbalho, D., Valette, R. *et al.* (1986) *From Petri Net*

*Based PLCs to Knowledge Based Control,* Proceedings of the IECON'86, pp. 817–22.

Bean, J.C., Birge, J.R., Mittenthal, J. *et al.* (1991) Matchup Scheduling with Multiple Resources, Release Dates and Disruptions, *Op Res,* **39**(3), pp. 470–83, May–June.

Beck, C.L. (1985) Modeling and Simulation of Flexible Control Structures for Automated Manufacturing Systems, Master's thesis, The Robotics Institute, Carnegie-Mellon University, Dec.

Beck, C.L. and Krogh, B.H. (1986) *Models for Simulation and Discrete Control of Manufacturing Systems,* The Proceedings of the 1986 IEEE International Conference on Robotics & Automation, San Francisco, Ca., pp. 305–10, Mar.

Ben Hadj-Alouane, N., Chaar, J.K. and Naylor, A.W. (1990) *The Design and Implementation of the Control and Integration Software of a Flexible Manufacturing System,* The Proceedings of The First International Conference on Systems Integration, Morristown, NJ, pp. 494–502, Apr.

Ben Hadj-Alouane, N., Chaar, J.K. and Naylor, A.W. (1991) Developing Control and Integration Software for Flexible Manufacturing Systems. *Jnl. of Sys Integ,* **2**(1), pp. 7–34, Jan.

Bispo, C.F.G. and Sentieiro, J.J. (1988) *An Extended Horizon Scheduling Algorithm for the Job-Shop Problem,* The Proceedings of the 1988 International Conference on Computer Integrated Manufacturing, Troy, NY, pp. 249–52.

Bollinger, J.G. and Duffie, N.A. (1988) *Computer Control of Machines and Processes,* Addison-Wesley Publishing Company, pp. 369–420.

Bonner, S. and Shin, K.G. (1982) A Comparative Study of Robot Languages, *IEEE Computer,* pp. 82–96, Dec.

Booch, G. (1986) Object-Oriented Development, *IEEE Trans on Sftwre Eng,* **SE-12**(2), pp. 211–21, Feb.

Bourne, D.A. and Fox, M.S. (1984) Autonomous Manufacturing: Automating the Job-Shop, *Computer,* **17**(9), pp. 76–86, Sept.

Bruno, G. (1984) Using Ada for Discrete Event Simulation, *Sftwre Prac and Exper,* **14**(7), pp. 685–95, July.

Bruno, G. and Balsamo, A. (1986a) *Petri Net-Based Object-Oriented Modeling of Distributed Systems,* OOPSLA '86. Conference Proceedings, Portland, Oreg, pp. 284–93.

Bruno, G. and Biglia, P. (1985a) *Performance Evaluation and Validation of Tool Handling in Flexible Manufacturing Systems Using Petri Nets,* 1985 International Workshop on Timed Petri Nets, pp. 64–71.

Bruno, G. and Elia, A. (1986b) Operational Specification of Process Control Systems: Execution of PROT Nets Using OPS5 (ed. H.J. Kugler), *Information Processing (IFIP)* **86**, pp. 35–40.

Bruno, G. and Marchetto, G. (1985b) *Rapid Prototyping of Control Systems Using High Level Petri Nets,* The Proceedings of the 8th International Conference on Software Engineering, pp. 230–35.

Bruno, G. and Marchetto, G. (1986c) Process-Translatable Petri Nets for the Rapid Prototyping of Process Control Systems, *IEEE Trans on Sftwre Eng,* **SE-12**(2), pp. 346–57.

Bruno, G. and Morisio, M. (1987a) *The Role of Rule Based Programming for Production Scheduling,* 1987 IEEE International Conference on Robotics and Automation, Raleigh, NC, pp. 545–50.

Bruno, G. and Morisio, M. (1987b) *Petri-Net Based Simulation of Manufacturing Cells,* 1987 IEEE International Conference on Robotics and Automation, Raleigh, NC, pp. 1174–79.

374     *Efficient and dependable manufacturing*

Camurri, A. and Franchi, P. (1990) *An Approach to The Design of The Hierarchical Control System of FMS, Combining Structured Knowledge Representation Formalisms and High-Level Petri Nets*, The Proceedings of the 1990 IEEE International Conference on Robotics & Automation, Cincinnati, O., pp. 520–25, May.

Chaar, J.K. and Davidson, E.S. (1990) *Cyclic Job Shop Scheduling Using Reservation Tables*, The Proceedings of the 1990 IEEE International Conference on Robotics and Automation, Cincinnati, O., pp. 2128–35, May.

Chaar, J.K. (1990) A Methodology for Developing Real-Time Control Software for Efficient and Dependable Manufacturing Systems, PhD thesis, University of Michigan, Ann Arbor, Mich., *Technical Report RSD-TR-20-90*.

Chaar, J.K., Teichroew, D. and Volz, R.A. (1993a) Developing Manufacturing Control Software: A Survey and Critique, *Int Jnl. of Flex Manuf Sys*, **5**(1), pp. 53–88.

Chaar, J.K., Teichroew, D. and Volz, R.A. (1993b) Real-Time Software Methodologies: Are They Suitable for Developing Manufacturing Control Software, *Int Jnl. of Flex Manuf Sys*, **5**(2), pp. 95–128.

Chaar, J.K. and Volz, R.A. (1993c) A Specification Language for Planning and Fault Recovery in Manufacturing Systems, *Int Jnl. of Flex Manuf Sys*, **5**(3), pp. 209–53.

Chang, S.J., DiCesare, F. and Goldbogen, G. (1989) *The Generation of Diagnostic Heuristics for Automated Error Recovery in Manufacturing Workstations*, The Proceedings of the 1989 IEEE International Conference on Robotics and Automation, Scottsdale, Ariz., pp. 522–27, May.

Chang, S.J., DiCesare, F. and Goldbogen, G. (1990) *Analysis of Diagnosability of Failure Knowledge in Manufacturing Systems*, The Proceedings of the 1990 IEEE International Conference on Robotics & Automation, Cincinnati, O., pp. 696–701, May.

Chintamaneni, P.R., Jalote, P., Shieh, Y.-B. *et al.* (1988) On Fault Tolerance in Manufacturing Systems, *IEEE Network*, **2**(3), pp. 32–9, May.

Corbeel, D., Gentina, J.G. and Vercauter, C. (1985) Application of An Extension of Petri Nets to Modelization of Control and Production Processes, *Advances in Petri Nets*, pp. 162–80.

Crockett, D., Desrochers, A., DiCesare, F. *et al.* (1987) *Implementation of a Petri Net Controller for a Machining Workstation*, The Proceedings of the 1987 IEEE International Conference on Robotics & Automation, Raleigh, NC, pp. 1861–67, March.

Cruise, A., Ennis, R., Finkel, A. *et al.* (1987) *YES/L1: Integrating Rule-based, Procedural and Real-Time Programming for Industrial Applications*, Proceedings of the Third Conference on Artificial Intelligence Applications, Kissimmee, Fl., pp. 134–39.

Davidson, E.S. and Chaar, J.K. (1993) Cyclic Job Shop Scheduling using Collision Vectors, University of Michigan, Ann Arbor, Mich., *Technical Report CSE-TR-169-93*.

Descotte, Y. and Latombe, J.-C. (1985) Making Compromises among Antagonist Constraints in a Planner, *Artificial Intelligence*, **27**, pp. 183–217.

Dove, R.K. (1988) *Process Design Automation: The Key to the 90's*, The Proceedings of AUTOFACT '88 Conference, Chicago, Ill., pp. 10.13–10.30 October.

Ekberg, G. and Krogh, B.H. (1987) Prototype Software for Automatic Generation of On-Line Control Programs for Discrete Manufacturing Processes, *Technical Report CMU-RI-TR-87-3*, The Robotics Institute, Carnegie-Mellon University.

Elsenbrown, B. (1988) Programmable Controllers Move to Systems Solutions, *Manuf Eng*, pp. 59–61. Jan.

Fikes, R.E. (1971a) *Monitored Execution of Robot Plans Produced by STRIPS*, The Proceedings of the IFIP Congress 71, Ljubljana, Yugoslavia, pp. 189–94.

Fikes, R.E., Hart, P.E. and Nilsson, N.J. (1972) Learning and Executing Generalized Robot Plans, *Artif Intell*, **3**, pp. 251–88.

Fikes, R.E. and Nilsson, N.J. (1971b) STRIPS: A New Approach to the Application of Theorem Proving to Problem Solving, *Artif Intell*, **2**, pp. 189–208.

Fisher, J.P. (1989) *Zone Logic–Increased Machine Productivity through Artificial Intelligence*, The Proceedings of the 18th Annual International Programmable Controllers (IPC) Conference, Apr.

Gayman, D.J. (1988) An Old Favorite Gets New Standards, *Manuf Eng*, pp. 55–8, Jan.

Gentina, J.C. and Corbeel, D. (1987) *Colored Adaptive Structured Petri Net: A Tool for the Automatic Synthesis of Hierarchical Control of Flexible Manufacturing Systems*, The Proceedings of the 1987 IEEE International Conference on Robotics Automation, Raleigh, NC, pp. 1166–73, Mar.

Goldsack, S.J., Holzoacker-Valero, A.A., Volz, R.A. *et al.* (1992) *AdaPT and Ada9X*, Tri-Ada '92, Orlando, Fl., Nov.

Georgeff, M.P. (1987) Planning, *Annual Review of Computer Sciences*, **2**, pp. 359–400.

Hasegawa, M., Takata, M., Temmyo, T. *et al.* (1990) *Modeling of Exception Handling in Manufacturing Cell Control and its Application to PLC Programming*, The Proceedings of the 1990 IEEE International Conference on Robotics & Automation, Cincinnati, O., pp. 514–19, May.

Huber, A. and Buenz, D. (1988) Using GRAI to Specify Expert Systems for the Control and Supervision of Flexible Flow Lines (ed. J. Browne), *Knowledge Based Production Management Systems (IFIP)*, **86**, pp. 295–308.

IBM, International Business Machines Corporation (1981) *IBM Robot System/1 AML Reference Manual*.

International Electrotechnical Commission (IEC) (1984) Preparation of Function Charts for Control Systems, Nov.

Jafari, M.A. (1990) *Petri Net Based Shop Floor Controller and Recovery Analysis*, The Proceedings of the 1990 IEEE International Conference on Robotics & Automation, Cincinnati, O., pp. 532–37, May.

Jones, J.E., White, D.R., Xiaoshu, X. *et al.* (1987) Development of an Off-Line Weld Planning System, *Texas Instr Tech Jnl*, pp. 47–53, Winter.

Kamath, M. and Viswanadham, N. (1986) *Applications of Petri Net Based Models in the Modelling and Analysis of Flexible Manufacturing Systems*, The Proceedings of the 1986 IEEE International Conference on Robotics & Automation, San Francisco, Ca., pp. 312–17, Mar.

Karmarkar, U.S. and Schrage, L. (1985) The Deterministic Dynamic Product Cycling Problem, *Op Res*, **33**(2), pp. 326–45, Mar–Apr.

Kasturia, E., DiCesare, F. and Desrochers, A. (1988) *Real Time Control of Multilevel Manufacturing Systems Using Colored Petri Nets*, The Proceedings of the 1988 IEEE International Conference on Robotics & Automation, Philadelphia, Pa, pp. 1114–19, Apr.

Kompass, E.J. (1987) Distributed Machine Control Uses Zoned Logic, Isolated Controllers, Fiber Optics, *Control Eng*, Aug.

Koren, Y. (1983) *Computer Control of Manufacturing Systems*, McGraw-Hill.

Krogh, B.H., Willson, R. and Pathak, D. (1988) *Automated Generation and Evaluation of Control Programs for Discrete Manufacturing Processes*, The Proceedings of the 1988 International Conference on Computer Integrated Manufacturing, Troy NY, pp. 92–99, May.

Le Pape, C. and Smith, S.F. (1987) Management of Temporal Constraints for Factory Scheduling, *Report CMU-RI-TR-87-13*, The Robotics Institute,

Carnegie-Mellon University, June.

Liu, Z. and Labetoulle, J. (1983) *A Heuristic Method for Loading and Scheduling Flexible Manufacturing Systems*, The Proceedings of the International Conference on Control' 88, Austin, Tex. pp. 195–200, Apr.

Martinez, J., Muro, P. and Silva, M. (1987) *Modeling Validation and Software Implementation of Production Systems Using High Level Petri Nets*, The Proceedings of the 1987 IEEE International Conference on Robotics & Automation, Raleigh, NC, pp. 1180–85, Mar.

Maxwell, W.L. and Singh, H. (1986) Scheduling Cyclic Production on Several Identical Machines, *Op Res*, **34**(3), pp. 460–3, May–June.

McDermott, D.V. (1987) Logic, Problem Solving, and Deduction, *Annual Review of Computer Sciences*, **2**, pp. 187–229.

Menga, G., Morisio, M., Picchiottino, P. *et al.* (1991) *A Framework for Object-Oriented Design and Prototyping of Manufacturing Systems*, The Proceedings of the 1991 IEEE International Conference on Robotics and Automation, Apr.

Meyer, W. (1986) *Knowledge-Based Realtime Supervision in CIM – The Workcell Controller*, ESPRIT' 86: Results and Achievements, pp. 33–52.

Nau, D.S. (1987) *Automated Process Planning Using Hierarchical Abstraction*, Texas Instr Tech Jnl., pp. 39–46, Winter.

Naylor, A.W. and Maletz, M.C. (1986) The Manufacturing Game: A Formal Approach to Manufacturing Software, *IEEE Transactions on Systems, Man, and Cybernetics*, **SMC-16**(3), pp. 321–34, May–June.

Naylor, A.W. and Volz, R.A. (1987) Design of Integrated Manufacturing System Control Software, *IEEE Transactions on Systems, Man, and Cybernetics*, **SMC-17**(6), pp. 881–97, Nov/Dec.

Nuutila, E., Kuusela, J., Tamminen, M. *et al.* (1987) XC-A Language for Embedded Rule Based Systems, *SIGPLAN Notices*, **22**(9), pp. 23–31, Sept.

Ow, P.S., Smith, S.F. and Thiriez, A. (1988) *Reactive Plan Revision*, Proceedings AAAI'88 Seventh National Conference on Artificial Intelligence, St. Paul, Minn, pp. 77–82, Aug. 21–6.

Parnas, D.L. (1972) On the Criteria to be Used in Decomposing Systems into Modules, *Comms. of the ACM*, **15**(12), pp. 1053–58, Dec.

Pathak, D.K. (1990) *Automated Development of Control Software for Discrete Manufacturing Systems*, The Robotics Institute, Carnegie-Mellon University, May.

Roundy, R. (1988) Cyclic Schedules for Job Shops with Identical Jobs, *Technical Report 766*, School of Operations Research and Industrial Engineering, Cornell University, July.

Russell, D. (1990) *Integration of PLCs and Databases for Factory Information Systems*, The Proceedings of The First International Conference on Systems Integration, Morristown-New Jersey, pp. 730–37, Apr.

Sacerdoti, E.D. (1974) Planning in a Hierarchy of Abstraction Spaces, *Artif Intell*, **5**, pp. 115–35.

Sahraoui, A., Courvoisier, M. and Valette, R, (1986) *Some Considerations on Monitoring in Distributed Real-Time Control of Flexible Manufacturing Systems*, Proceedings of the IECON '86, pp. 805–10.

Sahraoui, A., Atabakhche, H., Courvoisier, M. *et al.* (1987) *Joining Petri Nets and Knowledge Based Systems for Monitoring Purposes*, The Proceedings of the 1987 IEEE International Conference on Robotics & Automation, Raleigh, NC, pp. 1160–65, Mar.

Shaw, M.J. (1987) *Knowledge-Based Scheduling in Flexible Manufacturing Systems*, Tex Instr Tech Jnl, pp. 54–61, Winter.

Shaw, M.J. and Whinston, A.B. (1985a) *Automatic Planning and Flexible Scheduling:*

*A Knowledge-Based Approach*, The Proceedings of the 1985 International Conference on Automation & Robotics, St. Louis, M., pp. 890–94.

Shaw, M.J. and Whinston, A.B. (1985b) *Task Bidding and Distributed Planning in Flexible Manufacturing Systems*, Proceedings of the IEEE Second Conference on Artificial Intelligence Applications, Miami, Fl., pp. 184–89.

Shaw, M.J. and Whinston, A.B. (1986) *Application of Artificial Intelligence to Planning and Scheduling in Flexible Manufacturing*. (ed. A. Kusiak), Flex Manuf Sys: Methods and Studies, pp. 223–42.

Smith S.F. (1988) *A Constraint-Based Framework for Reactive Management of Factory Schedules* (ed. M.D. Oliff), Proceedings of the First International Conference on Expert Systems and the Leading Edge in Production Planning and Control, pp. 113–30.

Subramanyam, S. and Askin, R.G. (1986) *An Expert Systems Approach to Scheduling in Flexible Manufacturing Systems* (ed. A. Kusiak), Flex Manuf Sys: Methods and Studies, pp. 243–56.

Takata, S. and Sata, T. (1986) Model Referenced Monitoring and Diagnosis – Application to the Manufacturing System, *Computers in Industry*, **7**, pp. 31–43.

Thomas, B.H. and McLean, C. (1988) *Using GRAFCET to Design Generic Controllers*, The Proceedings of the 1988 International Conference on Computer Integrated Manufacturing, Troy, NY, pp. 110–19, May.

Unimation Incorporated (1982) *Users Guide to Val, Version II*, 2nd edn.

Van Brussel, H., Cottrez, F. and Valckenaers, P. (1990) *SESFAC: A Scheduling Expert System for Flexible Assembly Cells*, The Proceedings of the 1990 IEEE International Conference on Robotics & Automation, Cincinnati, O., pp. 1950–55, May.

Vasilash, G.S. (1987) *Rule-Based Breakthrough for Transfer Line Control*, Production, pp. 34–7, May.

Viswanadham, N. and Johnson, T.J. (1988) *Fault Detection and Diagnosis of Automated Manufacturing Systems*, The Proceedings of the 27th Conference on Decision and Control, Austin, Tex, pp. 2301–6, Dec.

Viswanadham, N. and Narahari, Y. (1987) *Coloured Petri Net Models for Automated Manufacturing Systems*, The Proceedings of the 1987 IEEE International Conference on Robotics & Automation, Raleigh, NC, pp. 1985–90, March.

Volz, R.A., Mudge, T.N. and Gal, D.A. (1984a) Using Ada as a Programming Language for Robot-Based Manufacturing Cells, *IEEE Transaction on Systems, Man, and Cybernetics*, **14**(6), pp. 863–77.

Volz, R.A. and Mudge, T.N. (1984b) *Robots are (nothing more than) Abstract Data Types*, Proceedings of the SME Conference on Robotics Research. The Next 5 Years and Beyond, Report #MS84–493, pp. 1–16.

Wilczynski, D. (1988) *A Common Device Control Architecture – The Savoir Actor*, The Proceedings of AUTOFACT'88 Conference, Chicago, Ill., pp. 10.31–10.40, Oct.

Willson, R.G. and Krogh, B.H. (1990) Petri Net Tools for the Specification and Analysis of Discrete Controllers, *IEEE Trans on Sftwre Eng*, **16**(1), pp. 39–50, Jan.

Wolter, J.D. (1989) *On The Automatic Generation of Assembly Plans*, The Proceedings of the 1989 IEEE International Conference on Robotics & Automation, Scottsdale, Ariz., pp. 62–8, May.

Zhou, M., DiCesare, F. and Desrochers, A.A. (1989) *A Top-Down Approach to Systematic Synthesis of Petri Net Models for Manufacturing Systems*, the Proceedings of the 1989 IEEE International Conference on Robotics & Automation, Scottsdale, Ariz., pp. 534–39, May.

Zhou, M.C. and DiCesare, F. (1990) *A Petri Net Design Method for Automated Manufacturing Systems with Shared Resources*, The Proceedings of the 1990 IEEE International Conference on Robotics & Automation, Cincinnati, O., pp. 526–31, May.

# CHAPTER 14

# Process plan representation for shop floor control

*A. Derebail, H. Cho and R. Wysk*

## 14.1 MOTIVATION FOR PROCESS PLAN REPRESENTATION MODELS

Product planning, design, manufacture and management are all elements of the engineering function (Webster's, 1985). Traditionally, the first three engineering elements are performed in a time-phased manner. A product engineer develops and details the design specifications for a product. The result of this activity is the **engineering product model**. Once the product design has been finalized, a process engineer determines which resources are required to manufacture the product. The result of this activity is the **process plan** for the product. Finally, a production engineer reconciles product demand with resource availability and manufactures the product. The production engineering activity results in an **execution plan**, which will be executed to manufacture the product. These activities are illustrated in Fig. 14.1. The product forms the link between these engineering activities (Wysk, 1992).

Today's engineer can process large quantities of data with computing power available for the analysis of complex, numerically intensive problems. The drafting boards used by manual drafters have been replaced by computer aided design (CAD) systems, which are capable of performing engineering analysis, in addition to drafting of product models. CAD systems allow designers to perturb and even optimize the performance of a design in a relatively easy manner. These systems also allow designers to extract geometric features directly from a design, so that engineering analysis can be performed automatically. It is this feature extraction capability that provides a process engineer with the capability of using computers to translate the product model into a process plan.

CAD systems are used increasingly for applications in all phases of

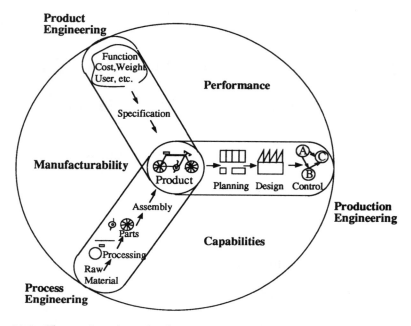

**Fig. 14.1**  The engineering wheel.

design, analysis and manufacture of a product. However, databases for CAD systems from different vendors are often incompatible with one another. A need was felt for a neutral data exchange format to facilitate the digital exchange of product data between different CAD systems. The initial graphics exchange specification (IGES), which defines this format, was adopted by the American National Standards Institute in 1981. Using IGES, the user can exchange product data models in the form of two- and three- dimensional wire frame and surface representations. Translators convert a vendor's proprietary internal database format into a neutral IGES format, and from the IGES format into another vendor's proprietary representation. Since the mid-1980s, another ISO standard called Standard for the Exchange of Product Model Data (STEP) has been under development. STEP is a neutral mechanism capable of completely representing product data throughout the life cycle of a product. The philosophy behind STEP is to make it suitable not only for neutral file exchange, but also to provide a basis for implementing and sharing product databases and archiving (ISO, 1991). Product Data Exchange using STEP (PDES) is the name given to the United States development activity in support of STEP.

During the last decade, a significant amount of work in the development of variant (computer-assisted) and generative (computer-generated) process planning systems has flourished. Generative process planning systems use the product model created by a CAD system to extract features from the model to create process plans. Research in the area of computer aided process planning (CAPP) has been simplified due to the availability of neutral formats (e.g. IGES, STEP) for the digital exchange of product data. The availability of these standard data representation formats has freed the process planner from having to deal with proprietary representations of specific CAD systems.

However, a similar statement cannot be made about the interface between process planning and shop floor control systems. Most of the research in CAPP has focused on the generation of process plans from the product model. This is a hard enough problem, so that most process planners have ignored the other aspect associated with process planning, namely shop floor control. Process plans form the basis for shop floor control, since they define the processing requirements for a product. Even though there are numerous CAPP systems available in the market today, none of them provides process plan data structured in a way that is compatible with SFCS designed by different vendors (Chryssolouris and Gruenig, 1990). This can be directly attributed to the fact that there is no neutral representation format for process plans analogous to the IGES/STEP format for the product model. This analogy is illustrated in Fig. 14.2. Interfaces between CAPP systems and SFCS today remain specific to the application at hand.

The importance of a neutral representation format for process planning system output cannot be overemphasized. In automated flexible manufacturing, parts can be produced using different resources depending on the system status. Traditional operation routing summaries used to represent a sequence of manufacturing activities fail to provide the detail necessary to represent resource and manufacturing process alternatives. It is highly desirable to develop a standard process plan representation model, which is capable of representing process planning information in a neutral format. This will provide a clean interface that

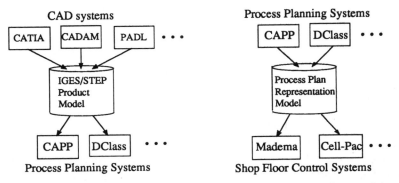

**Fig. 14.2** Analogy between IGES/STEP product and process plan models.

different CAPP systems can hook into and provide input for shop floor control.

With these issues at hand, the rest of this chapter is organized as follows. In section 14.2, we describe the functions performed by a shop floor controller, including planning, scheduling and execution. In section 14.3, we provide an introduction to the design of database systems for representing process planning and engineering data. In this section, we describe the role of conceptual data modeling in designing engineering databases, and the translation of conceptual schemas to database schemas. In section 14.4, we describe two conceptual schemas of process planning output, the International Standards Organization (ISO) schema, and A Language for Process Specification (ALPS) schema. Further, we propose the use of a process flow description capture based on the IDEF3 methodology (Mayer, 1991) to represent process planning output information. In section 14.5, we present a structured approach for the evolution of flexible process plans represented in a standard form into on-line execution plans. We describe this evolution in the context of a hierarchical control architecture of Joshi *et al.* (1990). In section 14.6, we provide a summary of the chapter.

## 14.2   A FUNCTIONAL MODEL OF SHOP FLOOR CONTROL

Davis and Jones (1989) suggest that the key to resolving integration issues in shop floor control lies in developing a better understanding of each manufacturing function and how it is related to other manufacturing functions. In order to understand the interface between process planning and SFCS, it is necessary to understand the functionality of an SFCS. In this section, we outline a functional model of an SFCS. It is assumed that the SFCS receives shop orders from a factory level control system and the process plans from a process planning system. The SFCS is responsible for manufacturing the product mix by the due-date specified in a shop order. In accordance with current shop conditions, the SFCS organizes and distributes, i.e. plans and schedules the activities necessary to manufacture the products specified in an efficient manner. It then downloads processing instructions and monitors the progress of the activities, i.e. executes shop floor activities.

In attempting to define the behavior of an SFCS, various control architectures that allocate decision-making responsibilities to control components of the SFCS have been defined. Dilts *et al.* (1991) trace the evolution of control architectures, where a number of approaches proposed for CIM, including centralized, proper hierarchical, heterarchical and modified hierarchical architectures are identified. However, the basic functionality of a shop floor controller is independent of the control architecture chosen for implementing the functions. The function of controlling a shop floor may be decomposed into three activities, namely planning for manufacture, scheduling the activities generated by the planner and executing the activities at the times specified by the scheduler.

In this section, we examine the planning and scheduling functions of a generic control entity and provide a brief introduction to the execution function. The functional decomposition of SFCS activities presented does not make any assumptions about any specific control architecture. A heterarchical control architecture may be obtained by extending the span of control of the control entity to most/all of the shop floor. A hierarchical control architecture may be obtained by limiting the span of control of the control entity to a section of the shop floor and by defining this control entity recursively over the number of levels in the hierarchy. Figure 14.3 depicts the major functions of a shop floor controller.

### 14.2.1  Planning

Planning may be informally described as the preparation of a set of activities to be scheduled in the future. Planning is initiated by an external order that identifies a set of products and provides due dates for a set of products to be manufactured. By external, we mean that the order is generated from outside the span of control of the control entity. Associated with each product is a process plan, which identifies alternative sets of form features that must be created and resource alternatives required to produce these form features.

The planning activity consists of a number of functions. The planner examines the current status of shop floor entities under its control and charts the future course of action over a planning horizon. In doing so, it takes into account the product mix to be produced, and the relative priorities of the products in the mix. The planner reconciles order priorities with the current status of shop floor entities and develops an initial routing specification of the products. This may involve splitting the order into batches of product, so that the load on all entities is balanced.

Planning also transforms the process plan with alternatives into a production plan that specifies:

- the form features to be produced on each part;
- the resources that will be used to produce each form feature;
- the relative priorities of the product batches.

The transformation may be performed over a number of stages. One possible method is to create intermediate plans by first choosing one of the form feature alternatives, and then choosing a resource alternative

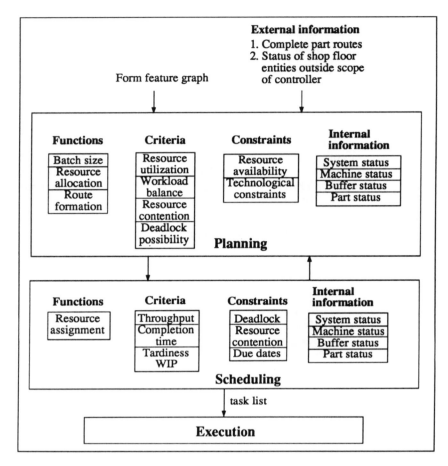

**Fig. 14.3**   Functions of a shop floor controller.

for each form feature. The production plan identifies the minimal set of form features that must be manufactured and related resources but does not necessarily specify the entire processing sequence.

### 14.2.2   Scheduling

If the flow in a facility is strictly linear in nature, and products are never sent back for rework, then developing a unique processing route for each product ensures that competition for resources will not arise. However, non-linear flow results in competition for resources, resulting in a need for scheduling. Scheduling is the assignment of resources to a batch or a part when competition arises, while possibly optimizing a performance measure and satisfying a set of constraints (Bona, 1990).

The current status of the system forms the basis for decision making in scheduling. Constraints on scheduling occur in the form of resource capacities and due dates. In CIM systems, there is also a need to take into account the problem of deadlocking when scheduling activities (Cho, 1992). The result of the scheduling activity is the imposition of a time window on the occurrence of all activities.

### 14.2.3 Execution of shop floor activities

The execution function plays an important role in:

1. monitoring and interpreting input messages;
2. making execution-based decisions;
3. broadcasting messages and downloading machine codes.

The execution function receives messages from the scheduling function that initiates action. The execution function communicates with planning and scheduling functions, and other controller's execution functions (if any). If the execution function is in direct command of a physical shop floor device, then it also communicates with the control unit connected to the physical device. The monitoring activity of an execution function plays the role of an input gate by collecting messages from other controllers and planning and scheduling functions. It also performs semantic interpretation for each message and forwards it to an execution-based decision-making activity. The execution-based decision-making activity may consist of a large number of detailed modules necessary to perform real-time analysis. These modules include manufacturing tasks including part movement, gripper changing on robots, re-fixturing, error handling, report generation, synchronization action and updating the system database. The output activity broadcasts control messages to other controllers and to planning and scheduling functions. At the physical device level, it also downloads device level instructions to the device control unit under its control.

### 14.3 CONCEPTUAL DATA MODELING
### FOR PROCESS PLAN REPRESENTATION

An SFCS uses process plan data, coupled with other information, in order to manufacture the product efficiently. This includes production requirements and inventory levels (demand related), factory configuration and resource status and availability (resource related) and product model (product related). Information related to these aspects is stored in different data structures, so that the shop controller actually uses a network of these data structures for decision making. The information required to drive a shop floor controller is shown in Fig. 14.4.

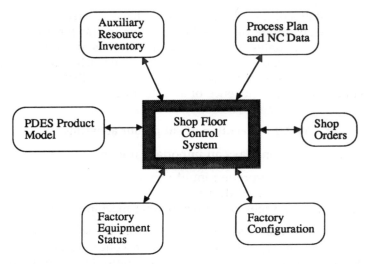

**Fig. 14.4** Information used by the shop floor controller for decision making.

Process plan data used for shop floor control is characterized by complex relationships between the structures that are used to represent the data. Process plans convey a large amount of processing information related to different aspects of controlling a shop floor. A set of form features must be manufactured in order to convert product from raw material form into finished form (product information). Each form feature is associated with one control entity that will execute instructions to create the form feature (primary resource information). Auxiliary resources including tools, jigs and fixtures are used in addition to the primary resource to produce the target form feature (secondary resource information). Process parameters (speed, feed, depth of cut) and machine level instructions (NC instruction, robot program) are associated with the production of a form feature with possible combinations of primary and secondary resources. There are a number of feasible precedence relationships between form features which can be specified using a form feature graph.

In traditional manufacturing systems, process plans have been represented by operation charts and process routing summaries. However, these representations do not meet the needs of flexible computer aided manufacturing, as follows:

1. They are incapable of representing processing sequence alternatives that are required for flexible manufacturing.
2. They are not capable of representing process plans at multiple levels of detail.
3. These representations are not all computer sensible.

4. They do not possess the power to express concepts of parallelism and concurrency of tasks.

There are a number of interrelationships between the data structures used to store the required process planning information. For example, the existence of a process plan instance is predicated on the existence of a product model of the product. A process plan may be represented at a detailed level in more than one way. A form-feature-based representation is independent of the facility configuration and the shop floor control architecture in which it is manufactured. A process plan may also be represented using a resource-based representation, where each node refers to the collection of form features that are produced using the resource as specified by the node. In this section, we introduce conceptual data modeling as a tool to create process plan representation models that will meet the needs of automated manufacturing. Process plan representation models may be converted into database schemas in order to create efficient computer sensible representations of process planning output.

Conceptual data modeling is a technology that has been used to address the issues involved in representing data for engineering applications such as process plan representation. It combines features of traditional database design with knowledge representation techniques (semantic networks) from the field of artificial intelligence (Brodie, 1984; Mylopoulos, 1989). There are a number of conceptual data modeling methodologies that may be employed including IDEF1-X, NIAM, and EXPRESS.

The result of this activity is an understanding of the requirements, expressed as a conceptual schema. The conceptual schema captures the semantics of the information involved in the processes being modeled. A conceptual schema is an abstraction of data about a system from the perspective of a systems designer, based on the synthesis of end user perspectives. The abstraction is independent of database implementation details, data access methods or application specific details (ISO TC184/SC4/WG7 – Document N23). A conceptual schema communicates the engineer's view of information about the world. It provides a method for managing the data model, which is independent from the physical organization of data. The flexibility provided by abstraction provides a basis for integration of different views of data. Conceptual data modeling provides the following mechanisms for managing the complex structure of information (Brodie, 1984):

- **Classification** is used to describe entities and their associations. Entities are abstractions of a system concept, about which we need to maintain information.
- **Association** is used to represent relationships between entities. Networks of relations are formed using association.

- **Aggregation** uses a single entity to represent the composition of several entities.
- **Generalization** represents the relationship between two entities, where one is a specialized kind of the other.

Conceptual data modeling also supports the specification of constraints on data. A number of constraint categories are supported by most conceptual data modeling methodologies (Morris, 1992):

- **Uniqueness**, where a group of attributes uniquely identify an entity.
- **Existence dependence**, when the existence of an entity without another is meaningless.
- **Cardinality** between entities, when it contributes to the meaning of entity relationships.
- **Optionality** of information associated with entities.
- **Domains** of attributes may be restricted.

An application designer translates the conceptual schema into a database schema that is suitable for computer implementation. Several database schemas can support the same conceptual schema. Both the conceptual schema and the database schema provide an abstraction of the information needed for one or more application uses in the real world. The database schema consists of data structures that define organization of data and operations on these structures to define the way in which information is accessed within a database management system. In Fig. 14.5, we illustrate the database design process graphically. Databases and physical files are alternative forms of representing data about process planning output.

In the next section, we describe two conceptual schemas for process plan representation that have emerged in recent years. We also propose a process plan representation method based on the IDEF3 process description capture methodology, and present a physical file representation method for capturing process plan information.


## 14.4   THE ISO PROCESS PLAN CONCEPTUAL SCHEMA

The main intent of the ISO process plan model development is to keep core components of the model as generic as possible so that any process can utilize the same components of the model. In the context of ISO 10303, a process consists of defined instructions required to complete an activity that defines or enhances the product definition. The ISO schema is intended to handle any type of process, and it provides for defining the following:

1. semantics for defining the decomposition of a process into a series of sub processes;

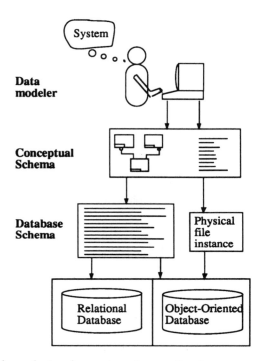

**Fig. 14.5** Database design for engineering applications.

2. control structure for defining the order of execution of the sub processes;
3. conditions which may alter the order of execution;
4. flexible means for defining resources.

The basic idea embedded in ISO process plan schema is the decomposition of a process plan. A process plan is defined recursively using a notion of a process plan activity which is defined as a 'decomposable unit of work'. The process plan activity has an attribute called activities which refers to a process plan set. Each element of the process plan set references another process plan activity, hence forming a recursive structure. In this way, a user may define as many different levels of decomposition as required. Figure 14.6 illustrates the recursive structure of the ISO schema. The ISO schema is developed using the EXPRESS conceptual modeling language. EXPRESS is a textual language by which a universe of discourse for an aspect of product data can be specified. EXPRESS is designed specially to deal with the information that is consumed or produced during product manufacturing.

In the ISO process plan schema, process planning form features are represented as sets of entities. Each entity *process_plan_activity* has 10

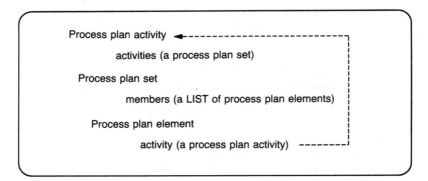

**Fig. 14.6** Recursive structure of ISO plan representation model.

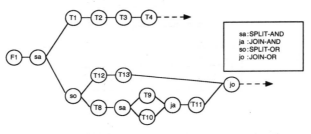

**Fig. 14.7** A form feature graph converted from a design feature graph.

different attributes to define a form feature. In Fig. 14.7, for example, the activity with identification 'T8' will be defined using other reference numbers filling the attributes such as creation date #50, creator #100, approval status #90, replacements '', resources #200, definition #300, parameters #350, shape reference #400.

A part of a STEP physical file representation (Fig. 14.8), to represent the form feature graph of Fig. 14.7 has been created on the basis of the document published in the National Institute of Standards and Technology (NIST) (Van Maanen, 1991). One of the attributes, **definition** may include NC code to process the form feature. Once each form feature has been defined, the precedence relationship among the form features is constructed using entity *process_plan_set*. This entity consists of three sub-types, *serial_process_plan_set*, *parallel_process_plan_set*, and *concurrent_process_plan_set*. The *serial_process_plan_set* further consists of six more sub-types, *sequential_set*, *iterative_set*, *serial_enumerative_set*, *binary_choice_set*, *serial_unordered_set*, and *serial_conditional_enumerative_set*. For example, step number 9 represents a sequential set including form features T12 and T13 which will be executed sequentially.

```
STEP;
HEADER;
ENDSEC;
DATA;
@1=PROCESS_PLAN_ACTIVITY('t8',#50,#100,#90,,#200,#300,#350,#400);
@2=PROCESS_PLAN_ACTIVITY('t9',#50,#100,#90,,#200,#302,#352,#402);
@3=PROCESS_PLAN_ACTIVITY('t10',#50,#100,#90,,#200,#304,#354,#404);
@4=PROCESS_PLAN_ACTIVITY('t11',#50,#100,#90,,#200,#306,#356,#406);
@5=PROCESS_PLAN_ACTIVITY('t12',#50,#100,#90,,#200,#308,#358,#408);
@6=PROCESS_PLAN_ACTIVITY('t13',#50,#100,#90,,#200,#310,#310,#410);
@7=PROCESS_PLAN_SET_ELEMENT(#5,,);
@8=PROCESS_PLAN_SET_ELEMENT(#6,,);
@9=SEQUENTIAL_SET(,2,(#7,#8));
@10=PROCESS_PLAN_ACTIVITY('t56',,,,,,,,#9);
@11=PROCESS_PLAN_SET_ELEMENT(#2,,);
@12=PROCESS_PLAN_SET_ELEMENT(#3,,);
@13=SERIAL_UNORDERED_SET(,2,(#11,#12));
@14=PROCESS_PLAN_ACTIVITY('t23',,,,,,,,#13);
@15=PROCESS_PLAN_SET_ELEMENT(#1,,);
@16=PROCESS_PLAN_SET_ELEMENT(#14,,);
@17=PROCESS_PLAN_SET_ELEMENT(#4,,);
@18=SEQUENTIAL_SET(,3,(#15,#16,#17));
@19=PROCESS_PLAN_ACTIVITY('t1234',,,,,,,,#18);
@20=PROCESS_PLAN_SET_ELEMENT(#10,special_tool?,);
@21=PROCESS_PLAN_SET_ELEMENT(#19,,);
@22=BINARY_CHOICE_SET(,1,(#20,#21));
@23=PROCESS_PLAN_ACTIVITY('t123456',,,,,,,,#2);
....
ENDSEC;
ENDSTEP;
```

**Fig. 14.8**   A STEP file to represent Fig. 14.7.

Step 10 defines the *process_plan_set* made of the two form features T12 and T13 to be a *process_plan_activity*.

While pure AND/OR graphs can be mapped into the ISO conceptual schema in a straightforward manner, the control flow of the schema cannot handle generalized AND/OR graphs, where sub-problem independence is not assumed. Consider the simple precedence graph shown in Fig. 14.9. The graph indicates that activity 5 cannot start before activity 2 and 4 are finished. The activities cannot be represented

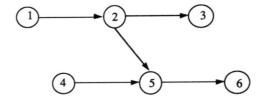

**Fig. 14.9**  Example precedence graph.

using a mutually exclusive set of parallel, serial or concurrent activities defined in the ISO process plan model. Other than this example, there are many cases that the control scheme of the ISO model does not cover. This drawback is due to the fact that the ISO model represents every activity by means of hierarchically structured mutually exclusive subsets whose control scheme is either serial, parallel or concurrent.

### 14.4.1  The ALPS conceptual schema

ALPS was designed as a generic process plan schema for discrete-process manufacturing at NIST. The development of ALPS took place in parallel with the ISO standardization efforts to define a standard process plan schema. However, unlike the ISO process plan conceptual schema, ALPS is intended to be an experimental language, not an international standard. The ALPS specification is built around a directed graph structure. Each node of ALPS has its own attributes as well as system defined attributes including node identifier, node name, node type, previous nodes and next nodes. The latest version of ALPS (Ray, 1992) allows variables, expressions and parameters in it.

Seven major classes of graph nodes are defined in ALPS: termination, task, split, join, synchronization, resource and information. Termination nodes consist of start node and end node. Task node defines primitive task or decomposable set of tasks. In conjunction with join nodes, split node defines branching specification of different processing paths. Resource nodes are for allocating and de-allocating resources to an activity. Synchronization nodes provide several means for coordinating multiple parallel tasks. Information nodes provide general purpose operations like database queries. More detailed description of ALPS can be found in Catron and Ray, 1991 and Ray, 1992.

ALPS is intended not only to represent process plans, but is also intended to capture the control perspective of SFCS by defining synchronization nodes and information nodes that carry control information. However, it does not possess the power to express arbitrary constraints on relationships between nodes, which is necessary for capturing the control perspective in a plan.

### 14.4.2 Process description capture and process plan representation

The primary requirements for any method that will capture flexible process plans for automated discrete part manufacture include:

- representation of alternative processes for processing a task;
- alternative resources that facilitate or support the processing of each task;
- text file representation that can be parsed by a computer program.

Here, we present a method for representing process plan information using the IDEF3 (Mayer, 1991) process description capture methodology, that meets the above requirements. The conceptual data modeling methodology presented earlier is designed to allow a description of the information that an analyst deems important to manage, in order to accomplish desired objectives. However, process plans can also be described from a process centered point of view. Using this method, a process plan can be described in terms of the tasks that need to be performed, the resources that participate in each of the tasks and any parameters associated with the tasks. Also, process flow descriptions are the most natural way for an expert process planner to model information about the process planning domain. The IDEF3 methodology provides a structured way to communicate this description. The structured process-centered description of a process plan can be constructed and saved in a neutral text file, which then provides input to an SFCS.

We now present a two-level hierarchical decomposition of a process plan using the IDEF3 concepts described. An important advantage of using this representation is the use of a flexible process plan kernel that can be transported between facilities and customized to suit the needs of a particular facility in an easy fashion. At the topmost level, the generic kernel of a process plan defines the alternative sets of form features that will be produced. This description is independent of the resources and resource parameters that will participate in each of the tasks. At the lower level in the decomposition, resource alternatives and process parameters associated with the production of each form feature are specified. The top level representation is generic in the sense that a physical file representation of this level is freely transportable between two facilities. The lower-level decomposition is customizable to the user's needs.

The IDEF3 process flow description method provides a structured method for expression of a domain expert's knowledge about how a particular system or organization works. A process plan is viewed as a collection of processes or tasks to be performed. In the context of process plan representation, an IDEF3 process represents the production of a form feature. The objects that participate in the process are the

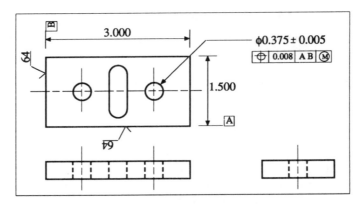

a) Bracket Design

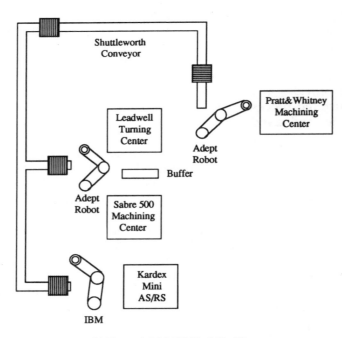

b) Texas A&M CIM Lab Facility

**Fig. 14.10**  Illustrative example of process plan representation.

| Oper # | | Feature | Time | Tooling |
|---|---|---|---|---|
| 1 | Side Mill | Loc surface | 0.6 min | 1.0" end mill |
| 2 | Side Mill | Loc surface B | 0.4 min | 1.0" end mill |
| 3 | Twist Drill | Hole #1 | 0.3 min | 0.3595" drill |
| 4 | Twist Drill | Hole #2 | 0.3 min | 0.3595" drill |
| 5 | Ream | Hole #1 | 0.3 min | 0.375" ream |
| 5a | Bore | Hole #1 | 0.3 min | 0.375" bore |
| 6 | Ream | Hole #2 | 0.3 min | 0.375" ream |
| 6a | Bore | Hole #2 | 0.3 min | 0.375" bore |
| 7 | Slot Mill | Slot #1 | 0.8 min | 0.5" end mill |

c) Operation routing summary

**Fig. 14.10**  *Continued*

primary and secondary resources used to manufacture the form feature. We provide an illustrative example of a process plan representation using the IDEF3 methodology in order to provide an introduction to IDEF3 and describe how process plans are represented. In Fig. 14.10(a) we provide an illustrative bracket. Its operation routing summary for the facility in Fig. 14.10(b) is shown in Fig. 14.10(c). The graphical IDEF3 representation of the top level process plan for this part is shown in Fig. 14.11(a).

The IDEF3 method is based on knowledge acquisition centered around process flow relations, with the identification of objects participating in the processes and object state transitions. It uses the notion of a scenario as the basic organizing structure for establishing boundary conditions and the focus for describing the process plan. Every process description has a scenario associated with it at the top level. A scenario binds the context of a description and provides the situation in which processes occur. Process flow description diagrams are the primary means used for capturing process planning output using IDEF3. In Table 14.1, we provide a description of each of the building blocks of IDEF3 used for process plan description. Detailed descriptions of each of the blocks are provided below.

### Unit of behavior (UOB)

In the context of process plan representation, a UOB represents an activity or an event. For example, in Fig. 14.11(b), 'mill locating surface

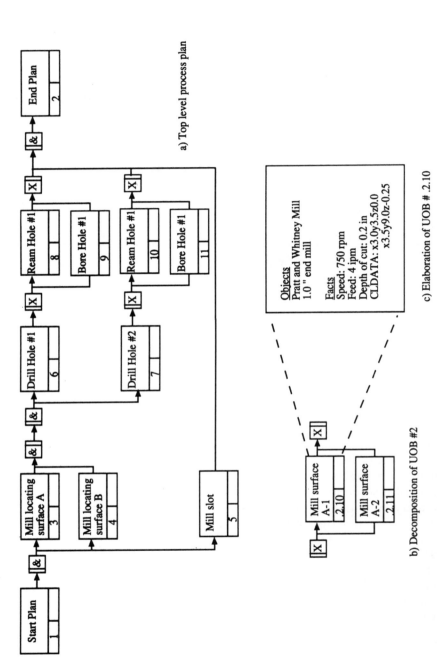

**Fig. 14.11** Process plan representation using IDEF3 description capture.

**Table 14.1**  Building blocks of IDEF3 for process plan representation

| name | description | symbol |
|------|-------------|--------|
| Unit of behavior (UOB) | Captures a description of 'what's going on'. | Start Plan / 1 |
| Elaboration of a UOB | Describes a UOB in terms of participating objects and relations. | Objects / Facts |
| Decomposition of a UOB | Describes a UOB in terms of other UOBs. | |
| Object | Agents and actors that participate in the activation of a UOB. | |
| Fact | Describes properties of objects and relations between objects. | |
| Link | Denotes significant relation between UOBs. | → |
| Junction | Highlight constraints on sequencing relations among UOBs. | |

A' and 'mill slot' are activities that are self explanatory in nature. The first and the last UOBs, namely 'start plan' and 'end plan' are events that represent the beginning and ending of plan execution.

### Elaboration of a UOB

Figure 14.11(c) shows the elaboration of the UOB 'mill surface A'. This elaboration identifies two resources, 'Pratt and Whitney mill' and '1.0" end mill' as the objects that participate in this activity. Objects participating in the UOB can be tagged differently. In process plan representation, an object is either: a) an agent or primary resource (Pratt and Whitney mill), which is considered the primary effector of the UOB, or b) affected or secondary resource (1.0" end mill), wherein relations to that object are changed by or during the UOB. Facts about the activity include process parameters for the combination of the primary and secondary resources.

### Decomposition of a UOB

The decomposition of a UOB provides a description in terms of other UOBs. As mentioned earlier, a process plan is represented using a two-level decomposition. At the top level, there are no references to resource alternatives, if they exist. The top level in the plan provides the possible form feature alternatives as shown Fig. 14.11(a). Activities that have a number of resource alternatives are decomposed at a second level. For instance, Fig. 14.11(b), the activity 'mill locating surface A' is decomposed into two activities that are linked by exclusive OR junctions. This means that either of the two activities in the decomposition must be completed in order to perform the activity at the top level. Resources and process parameters are provided for each combination of primary and secondary resources at the lower level. Thus decomposition of a UOB provides a means to describe resource alternatives for producing form features.

### Object

Objects in a process plan refer to resources that take part in the activation of a UOB. As mentioned earlier, objects are of two types, namely primary (agent) and secondary (affected) resources, depending on whether they are considered effectors of a UOB.

### Fact

The facts listed in the elaboration of a UOB provide the characteristics of the resources in the UOB, including process parameters (feed, speed, depth of cut, CLDATA) associated with the UOB.

### Precedence link

Precedence links express simple temporal precedences between the instances of a UOB type with another UOB type.

### Junction

Junctions are used to highlight constraints on possible precedence relations between UOBs. Two major types of junctions are used, namely AND and XOR junctions. The AND junctions may have synchronous and non-synchronous interpretation, allowing for paths to start or end together as the case may require. The junction types and their descriptions are shown in Table 14.2.

**Table 14.2** Junction types and their semantics

| Junction type | symbol | meaning |
|---|---|---|
| Asynchronous AND | ‖&⌉ | Each of the following paths will start and all UOBs will eventually happen. |
| Synchronous AND | ‖&‖ | Each of the following paths will start together and all UOBs will happen. |
| XOR | ‖X⌉ | Exactly one of the following paths will start and only one of the UOBs will happen. |

a) Front of link semantics for junction types.

| junction type | symbol | meaning |
|---|---|---|
| Asynchronous AND | &‖ | Each of the preceding paths must finish |
| Synchronous AND | ‖&‖ | Each of the preceding paths must finish at the same time |
| XOR | X‖ | Exactly one of the preceding paths will have completed |

b) Back of link semantics for junction types.

### 14.4.3 Physical file representation of the process plan

We use Standard Generalized Markup Language (SGML), a like language, as a neutral text file representation to capture process planning information described using IDEF3. In general, the text file can be viewed as a collection of attribute/value pairs, where attribute is the name of a field and value is the value of the field. The text file is broken into the following major sections:

- file descriptor;
- project summary information;
- pool information;
- process decomposition information.

*File descriptor*

This is a single line of text appearing at the beginning of the file. It consists of a descriptive label, version information and date/time of last modification.

### Project summary information

This section defines the purpose, scope and need of the project.

### Pool information

Pools include the process pool, object pool, fact pool, link pool and junction pool. Each pool is listed separately, and all pool items are contained in the text file. Processes belonging to the process pool also have an elaboration field which consists of a list of objects participating in the process and facts about the process. The link pool contains all the link IDs occurring in the diagram, the cardinality and type of each link and its origin and destination.

### Process decomposition information

Process decompositions have a parent field which identifies the process that owns the decomposition.

## 14.5   EVOLUTION OF PROCESS PLANS IN THE SHOP FLOOR

In this section we describe the evolution of process plans from an off-line representation to an execution plan that is downloaded to equipment for execution (Cho, 1994). For illustrative purposes, a form feature graph created by off-line process planning forms the principal input to the SFCS. It is assumed that implementation of the process plan occurs in a facility that is hierarchically structured into three levels (shop, workstation, equipment) for control purposes (Joshi, 1990). The process plan can also be structured accordingly to reflect the processing steps required at each level in the control hierarchy. The generated form feature graph is converted into a set of hierarchical process plans. Even though the form feature graph describes alternative resources to finish the form features and form feature precedence relationships, it is not tailored to take into account the current facility status and nor does it take into account the shop configuration, (for example, which piece of equipment belongs to a particular workstation). In other words, the form feature graph has been created on the basis of primitive features that do not reflect the hierarchical nature of the shop floor. The shop floor is structured based on the resources used to produce form features. The planning functions of each level in the hierarchy are responsible for decomposing the form feature graph into a set of hierarchical process plans that correspond to the three levels of the hierarchy. This decomposition is resource based. The overall architecture of this conversion mechanism is given in Fig. 14.12 (Cho, 1994).

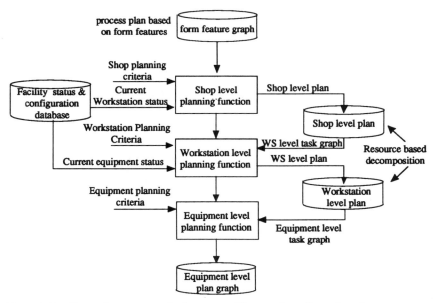

**Fig. 14.12** Form feature graph decomposition and on-line planning.

The shop level planning fuction receives a series of job orders associated with the form feature graphs from the factory level controller. It then creates a shop level plan graph in which a node represents a workstation form feature graph that defines the set of activities performed by a workstation controller. Finally it generates the best workstation sequence in terms of workstation availability, resource availability and travel times between workstations. Table 14.3 depicts the taxonomy of planning functions at each level in the control hierarchy.

The workstation controller receives an order and its related form feature graphs from the shop level controller. The workstation planning function then prepares the activities of the various equipment and generates a workstation level plan graph in which a node represents an equipment form feature graph that specifies equipment activities and an edge represents precedence relationships. The best workstation level route from the workstation level plan graph is obtained by considering machine breakdown, resource availability and machine utilization.

The equipment form feature graph equivalent to an equipment level plan graph can be serialized into a large number of operation sequences. Heuristics, such as number of tool changes required are used to prune the multiple sequences that do not seem promising or are unrealistic. The node and edge costs for each operation sequence are computed to select the best sequence. Node cost computation consists of optimizing and/or selecting appropriate machining parameters for each form fea-

**Table 14.3**  Evolution of process plan from the shop down to the equipment

| level | form feature graph | plan graph | route graph |
|---|---|---|---|

**Shop**

form feature graph

w1 — t0-sa, t1-sa, t2, t3, ja-t4, t5-t6, ja-t7-t8 — w2

plan graph: W1 — W2

route graph: workstation2 | W1 — W2 | workstation1

**Work-station**

Workstation task graph w1

o1: t0-sa — o2: t1-sa, t2, t3, ja-t4, t5-t6, ja — o3

plan graph: o1-sa, o2, o3, ja

route graph: machine1 | o1 o2 o3 | machine1 machine2

**Equip-ment**

Equipment task graph o2: t1-sa, t2, t3, ja-t4

Equipment plan graph: t1-sa, t2, t3, ja-t4

route graph: tool1 tool4 | t1 t2 t3 t4 | tool1 tool2

ture and computing the machining time and cost associated with each form feature based on the selected parameters. Edge costs represent non-cutting times, including tool change and rapid travel between form features. An equipment level route is selected using the node and edge costs. NC code is generated using the CLDATA information between form features for each edge in the sequence and a single file consisting of G and M codes is created. The file may now be downloaded to the equipment control unit for execution.

## 14.6   CONCLUSION

In addressing the integration of CAD, CAPP and CAM, this chapter provides a structured approach to the integration of process plans and shop floor control. We have emphasized the importance of process planning output representations for integration of process planning and shop floor control. A description of the current approaches to process plan representation has been provided. The drawbacks of the current approaches have been highlighted, and a representation method based on process description capture provided. We have provided an illustrative example of process plan representation using process description capture methodology. In the last section, an approach to planning for hierarchical shop floor control, based on the process plan representation is provided. The approach briefly describes the integration issues that arise when an interface between process planning and shop floor control is considered in a hierarchical control architecture. A form feature graph is decomposed into a hierarchical set of process plans, that reflect the resource requirements at each level in the control hierarchy. The output is a sequence of NC instructions that may be directly downloaded to the relevant piece of shop floor equipment. It is hoped that further research into process plan representation will result in the evolution of a standard representation method that will promote integration of process planning and shop floor control.

## REFERENCES

Bona, B., Brandimarte, P., Greco, C. *et al.* (1990) Hybrid hierarchical scheduling and control systems in manufacturing. *IEEE Trans on Rob and Autom*, **6**(6), 673–86.

Brodie, M.L., Mylopoulos, J. and Schmidt, J. (eds) (1984) *On Conceptual Modeling: Perspectives from Artificial Intelligence, Databases and Programming Languages*, Springer Verlag, NY.

Catron, B.A. and Ray, S. (1991) ALPS: A language for process specification. *Int Jnl. of Comp Integ Manu*, **4**(2), pp. 105–13.

Cho, H., Derebail, A., Hale, T. and Wysk, R.A. (1994) A formal approach for integrating process planning and shop floor control. *Jnl. of Eng for Ind*, **116**(1), pp. 108–16.

Cho, H., Kumaran, T.K. and Wysk, R.A. (1992) A graph theoretic approach to deadlock detection and avoidance in manufacturing systems Texas A&M University, Industrial Engineering Working Paper No. 92–21.

Chryssolouris, G. and Gruenig, I. (1990) *Process Planning Interface for Intelligent Manufacturing Systems*, MIT Laboratory for Manufacturing and Productivity.

Davis, W.J. and Jones, A.T. (1989) A functional approach to designing architectures for CIM. *IEEE Transactions on Systems, Man and Cybernetics*, **19**(2), pp. 164–74.

DeFazio, T.L. and Whitney, D.E. (1987) Simplified generation of all mechanical assembly sequences. *IEEE Jnl. on Rob and Autom*, **RA-3**(6), pp. 640–58.

Dilts, D.M., Boyd, N.P. and Whorms, H.H. (1991) The evolution of control architectures for automated manufacturing systems. *Jnl. of Manuf Sys*, **10**(1), 79–93.

Homen DeMello, L.S. and Sanderson, A.C. (1991) A correct and complete algorithm for the generation of mechanical assembly sequences. *IEEE Trans on Rob and Autom*, **7**(2).

Homem DeMello, L.S. and Sanderson, A.C. (1990) AND/OR graph representation of assembly plans. *IEEE Trans on Rob and Autom*, **6**(2).

Homen DeMello, L.S. and Sanderson, A.C. (1991) Two criteria for the selection of assembly plans, Maximizing the flexibility of sequencing the assembly tasks and minimizing the assembly time through parallel execution of assembly tasks. *IEEE Trans on Rob and Autom*, **7**(5).

ISO TC184/SC4/WG7 Document N23, Release 3 (1991) *Functional Requirements for a STEP Data Access Interface*.

Joshi, S., Wysk, R.A. and Jones, A. (1990) *A scalable architecture for CIM shop floor control*. Proceedings of CIMCON'90 pp. 21–33.

Mayer, R.J. *et al.* (1991) *The IDEF3 process flow description method practice and use manual*, Knowledge Based Systems, Inc., College Station, Tex.

Morris, K.C., Mitchell, M. and Dabrowski, C. (1992) *Database Systems in Engineering*, NISTIR 4987, National Institute of Standards and Technology, Gaithersburg, Md.

Mylopoulos, J. and Brodie, M.L. (eds) (1989) *Readings in Artificial Intelligence and Databases*, Morgan Kaufmann Publishers Inc, San Mateo, Ca.

Nixon, B. and Mylopoulos, J. (1990) *Integration Issues in Implementing Semantic Data Models*, in Bancilhon.

Ray, S.R. (1992) Using the ALPS process plan model. *Factory Automation systems Division, NIST*, Gaithersburg, Md.

Sanderson, A.C., Homem De Mello, L.S. and Zhang, H. (Spring 1990) Assembly sequence planning, *AI Magazine*, pp. 62–81.

Van Maanen, J. (1991) STEP Part 21-*Clear text encoding of the exchange structure*.

*Webster's New World Dictionary* (1985) School and Office Edition, The World Publishing Company, Cleveland, Ohio.

Wysk, R.A. (1992) *Integration requirements for the engineering of a product.* AUTOFACT92 Proceedings, Detroit, Mich.

# CHAPTER 15

# Integration of cutting-tool management with shop-floor control in flexible machining systems

*D. Veeramani*

## 15.1 INTRODUCTION

The problem of controlling manufacturing systems has existed ever since the first time that a collection of machines and operators were pooled together to form a factory. The fundamental issue underlying the control of any manufacturing system is that of organizing the resources and orchestrating the activities on the shop-floor in a manner that enables desired objectives to be met. Over the decades, the manufacturing system has undergone major structural changes, evolving from a congregation of manually operated machine tools into an integrated system of computer numerically controlled (CNC) machines. This metamorphosis of manufacturing systems has also been accompanied by marked changes in the philosophy that has guided the operation of these systems, namely from that of mass production to one of being agile and responsive to customer demands.

The emphasis in manufacturing system control has traditionally focused almost exclusively on issues related to workpieces (such as due dates) and machine tools (such as machine utilization). The validity of such parochial approaches to shop-floor control is questionable in the context of modern flexible production systems. This is due to the significant differences between the machines that comprise conventional factories and those in modern factories. Conventional machine tools, such as a drill press or a shaping machine, are capable of performing a limited number of similar operations. In essence, if a workpiece required a variety of operations, it often mandated that the workpiece be processed on a number of different conventional machine tools, each performing a specific operation on the workpiece. Modern CNC machines, on the other hand, are capable of performing a wide variety of operations such as milling, drilling, boring, etc. Some machines, such as mill/turn centers, are even capable of performing traditionally 'disjoint' operations, such as turning of cylindrical surfaces and milling of flat surfaces, on the same workpiece in one set-up. The key point to note here is that the capabilities of these modern machining centers are defined to a great extent by the cutting-tools that are made available to these machines. Hence, the manner in which cutting-tools are distributed among the machines has a significant impact on the capability of the manufacturing system. Therefore, in modern manufacturing systems, it is no longer possible to ignore the role of cutting-tool management (CTM) in controlling the manufacturing system. (Some of the early FMSs unfortunately learned this lesson the hard way.) Lack of attention to cutting-tool related issues can prevent an FMS from reaching its fullest potential and can make it 'inflexible' in practice.

Fortunately, advances in manufacturing hardware and software technologies now enable the consideration of CTM as an integral part of shop-floor control. For instance, the ability to transport tools automatically (from the tool-crib to the machines and between machines), and to maintain tool-specific information on a chip in the tool holder and at the machines, now allow the shop-floor control system to coordinate the flow of each cutting-tool within the manufacturing system. These technologies now allow the promotion of cutting-tools to the same cadre as workpieces and machines from the perspective of shop-floor control. Indeed, in modern flexible machining systems, the shop-floor control system needs to ensure that the correct workpiece **and** the correct cutting-tools are made available at the right machine and at the right time.

The integration of CTM with shop-floor control requires the establishment of an effective system to plan, control and monitor cutting-tool usage. CTM is multi-faceted and affects various functions within an organization, such as design, purchasing, process engineering, production planning, tool preparation and shop-floor operation. Many companies have already realized significant benefits in terms of reduced cutting-tool inventory and costs, reduced production delays (caused by unavailability of tools), improved productivity and system flexibility and better shop-floor control. CTM is therefore attracting increasing attention within manufacturing organizations (Veeramani, Upton, and Barash, 1992).

This chapter examines some key aspects of CTM and provides insight into how a CTM system can be integrated with shop-floor control in FMSs. The chapter is organized as follows. The next section describes a structured approach to rationalization of cutting-tool inventory, the first step towards the establishment of an effective CTM system. The following section examines the design of an information system that forms the backbone of the CTM system. The next three sections in the chapter discuss the role of CTM in the context of shop-floor control, with emphasis on policies for tool allocation, tool replenishment and tool flow control. Finally, the impact of the shop-floor control architecture on CTM is examined.

## 15.2   RATIONALIZATION OF CUTTING-TOOL INVENTORY

The first step towards a successful implementation of a CTM system is to rationalize the existing cutting-tool inventory. This requires a critical examination of the tool inventory with the objective of reducing it to a 'standardized' set of tools of manageable size. In most companies involved with machining, there is often no mechanism for finding out tool-related information such as the total number of tools in the factory, the amount of money spent each year in perishable tooling and the utilization of each type of tool. Even if there exists some system that can provide this information, it is rarely available in a form that enables its seamless integration with the shop-floor operation and control system. Thus, companies usually do not have in place a system that readily facilitates rationalization of cutting tools. This often forms a major deterrent for initiating the implementation of a comprehensive CTM system.

The need for cutting-tool rationalization stems from the high varieties and numbers of cutting tools that are typically found in machining systems. This has been primarily due to the fundamental differences among the machining operations that necessitate different types of tools (for instance, drilling and milling operations require different types of tools). Another reason that has led to the proliferation of cutting tools in machining systems is the sequential approach to product and process development that has been traditionally followed. This 'over the wall' approach discourages communication among the design and manufacturing departments, thereby requiring the purchase of new cutting tools which could have been avoided by minor modifications to the design. Indeed, the cutting tool manufacturers have also indirectly contributed to this matter by making available for purchase a large variety of cutting tools. For instance, a study of various cutting-tool manufacturers indicated that it is possible to purchase around 85 different sizes of standard, jobbers length, straight shank, HSS twist drills in the 0.25″ to 0.50″ range. Clearly, the cutting tool manufacturers are producing these standard tool sizes because some machining company requires them. If a conscious effort is made during the design process towards using only the tools that are available in the shop-floor, the increase in tool inventory due to purchase of new tools can be reduced. The growing appreciation of the benefits of concurrent engineering is now facilitating this process.

The trend towards increased product variety and smaller product life cycles has resulted in a growing pressure on manufacturing systems to be flexible and responsive to dynamic production requirements. Many companies fail to realize the contribution of cutting tools towards cost; in some machining systems, tools can account for as much as 25% of the manufacturing operating costs (Tomek, 1986). In addition, some of the costs, for instance due to production delays caused by unavailability of proper tools, are not quantifiable. Yet they can have significant economic ramifications. In these days of time-based competition, companies want to prevent situations in which an order is lost due to the lack of a tool. This can promote the maintenance of higher tool inventories to guard against unexpected scenarios. Such increased and unnecessary levels of tool inventory can be avoided by proper CTM.

Developing a method for rationalization of cutting-tool inventory is not a trivial task. Not only must this method provide for 'standardization' of existing tool inventory, but it should also provide a means by which the tool inventory can be continuously upgraded as the production requirements change and with changes in cutting-tool technology. The difficulties associated with rationalization of cutting-tool inventory has caused some companies to make 'half-hearted' attempts at establishing CTM systems that eventually proved to be far less effective than anticipated.

The task of cutting-tool rationalization is further complicated by the fact that a cutting-tool is often an assembly of a number of components, such as tool holder, adapter, body and insert. Many companies have adopted *ad hoc* approaches to tool rationalization. Part of the difficulty in developing a strategy for tool rationalization arises due to differences in the production environment within which a manufacturing system functions. Some companies have based their tool rationalization strategy by focusing on tool requirements for each part family. For instance, a Japanese company reportedly had a team of designers evaluate the tool requirements for a family of housings, and were successful in reducing the tool requirements from 600 to 60 (Hartley, 1984). Others have chosen to focus on specific cutting-tool components such as inserts. One automotive company reportedly reduced the number of square inserts from 3280 to 270 (Gayman, 1987). Indeed, some studies suggest that about 50% of the average tool inventory consists of obsolete tooling (Boyle, 1986). Hence, it is not surprising that significant reductions in cutting-tool inventory are possible through rationalization of tool requirements.

An approach to rationalization of cutting-tool inventory is now presented. Indeed, the proposed methodology would require modifications depending on the nature of the production system at hand. The key to success in this task is discipline, information sharing and cross-functional team involvement. The proposed approach follows three stages: information gathering, rationalization of tool inventory and continuous improvement of tool inventory.

The first step in initiating a cutting-tool rationalization exercise is to form a 'task force' with representatives from all areas within the manufacturing company that are affected by cutting-tool related issues. A typical task force would consist of members from design, manufacturing engineering, purchasing, tool preparation, production planning and shop-floor areas. The first task of this team would be to organize an information system that would integrate and cater to the needs of the various departments.

Armed with information about the cutting-tool inventory, the next step is to examine the tooling requirements of the products fabricated at the facility. The manner in which this evaluation is conducted would depend to a certain degree on the nature of the product mix.

If the company manufactures a relatively stable, well-defined set of products, then the team should first classify the parts into families on the basis of their cutting-tool requirements. (Parts having similar cutting-tool requirements would qualify to gain membership in the same family.) The next step is to examine the tool requirements for parts **within** each family and to identify opportunities for reduction of the number of tools required by substituting tools and eliminating redundant tools. Subsequently, opportunities for trimming the tool inventory further need to be identified by rationalizing tool requirements **across** part families (Veeramani, 1992a). Since the team is cross-functional, the process of cutting-tool rationalization can take into account the ramifications of the decisions that are made including those related to design changes to facilitate tool inventory reduction.

A more general approach is to rationalize the cutting-tool inventory on the basis of frequency of tool usage. To perform this analysis, the team needs to examine the products that were manufactured over the past year or two and those scheduled for manufacture over the next few months (if known). Based on the tool requirements for these parts, the tool inventory can be classified into groups according to the number of times that tools in each group were used. For instance, a coarse classification might be:

1. frequent use;
2. average use;
3. infrequent use;
4. never used over the past two years.

The tools that have a history of low usage are potential targets for elimination. All ramifications of elimination of these tools have to be considered before any irrevocable decision is made. For instance, there may be opportunities for utilizing some component of the tool (such as the insert) for other tools. Thus, some parts of these tools can be salvaged for continued use in the system. On the other hand, the availability of new, high-performance inserts in the tool inventory may have made the older inserts obsolete. In such a case, there is no risk associated with eliminating the older inserts. The rationalization of cutting-tools on the basis of frequency of usage has to be done with caution. The usage relates primarily to the number of parts on which it was used and to a secondary extent to the number of tools of a type that was utilized. The total consumption of a particular tool type is not necessarily a good indicator of its 'goodness'. For instance, in an aerospace company, the poor performance (tool life) of a particular type of insert resulted in a high consumption rate for it. The purchasing department misinterpreted this high consumption rate as a sign of approval for the insert and therefore ordered large quantities of the insert. Clearly, this situation arose because of a lack of communication among the parties concerned. This example serves to emphasize the importance of adopting a cross-functional team to perform the tool rationalization exercise.

Another aspect of tool rationalization relates to the number of duplicates of different tool types that are maintained in the system. By tracking the tool requirements for jobs in the production horizon, and by controlling the distribution and flow of tools in the shop-floor and their availability in the tool preparation and maintenance areas, it is possible to reduce the number of duplicate tooling that needs to be maintained. For instance, the availability of an automated tool handling system integrated with the shop-floor control system would enable the sharing of cutting-tools between machines, thereby reducing the need for redundant copies of tools on each machine.

A difficult aspect of tool rationalization is that relating to design modifications which would enable a spectrum of parts to be made with a relatively small set of tools. Opportunities for such savings exist when designs have been made without consideration to manufacturability (specifically, tool availability) issues. Such situations are common in most organizations where product designs have been developed over a period of time by different groups of people. In such cases, it may be possible to redesign parts, without compromising their functionality, so that the modified designs result in a smaller variety of cutting-tools and prevent the purchase of new tools. For instance, if it is found that the tool requirements for parts to be made in the forecasted future require tools that are not available in the current tool inventory, then the team should re-examine the design with the aim of modifying those features that necessitate the purchase of new cutting-tools.

At the end of the second phase of the tool rationalization exercise, the team should have arrived at a preferred set of tools for use in machining products on the shop-floor. (The design of new products and the process planning for these products should attempt to utilize this preferred set of tools.) In order to take advantage of developments in cutting-tool technology and to accommodate changes in the production requirements for the manufacturing system, a method for continuous evaluation and improvement of the tool inventory needs to be put in place. This aspect of the tool inventory rationalization process is perhaps the most difficult one to sustain. It requires that the organizational culture be transformed into one that adopts a concurrent engineering approach to product and process design in place of the traditional sequential approach. Continuous evaluation of the tool inventory, specifically the preferred tool set, would enable the company to further trim the tool inventory (without sacrificing operational flexibility) by incorporating multi-functional tools (for instance, the CUT-GRIP turning tools available from ISCAR) and modular tooling systems.

Rationalization of cutting-tool inventory also helps improve the organization and operation of the tool crib. The cutting-tools can be arranged in a manner that enables frequently used tools to be easily accessible to the tool crib attendant. This will not only make the tool crib operation more efficient but will also assist in the continuous improvement of the tool crib inventory.

Rationalization of the cutting-tool inventory is achieved through a process of evolution. With time, through continuous evaluation of the cutting-tool inventory, the company can not only eliminate redundant and obsolete tooling, but maintain the most effective set of cutting tools that would enable the production of parts to a high quality and at a low cost while maintaining a high level of system performance and flexibility.

Cutting-tool rationalization offers many challenging issues that require further investigation. The first relates to the development of a systematic procedure for evaluation of designs (both old and new) from the perspective of cutting-tool rationalization. The second involves the development of an effective means of continuous evaluation and improvement of the cutting-tool inventory. Both these issues require a team-oriented approach. The dynamics of the team effort in this exercise is yet another area for further research. In addition, analytical models that capture the ramifications (economic and otherwise) of cutting-tool rationalization need to be developed.

## 15.3 TOOL INFORMATION MANAGEMENT

At the foundation of every effective CTM system is an information system that encompasses all facets of tool-related information. Indeed, many companies have equated tool information management to CTM.

It is important, however, to recognize that CTM goes beyond tool information management and covers a variety of activities that range from rationalization of cutting-tool requirements at the product design phase to orchestration of tool flow on the shop-floor.

The key to implementing an effective tool information management system is to recognize the various departments and functions within the organization that are impacted by cutting-tool related issues, identifying the information needs of each of these areas, integrating the information available on cutting-tools and making it available in a manner that facilitates decision-making in accord with the tool management objectives. In many organizations, information related to cutting-tools is typically collected and stored in a distributed fashion within various departments in the company. For instance, tool information might be maintained in the tool crib using a simple spreadsheet software package, in the purchasing department using a database package, and in the process engineering department on a mainframe computer. Integrating and sharing information from these various sources is not a trivial task.

To enable a better understanding of the requirements of a tool information management system, we now examine the tool-related information needs of various areas within an organization and the interaction between them. Cutting-tool related information is relevant primarily for performing the following functions: product design, purchasing, process planning, tool crib operation, tool preparation and maintenance and tool control on the shop-floor (including tool allocation, tool handling, tool monitoring and tool replenishment).

At the product design phase, it is desirable to provide the designer with the ability to evaluate the manufacturability of the part, specifically in terms of the cutting-tools available in the tool crib. There are different mechanisms (dependent on the sophistication of the design tool) by which the designer could be advised about the compatibility of the shape features in the design to those that can be generated by a preferred set of tools. Modern CAD/CAM systems are striving towards providing the ability to generate alternative process plans at the design stage itself, thereby enabling the analysis of the design from the perspective of manufacturing. Such systems would assist in not only identifying 'non-standard' features that require the purchase of new tools or form tools, but could also provide feedback to the designer on how to modify the design so as to conform to the cutting-tools available in the tool crib. Till such truly integrated, comprehensive CAD/CAM systems become available, development of process plans will continue to be done by the manufacturing engineering department (or possibly in a team-based approach that includes representatives from other areas, particularly design, in keeping with the concurrent engineering philosophy). In developing the detailed process plan, information regarding the types of cutting-tools available (including geometry, identification number, material and tool life), the suitability of various cutting-tools for the desired operations, performance history, etc. needs to be made available.

The purchasing department needs to maintain information about in-house requirements for cutting-tools and cutting-tool suppliers. To be effective, the information system should facilitate interaction between the purchasing department and other areas such as the tool crib and the process planning department. This would enable the purchasing department to inform the other areas about new cutting-tools or good purchase opportunities, and on the other hand, to receive input on or requests for specific cutting-tools. The availability of a seamless tool information system would allow the purchasing department to assume the role of an active intermediary between the company and the cutting-tool manufacturers as opposed to being a passive procurer of cutting-tools.

The tool crib, including the tool preparation and maintenance areas, forms the nucleus of the CTM system on the shop-floor. Information on each type of cutting-tool is maintained at the tool crib including a tool identification number, nominal dimensions and assembly components (including tool holder, adapters, and inserts). This information facilitates the preparation of cutting-tools for use on the shop-floor, retrieval of tools from the tool crib, tracking of tool use, requisition of tools through purchasing, preparation of tool maintenance (regrinding) tasks and interaction with the shop-floor control system.

It is important to recognize that part of the complexity of cutting-tool information management stems from the heterogeneity of cutting-tool data types. For instance, some types of information are best expressed in terms of geometry, others in terms of numbers and letters, and yet others in terms of structures for knowledge representation (such as frames). Ideally, the information system should be based on a schema that enables all these different types of data to be represented and retrieved in a unified manner. The development of such an integrated cutting-tool information system is essential for implementing a comprehensive CTM. However, the difficulty associated with developing such a system has proven to be a major hurdle in developing comprehensive CTM systems.

In addition, cutting-tool information changes with time, some more rapidly than others. Therefore, different mechanisms for information collection, storage, and retrieval need to be considered in designing the cutting-tool information management system. For instance, information pertaining to the life remaining on a cutting-tool changes continuously during the machining operation. The contents of a tool magazine at a machine changes more intermittently, and changes to information on suppliers occurs even less frequently. The challenge lies in being able to implement a mechanism by which the cutting-tool information system is capable of maintaining up-to-date and accurate information that is consistent with physical reality. For practical reasons, it might therefore be prudent to have a distributed information system (ranging from a read-write memory chip on a tool holder to a database system in the purchasing department). However, the consistency of the information in these different sources of tooling information within the organization has to be maintained, and mechanisms for abstraction and reasoning with information from one source for use at another source should be developed.

The significance of cutting-tool information management is further highlighted by its intimate relationship with the shop-floor control system. The shop-floor control system is responsible for orchestration of all shop-floor level activities, including allocation of jobs to machines and coordination of material flow. In machining systems, cutting-tool related information is an integral part of such decision-making processes. The shop-floor controller has to ensure the availability of the requisite cutting-tools (types, numbers, and tool-lives) at a machine to which a job is assigned. The control system also needs to coordinate the flow of cutting-tools in the manufacturing system, including deciding when tools at a machine need to be replenished, which transport entity to assign the task to, and ensuring that the necessary tools are made available at the tool preparation area. Thus the shop-floor control system has to interact with the various entities in the manufacturing system, including machines and transport vehicles, and other areas such as the tool crib and the tool preparation room, to ensure the availability of cutting-tools on the shop-floor. To enable the shop-floor controller to perform this function effectively, it needs to depend heavily on the tool information management system.

One method by which the cutting-tool information management system can be integrated with planning and control decisions is through a tool requirements planning system (TRP) (Wassweiler, 1982). For instance, in an MRP-based system, TRP can be used to determine the schedules for cutting-tools by time bucket and to maintain the schedule dates as MRP responds to changes in the manufacturing production schedule. The information system supporting TRP would include information on tool masters, tool bills, routings, etc. that would enable the assurance of the requisite tools at a machine when needed by a workpiece undergoing processing at that machine.

A variety of software packages are available to facilitate cutting-tool information management at various levels (Mason, 1986). However, for effective cutting-tool management, the information system must be capable of collecting, integrating and providing all tool-related information as needed within all functions within an organization including design, purchasing and the shop-floor.

A number of interesting research issues need to be addressed with regard to cutting-tool information management. The first one relates to the development of methods for integration of cutting-tool related information with manufacturing planning and control decision making. The second relates to the development of a unified schema for representation of various types of cutting-tool information. The third issue involves the development of methods to ensure consistency and concurrency control in a distributed tool information system. Fourth, a convenient method for checking the validity of the information stored in the cutting-tool information system needs to be developed. Fifth, effective methods for automatic collection and analysis of cutting-tool data from the shop-floor need to be designed. Finally, a method for tracking cutting-tool related costs in the system through the tool information management system needs to be developed. This would not only help assess the economic impact of cutting-tools but would also enable appropriate tool-related cost allocation to jobs.

## 15.4 TOOL MANAGEMENT ON THE SHOP-FLOOR

In this section, we examine in greater detail the role played by cutting-tool management on shop-floor control. Specifically, we examine the three primary functions of tool allocation, tool replenishment, and tool transport.

### 15.4.1 Tool allocation

The operational capabilities of an advanced CNC machine are determined and/or constrained to a great degree by the set of cutting-tools assigned to the machine. Maximizing the operational flexibility of a machining center would be easy if each machine had a tool magazine of infinite (or very large) size. This would enable each machine to be loaded with all the different types of tools available in the tool crib. However, this option is not economically viable, since it would require a very high investment in cutting-tools. In addition, the average utilization of cutting-tools in the system would be poor. In practice, therefore, cutting-tool magazines at machines are of limited size, typically having a capacity of 20 to 60 tools.

A number of different methods can be utilized to allocate tools to machines so that cutting-tool availability is ensured at the machine while satisfying tool magazine constraints. One approach is to first cluster the parts to be manufactured in the manufacturing system into families and to dedicate one or more machines within the system to the machining of each family. This forms the basis for cellular manufacturing. Tool allocation within a cellular manufacturing framework begins with the identification of the tool requirements for each family of parts. Subsequently, allocation of cutting-tools to the machines within each cell can be done in a variety of ways (Kusiak, 1986). If all the machines within a cell are identical, one strategy is to provide each machine with an identical set of cutting-tools. Thus, parts being routed to the cell can be machined on any machine in the cell. This approach to tool allocation is simple; however, it requires a high level or duplicate cutting-tool inventory in the system. The second approach is to equip each machine within the cell with a distinct set of tools, thereby requiring parts to move from one machine to another in order to undergo operations. This approach allows a lower tool inventory and requires a smaller tool magazine capacity, but can create problems associated with work load imbalance and ineffective use of the flexibility of the machines.

The third approach allows each machine to maintain a limited set of (possibly unique) tools but enables access to additional tools when needed from a tool storage area via a tool handling system. This would allow each machine to be dedicated for a specific set of operations but still provide the flexibility of performing other operations by accessing tools from a common storage area (such as a common tool magazine or the tool crib). This ability to change or add to the operational capability of the machine is useful because it allows the cell to cope gracefully with the failure of another machine within the cell.

The cellular manufacturing approach is most suited to production environments where the part families are clearly identifiable (in terms of the commonalty in operational requirements) and the demand is relatively stable. In such cases, it is possible to dedicate each cell for the machining of a specific family of parts. This approach is, however, not suitable for a low volume, high variety production environment because it does not take advantage of the true operational flexibility of the machines. In many cases though, companies have adopted this constrained approach to operating the manufacturing system because it enables easier shop-floor control. In situations where the product mix changes with each production window, it is necessary to change the composition of the tool magazines of the machines with each production window. This approach takes advantage of the operational flexibility of the machines to a greater extent than does the cellular approach. Tool allocation to machines in this approach is, however, a more complex problem. Several different strategies for tool allocation are possible under this approach (Hankins and Rovito, 1984). The first strategy, known as the bulk exchange strategy, provides a machine with a copy of each tool needed for each job visiting it. A complete set of tools is delivered to the machine for each different part being run at the machine. This strategy results in high cutting-tool inventories, a high level of tool handling, and consequently high set-up times.

The second strategy allows sharing of tools at a machine between different parts being machined within a production window. It takes advantage of the commonalties within the tool requirements for parts being run at a machine during a production window. Each machine is therefore loaded only once at the beginning of each production window with all the tools needed during that period; common tools are not duplicated for each part. This strategy results in lower tool inventory relative to the bulk exchange strategy. However, it requires a large tool magazine capacity at the machine and also limits the level of routing flexibility in the system.

The third strategy takes advantage of commonalties in tool requirements between production windows. In this strategy, at the end of a production window, the tools that are unique to that production period are removed from the tool magazine, thereby creating room for the unique tools of the next production window to be loaded at the machine. This scheme allows further reduction in the cutting-tool inventory by enabling sharing of common tools from one production window to the next.

The fourth strategy maintains a set of high usage tools permanently on the tool magazine. Provision of such a set of resident tools at each machine enhances the routing flexibility in the system. However, it typically results in a higher tool inventory. This strategy is particularly effective when coupled with rationalization of part designs with respect to their cutting-tool requirements.

The selection of a suitable tool allocation strategy is contingent upon the characteristics of the production environment in the manufacturing system. If there exist distinct similarities among the tool requirements of various parts, it would allow the formation of part families and adoption of a cellular approach to manufacturing and tool allocation. If, on the other hand, the product mix is constantly in a state of flux, then the tool allocation strategy chosen should be one that promotes routing flexibility in the system. The tool allocation problem is compounded by the finite lives of cutting-tools. It is therefore necessary to consider not only the types of tools that are needed at a machine but also the number of duplicates that need to be allocated so as to allow uninterrupted operation of the machine. Since tool magazines have a limited capacity, increasing the level of tool duplication in a tool magazine reduces the number of different types of tools that can be stored at the machine. Thus there exists an interplay between tool allocation decisions on the one hand and tool replenishment needs and tool transportation capabilities on the other. Another issue relates to the physical size of cutting tools. Some tools such as boring bars or large milling cutters may occupy the space corresponding to more than one slot in a tool magazine, thereby reducing the overall tool magazine capacity. In addition, the manner in which tools are located in a tool magazine affects its weight distribution and can affect the operation of the tool magazine.

A number of optimization models have been reported in the literature to address the tool allocation problem. These models typically consider tool allocation as a part of the machine loading problem in FMSs in which the objective is to find the 'best' allocation of jobs to machines in the system (Stecke, 1989). The problem of tool allocation in dynamic production environments has not been investigated in great detail. This area offers great challenges because it requires the concurrent evaluation of the tool allocation strategy with the capabilities of the tool handling system and the state of the manufacturing system (in terms of the types of parts currently being processed in the system and the contents of the tool magazines).

### 15.4.2 Tool replenishment

Cutting-tools periodically need to be replaced at the machine due to wear and breakage. Tool breakage should be avoided because it often results in damage to the workpiece that necessitates rework or scrapping of the workpiece. Tool breakage therefore typically results in increased cost and affects the productivity at the machine. One mechanism to avoid tool breakage is to replace it after a 'safe' period of operation. Cutting-tool manufacturers often prescribe a 'standard tool life' for a particular tool under specific machining conditions. Many companies use this as a basis for making tool replacement decisions at the machine. One method for detecting tool replacement time uses the machine tool controller to track the actual usage of each cutting-tool at the machine. When a particular tool has been used for a period of time equaling its 'standard tool life' it is ready for replacement. The actual tool life is stochastic in nature and cannot be predicted in advance. Hence, this conservative approach to tool replacement is often adopted in practice. The disadvantage with this approach is that tools are often under utilized thereby resulting in increased costs and tool inventories. In addition, this approach does not guard against premature tool failure. An alternative approach is to utilize real-time tool monitoring systems that track the performance of the cutting-tool during the machining operation. When abnormal tool behavior is noticed, the monitoring system triggers a tool replacement request. The difficulty with this approach is the lack of robust tool monitoring systems that can accurately predict tool wear and prevent tool breakage. In addition, it is important to ensure that a tool does not wear out in the middle of an operation as it can damage the workpiece. Therefore, tools need to be replaced when they lack sufficient tool life to perform a machining operation.

Keeping worn tools in a machine's tool magazine is an unproductive use of space on the magazine. Worn tools need to be removed to create room for replacement tools. An effective tool replenishment scheme would ensure that the machine is never starved for requisite tools in good condition. There are several approaches to replacement of tools at a machine. The suitability of each strategy depends on the type of tool handling system, the tool allocation strategy, the tool monitoring system and the production environment in the manufacturing system.

One approach is to send a replacement tool as soon as a machine signals that a tool has been worn out. While this strategy would enable the minimization of duplicate tools in the tool magazine, it requires an efficient tool handling system that can quickly transport tools from the tool crib to the tool magazine for exchange with the worn tool. Since such tool handling systems are uncommon, the typical strategy is to provide sufficient number of tools at a tool magazine so that the tool replenishment can be done periodically (for instance at the beginning of every shift). Another aspect of the tool replenishment strategy involves the determination of the number of tools that need to be replaced. For instance, one approach is to replace all the tools in the tool magazine whenever tool replenishment is done (analogous to the bulk exchange process). Although this approach results in inefficient use of tools, it reduces the number of times that the tools have to be replaced in the system. The second approach is to replace only those tools that are worn out. This strategy allows each tool to be used to its 'fullest' extent. However, it increases the number of times that tools need to be delivered to the machine and consequently the complexity of coordinating the tool replenishment process.

There exists a variety of research issues related to tool replenishment that warrants further investigation. The areas of adaptive machining systems and tool-life monitoring systems continue to offer opportunities for research. In terms of tool replenishment strategies, one issue that requires investigation is the evaluation of different approaches to tool storage in the manufacturing system. One approach is to store all additional tools in the tool crib and transport them to machines when necessary. The other approach is to have several secondary tool stores on the shop-floor. The benefits of this alternative approach need to be investigated. In addition, how should these tool stores be distributed in the system? Should local 'hubs' be formed such that each hub is responsible for tool provision to a set of machines in its neighborhood? An additional issue to investigate is the impact of tool sharing among machines.

### 15.4.3 Tool handling

One function of the shop-floor control system is the coordination of material flow in the manufacturing system. When material is moved on the shop-floor using automated systems, the controller has to make a number of decisions related to the selection of a transport vehicle for a task, determination of the path that the transport vehicle needs to take and coordination of the traffic in the system. Two types of entities need to be moved on a shop-floor, namely workpieces and cutting-tools. Requests for transportation of cutting-tools arise because of tool allocation, tool replenishment and tool sharing needs. Tool transportation typically takes place between the tool crib or preparation area and the machines or between machines. Tools can be moved on the shop-floor using a number of different methods. At one end of the spectrum is manual handling of tools. This method is perhaps the most common one, but has some inherent disadvantages. Manual handling of tools often limits the frequency with which tools can be transported in the system (particularly in the case when the cutting-tools are heavy) because of the potential for worker injury and fatigue. It may also require more time and be more prone to error in comparison to automated tool handling.

Automated tool transport can be done using either the workpiece handling system or a dedicated tool handling system. The workpiece handling system can be used for transportation of tools in manufacturing systems where the processing times are high thereby resulting in low utilization of the vehicles for transportation of workpieces. It is also suitable in situations where a large number of tools need to be delivered to a machine. In such cases, the workpiece transport vehicle can deliver a secondary magazine of tools to the machine where the machine's automatic tool changer can move the tools between the two magazines. In situations when special tools are needed for machining a part, the workpiece transporter can be used to deliver both the workpiece and its special tools to the machine.

There exist a number of different types of automated material handling systems dedicated to the transportation of cutting-tools. One method employs a large secondary tool magazine that is accessible to all machines in the system. The typical layout follows a straight line configuration. When a machine requests a tool, the secondary tool magazine (chain) moves so as to position the slot with the requested tool next to the machine. Then a tool exchanging mechanism picks up the tool from the secondary magazine and places it on to the machine's magazine. This layout limits the number of machines that can be served by this tool storage unit. In addition, the total time needed for exchanging tools may be unsatisfactory particularly in cases where a large number of tools needs to be transferred between the machine and the secondary magazine.

Another method that is commonly used is based on a gantry tool delivery system. Tools are transferred individually in this case from the storage area (possibly a secondary tool storage rack or a machine's tool magazine) using a gripper on a gantry type configuration. This approach has the disadvantage of limited tool carrier capacity and potentially high transportation time. Having multiple tool carriers in this system can create deadlock or traffic coordination problems. The tool grippers can also be installed on an overhead conveyor type system. While such a system increases the number of tool carrier entities and the speed of tool delivery, there are limitations to the routing flexibility of the tool carriers. Another option is to utilize higher capacity tool carriers, such as tool cassettes or racks, capable of carrying between four and eight tools. This would allow simultaneous exchange of multiple tools and enhance the efficiency of the tool handling process.

The selection of a tool handling system depends on a variety of factors including the rate with which the product mix changes on the shop-floor, the typical processing times and tool lives, the tool magazine capacity at each machine, the average number of tools required to process a part at a machine, the weight of a typical tool, the distribution of tool stores in the system, the level of similarity among the tool requirements for the various part types and the capabilities of the information management system and the shop-floor control system. In general, however, a tool handling system that is capable of operating at high speeds, but has a low capacity per carrier would result in lower tool inventory and more effective shop-floor performance than one that has a large tool carrier capacity and travels at a low speed. Research is ongoing at the University of Wisconsin-Madison towards the development of a high-speed tool delivery system capable of reaching a maximum speed of 33 meters per second. Preliminary studies on the performance of manufacturing systems having such high-speed tool delivery systems have been encouraging.

The coordination of tool flow within a manufacturing system has serious implications on shop-floor performance. While the strategy of intermittent change of tools (such as between shifts) within the system is simpler to implement, it restricts the operational flexibility of machines and the routing flexibility of parts in the system. The availability of a tool handling system would enable the manufacturing system to respond more effectively to unexpected changes such as the failure of a machine tool or the arrival of 'urgent' jobs. The challenge lies in developing robust shop-floor control systems that are capable of integrating tool flow and workpiece flow so as to achieve desired production objectives and a high level of system performance.

As alluded to earlier, there exists strong interdependence between the strategies for tool allocation and replenishment and the capabilities of the tool handling system. The development of enabling technologies in the areas of tool handling system hardware and control software are now making it possible to promote tool flow control as an integral part of shop-floor control. This would enable the scheduling and control of not only workpieces on the shop-floor but also individual cutting-tools. Manufacturing systems with such capabilities would be truly flexible and be capable of responding effectively to external and internal disturbances.

## 15.5  IMPACT OF SHOP-FLOOR CONTROL ARCHITECTURE ON CTM

The architecture of a shop-floor control system dictates the manner in which decisions related to task and resource allocation are made in the manufacturing system. Most shop-floor control systems follow either a centralized architecture or a hierarchical architecture. In the centralized scheme, a central controller oversees all the activities in the manufacturing system. This controller has global knowledge of the manufacturing system and generates commands for subordinate equipment controllers (in machine tools and transport vehicles) to execute desired actions. This central controller also receives feedback from the shop-floor equipment regarding their status and upon completion of the specified tasks. The advantage of a centralized control architecture lies in its ability to base its decisions on the global knowledge of the shop-floor. The disadvantage with this approach to shop-floor control stems from the constraint on the size of the system that can be effectively controlled by a central controller. In addition, the reliability of such systems is also a matter of concern.

The other approach, namely the hierarchically structured control system, partitions the control system into different levels of responsibility and authority. The interaction is typically limited to controllers in neighboring levels and follows a master/slave relationship. The delegation of responsibilities allows hierarchically structured control systems to control larger systems. However, the hierarchical structure also imposes rigidity on the control mechanism and limits the flexibility of the manufacturing system. In addition, the complexity of these control systems grows steeply with the size of the manufacturing system.

Tool flow control on the shop-floor under the centralized and hierarchical paradigms can be achieved in a number of ways and depends on the nature of the tool handling system and the strategies being adopted for tool allocation and replenishment. When tool handling is achieved through the workpiece handling system, the shop-floor control system has to coordinate the demands on the material handling system and assign tasks to it in a manner that enables both workpiece transport and tool transport requests to be met effectively. Complications arise due to the potential for conflicting demands on the transport system. The shop-floor control system therefore has to delegate tasks to the material handling system based on a well-defined set of priorities for the various types of transport requests that need to be satisfied.

Having a dedicated tool handling system allows segregation of tool transport and workpiece transport requests. However, the shop-floor controller would now need to oversee the operation of two material handling systems. This increases the complexity of the shop-floor controller. To control the tool handling system, the controller has to track events (such as part type change) that necessitate tool transportation. If multiple tool carriers are available, the controller needs to select a suitable carrier and assign the tool transportation task to it. In addition, the controller needs to ensure the availability of the requisite cutting-tool through communication with the tool crib or the tool preparation area. The controller also needs to coordinate the actions of the tool carriers to prevent deadlock situations. Such a situation can arise when one tool carrier reaches a machine to deliver a tool but finds no empty tool slot while another tool carrier that was assigned the task of removal of a worn-out tool from the magazine is unable to reach the machine because of the presence of the first tool carrier. Even though the hierarchical approach allows the delegation of the tool flow control to a tool handling system controller, the overall complexity of shop-floor control resulting from the consideration of cutting-tool management on the shop-floor makes this approach unwieldy in large systems.

The difficulties associated with hierarchical control of large systems has promoted a growing level of interest on a decentralized (or heterarchical) approach to shop-floor control. Under this paradigm, shop-floor control is achieved through message passing and cooperative decision-making among autonomous manufacturing system entities (such as machines, transport vehicles, workpieces and cutting-tools). No entity in the manufacturing system has complete knowledge of the state of the manufacturing system. Decisions are made by each autonomous entity on the basis of local knowledge and information obtained through message passing. This approach is motivated by the success of distributed computing systems. Research to date has shown that a heterarchically structured shop-floor control system offers multiple advantages including simplicity, modularity, robustness, adaptability and extendibility (Dilts, Boyd and Whorms, 1991). One model for implementation of the decentralized paradigm is the contract net model. Task and resource allocation under this approach is achieved through an auction-based mechanism. In the simplest scenario, workpieces enter the system as customers and announce their processing needs to the machines. Each machine acts as a contractor and submits a bid to each workpiece. In constructing the bid, the machine takes into account its processing capabilities, the load on the machine and the impact of other auctions in progress. The workpiece selects a winner of the auction on the basis of the bids submitted. A material handling vehicle for transporting the workpiece to the chosen machine is also similarly selected through an auctioning process. The workpiece is then transported to the machine where it subsequently undergoes processing.

The consideration of cutting-tools explicitly in the auctioning process outlined above allows integration of cutting-tool management with shop-floor control (Veeramani, 1992b). It does, however, require the careful formulation of protocols for the behavior of entities so that conflicts and deadlocks relating to cutting-tools do not arise. The inclusion of cutting-tool related factors in the bid-construction process would require a machine to check for availability of requisite cutting-tools in its tool magazine. If some cutting-tools are lacking, then the machine has to enter into a secondary level of negotiation with other machines and the tool crib to identify donors for the necessary tool cribs. In preparing the bid, the machine has to estimate the amount of time needed to transport the requisite tools from the donors to its tool magazine. Subsequently, if the machine wins the auction, it needs to conduct an auction to identify the best tool carrier(s) to perform the tool transportation tasks.

While a decentralized approach to shop-floor control offers a great level of flexibility to system operation, it presumes a high level of intelligence inherent in the system entities and the existence of a well-defined set of protocols that prevents chaos in the system. Advances in manufacturing system hardware and software technologies are now enabling the development of heterarchical control systems. Research on decentralized shop-floor control systems is still in its infancy. This paradigm requires a rethinking of the roles played by manufacturing system entities in traditional hierarchically controlled systems. A number of issues need to be resolved before a comprehensive shop-floor control system can be implemented. Some of these issues include the identification of the types of information that need to be maintained by each system entity, protocols for participation in auctions, strategies for bid-construction, rules for inter-entity negotiation for sharing of or access to resources and mechanisms for reacting to unexpected changes in the manufacturing system. These research issues assume a higher level of complexity when cutting-tools are allowed to participate in the shop-floor control process as active, autonomous entities. The implications of such a high level of decentralization are very significant to the manner in which the communication, information and control systems are designed.

## 15.6 CONCLUSION

Cutting-tool management plays a crucial role in the control of modern flexible machining systems. It spans a large variety of functions including design, purchasing, tool crib operation and shop-floor control. The implementation of an effective cutting-tool management hinges on the availability of a well designed information management system. There exists an intimate relationship between cutting-tool management and shop-floor control. To maintain a high level of performance of the manufacturing system, the shop-floor control needs to coordinate both workpiece and tool flow within the shop-floor. However, this imposes a higher level of complexity on the shop-floor control function. The heterarchical paradigm offers many advantages that make the development of decentralized shop-floor control systems attractive. The integration of a comprehensive cutting-tool management system with a heterarchical shop-floor control system would enable the evolution of truly flexible machining systems.

## REFERENCES

Boyle, C. (1986) *Using group technology for tool inventory control*, Proceedings of the 3rd Biennial International Machine Tool Technical Conference, Chicago, Ill., September 3–11, National Machine Tool Builders' Association, pp. 111–20.

Dilts, D.M., Boyd, N.P. and Whorms, H.H. (1991) The evolution of control architectures for automated manufacturing systems, *Jnl. of Manuf Sys*, **10**(1), pp. 79–93.

Gayman, D.J. (1987) Meeting production needs with tool management, *Manuf Eng*, Sept. pp. 41–7.

Hankins, S.L. and Rovito, V.P. (1984) *The impact of tooling in Flexible Manufacturing Systems*, Proceedings of the 2nd Biennial International Machine Tool Technical Conference, Chicago, Ill., September 5–13, National Machine Tool Builders' Association, pp. 175–98.

Hartley, J.R. (1984) *FMS at Work*, IFS Publications, Bedford, UK.

Kusiak, A. (1986) Parts and tools handling systems, in *Modelling and Design of Flexible Manufacturing Systems* (ed. A. Kusiak), Elsevier Science Publications, Amsterdam, pp. 99–109.

Mason, F. (1986) Computerized cutting-tool management, *Amer Mach & Autom Manuf*, May, pp. 106–20.

Stecke, K.E. (1989) Algorithm for efficient planning and operation of a particular FMS, *Int Jnl. of Flex Manuf Sys*, **1**, pp. 287–324.

Tomek, P. (1986) Tooling strategies related to FMS management, *The FMS Mag*, April, pp. 102–7.

Veeramani, D., Upton, D.M. and Barash, M.M. (1992) Cutting-tool management in computer-integrated manufacturing. *Int Jnl. of Flex Manuf Sys*, **3/4**, pp. 237–65.

Veeramani, D. (1992a) Rationalization of cutting-tool requirements, in *Concurrent Engineering*, (eds D. Dutta, A.C. Woo, S. Chandrashekhar, S. Bailey and M. Allen), PED-Vol. 59, The American Society of Mechanical Engineers, NY., pp. 307–13.

Veeramani, D. (1992b) *Task and resource allocation via auctioning*. Proceedings of the 1992 Winter Simulation Conference (eds J.J. Swain, D. Goldsman, R.C. Crain and J.R. Wilson), Arlington, Va, December 13–16, pp. 945–54.

Wassweiler, W.R. (1982) *Tool requirements planning*, Proceedings of the 1982 Conference of American Production and Inventory Control Society, pp. 160–2.

# CHAPTER 16

# An object-oriented control architecture for flexible manufacturing cells

*J.M. Hopkins, R.E. King and C.T. Culbreth*

## 16.1 INTRODUCTION

In response to foreign competition and rapidly changing customer demands, domestic manufacturers in many industrial sectors have developed a strong interest in flexible automation. Decreased costs and improved capabilities in robotics and computer-integrated systems have increased the feasibility of such systems, particularly for small to medium-sized companies. Their primary benefit is relaxation of the restrictions on the range of manufacturable products associated with 'hard' automation systems.

In concert with the 'just in time' (JIT) manufacturing philosophy, companies are seeking to move from large volume/low product mix production capabilities towards reduced lot sizes. Ideally, systems should provide the ability to produce a significant set of parts, in lot sizes as small as one, without consuming appreciable time in setting up or adjusting production equipment for part changeovers. Flexible automation is a vehicle for the reduction of both manufacturing lead times and lot size. Additional goals of flexible automation are the reduction of direct labor costs, improved quality and faster delivery. However, experience has shown that the implementation of full-blown flexible manufacturing systems has not, in general, provided the anticipated rewards.

Consequently, many companies have pursued the use of flexible cells as a means to learn and gain necessary experience in flexible automation (O'Grady, 1989). One critical task in the successful implementation of flexible cells is the development of the cell control system. To achieve its potential, flexible automation must have robust control software which can run on a number of hardware platforms and provide the following capabilities.

1. The software must be able to communicate with all devices in the cell and control cell action, minimizing delays between signals and corresponding actions.
2. Errors in the system must be detected and handled within strict time limits.
3. The cell control software must easily adapt to different cell configurations.
4. The software must be extensible to accommodate the addition of new components (machinery) to the cell.

A scheduling module should also be included in this software specification unless scheduling will be handled by a separate, but interfacing, program at the 'shop level' of the manufacturing system control hierarchy as described by Weber and Moodie (1989).

In this chapter, a cell control system is presented which embodies the concepts of object-oriented design and programming and therefore has the following characteristics:

• data encapsulation and information hiding;
• structured, hierarchical knowledge representation;
• inheritance of values; and,
• flexible and reusable code.

To provide context, a review of the recent literature discussing current approaches to cell control and modeling is provided. Then a generalized model for cell control is developed using the concepts of object-oriented design and programming. This model is further refined into a generic cell control architecture and class hierarchy. The architecture is then applied to a flexible manufacturing cell for furniture part production (FFMC) as described by Culbreth, King and Sanii (1989). The system consists of the following components (Fig. 16.1):

1. Three-axis CNC work center.
2. Six-axis robot with force/torque sensing.
3. Two-dimensional gray scale vision system.
4. Custom interchangeable robot end-effector system.
5. Programmable logic controller.
6. Cell control computer.
7. Input and output conveyors.

A real-time, multi-tasking, state-transition table control system was originally implemented to control the prototype cell using message passing for inter-task communication (Fig. 16.2). Discrete, low-level I/O signals from cell sensors are managed by a programmable logic controller (PLC) which communicates with the cell control computer. A vision system identifies the type and location of an entering part upon

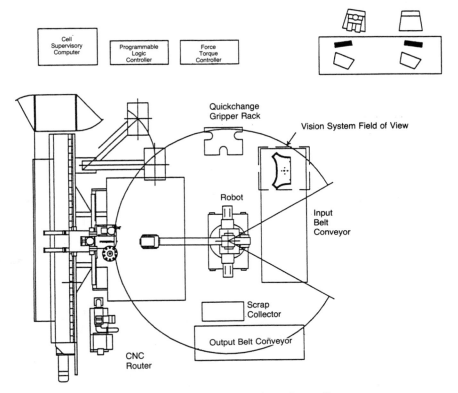

**Fig. 16.1** Prototype flexible furniture manufacturing cell.

detection by the PLC. Based on this information, the cell proceeds with processing the part as directed by the control software.

The characteristics and limitations of the original cell control software are discussed in relation to the proposed object-oriented approach. In this way, a meaningful comparison of the capabilities can be constructed and a more concrete understanding of the new paradigm can be gained. Testing of the object-oriented architecture in the context of the proto-type cell provides a basis for defining appropriate boundaries of applicability and for recommending system extensions.

## 16.2 REVIEW OF CELL CONTROL AND MODELING SYSTEMS RESEARCH

There has been a variety of cell control and modeling systems devel-oped to date. Research emphasis in flexible manufacturing can be roughly grouped into the following areas: scheduling; simulation/ modeling; fault diagnosis and control. There has also been a particularly

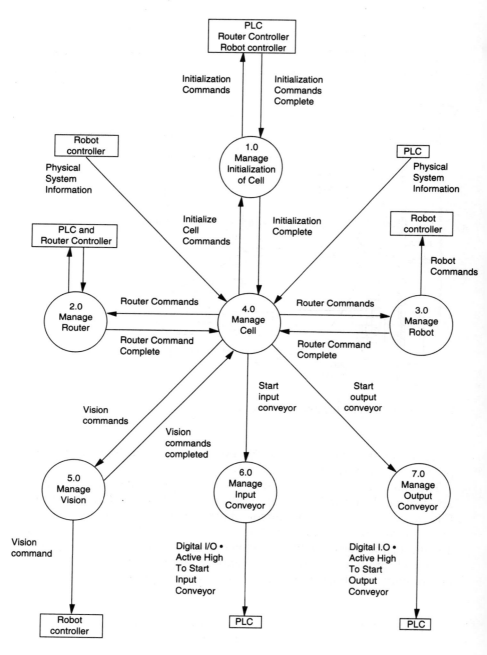

**Fig. 16.2** Context level data flow diagram of the original FFMC control software.

high level of interest in artificial intelligence techniques for manufacturing modeling and control over the past decade. While not exhaustive, the following describes some of the significant efforts in this area.

Gliviak *et al.* (1984) designed an expert system shell (CEMAS) to supervise a manufacturing cell. It consists of a database, a knowledge base, a knowledge acquisition block and an explanation block. The main activities of the system involve the recording of dynamically changing situations, the scheduling of tasks to particular workplaces and detection and resolution of critical situations. Each machine tool requires control by an individual microcomputer making the system hardware intensive.

Bourne and Fox (1984) and Wright and Bourne (1987) describe the Transportable Cell (Transcell) system which was developed and partially implemented for a swaging operation at Westinghouse Inc. The control software was written in a programming language designed for cell control called Cell Management Language (CML). It uses a rule-based system at its top level that communicates with a database and a device driver. Information is stored in state tables in the database. Rules firing on the upper level lead to changes in the database, which in turn leads to the firing of more rules that involve communication with the device drivers.

Sepulveda and Sullivan (1988) developed a prototype knowledge-based system for real-time control and short-term planning of an automated manufacturing facility called Cell Manager. The system receives a schedule from a higher level scheduler and then controls local scheduling and routing. The knowledge base and inference engine are separated such that rules can be modified or added as needed.

The flexible automated modeling and scheduling (FAMS) system developed by Buckley *et al.* (1988) was designed with the objectives of:

- assisting shop managers in planning resource requirements and utilization;
- providing accurate estimation of work order completion dates;
- determining dynamic optimal resource allocation, and
- providing capability for users to experiment with changing system parameters.

The FAMS system consists of a scheduler module, a modeler module and a shop monitor module which collectively make up a 'kernel'. System communications are accomplished using a database interface module and a user interface module which make up the system's 'shell'. The system can schedule and monitor and model usage of system resources. It is a effective modeling tool, but cannot support real-time device communications and thus cannot be used for cell control.

O'Grady and Lee (1988) proposed the use of a blackboard and actor-based intelligent cell control system (ICCS). The framework for the ICCS is called the production logistics and timing organizer (PLATO-Z). The cell control system is structured as the cell level of a combined computer aided manufacturing international (CAM-I) and NBS (NIST) hierarchical representation of the factory including factory, shop, cell and equipment levels of control. The system consists of four black-boards for scheduling, operation-dispatching, monitoring and error-handling. There are also support functions for initialization and termination, communication and networking, and interfacing with the user. Production rules and procedures govern the operations of the control system. System shortcomings include inefficiencies due to redundant message passing and rule consultation for repetitive tasks.

MPECS, or multi-pass expert control system (Wu and Wysk, 1988), is a supervisory control system for manufacturing cells. The control structure uses both expert system technology and discrete event simulation. The primary purpose of the system is to utilize all available data in the cell for 'good' strategies to guide the system, and to make real-time control decisions. The major drawback of MPECS is the difficulty of controlling the length of the simulation period in which scheduling is being performed; a long period can generate delays in the cell.

Manivannan and Banks (1991) present a real-time knowledge-based simulation (RTKBS) framework for accomplishing real-time control that integrates data collection devices within a dynamic knowledge base and a simulation model. In order to accomplish real-time control the framework synchronizes event times in the cell and in the simulation using a temporal knowledge base. This event/time synchronization is carried out using past, future and current event lists. As data is obtained from the cell these event lists are compared and updated. The RTKBS also features a dynamic knowledge base in which outputs from each simulation run are stored. This reduces response time delays often encountered in systems which employ simulation-based scheduling.

Menon and Ferreira (1988) developed a Petri net-based system for reconfiguring systems controllers for easy adaptation to system changes. A coordination controller was structured in three layers: a planning level that processes orders arriving from a scheduler, an intermediate level that converts the output of the planning level into actions and an execution level that interfaces with the hardware in the FMS.

Weber and Moodie (1989) produced a Prolog-based prototype intelligent information system for an automated, integrated manufacturing system which is a distributed system of cooperating, consistent knowledge base sites. A semi-intelligent mechanism predicts user process timing and data needs and then plans and schedules the operation of the site and prepares required information before the retrieval request arrives. The prepared information is stored in object frames for rapid access by the user. The major drawback of the system is that although Prolog allows rapid prototyping, the linear search procedures make them inherently inefficient.

In order to decrease the amount of information storage required in a conventional state-table driven control system, Drolet and Moodie (1989) describe the use of a relational database for storage of part descriptions and other data not directly related to system state transitions or fault recognition. A fully functional manufacturing cell replica was constructed to test the software developed, which included a database main module, a compiled Basic robot module and a ladder logic module. A sequentially-driven software system was coded such that the main program was used to poll input ports and communication channels in a predetermined order. The software was tested and performed well, but was not designed to handle system interrupts or multi-tasking, which prevents the system from reacting to signals from the cell as they occur.

The National Institute of Standards and Technology built an automated research manufacturing facility (AMRF) designed to reflect the 'factory of the future'. The AMRF allows investigation of the manufacturing enterprise as a system of intelligent machines. A real-time control system architecture is employed using sensory feedback (Swyt, 1989). The system communicates among hierarchical levels using message passing through mailboxes. Lisp is used in the construction of semi-intelligent process planning modules utilizing AI concepts of knowledge representation for process plan work elements. The system is still in the development stages and is more a research tool than a directly applicable system for cell control.

Rohr Industries of California has developed a flexible robotic workcell for the assembly of airframe components for the air force which has been fully implemented (Olsen, 1990). The program is called flexible assembly subsystems (FAS), and incorporates control of a gantry robot, an articulated arm, a flexible fixture, a 3D metrology system, a parts presenter and an automatic fastening machine. Engineers enter a detailed description of the assembly procedures through an interpretive program called MAS (master assembly sequence), and using this information and CAD drawings the software constructs the procedures to be performed for a given assembly. Low-level coding such as robot path motions are stored in preprogrammed routines that are accessed as needed for a specific assembly task, negating the need for explicit programming of all robot moves.

Camurri and Franchi (1990) designed a coordination subsystem and real-time scheduling subsystem for the simulation and real-time control of flexible manufacturing systems (FMS). They combined structured object-oriented paradigms and an extension of high-level Petri nets. Frame-based multiple inheritance semantic networks were used for the knowledge base, which were in turn used for automatic synthesis of Petri net models of the coordination subsystem. Advanced software development is currently underway using the PETREX C Tools software library.

Jafari (1990) developed a modular shop floor controller using adaptive Petri nets. Adaptive Petri nets make possible the modeling of dynamic structural changes of a system, such as machine breakdowns, which cannot be modeled with conventional Petri nets. The model itself was analyzed and validated with Petri nets. Work was also done on error recovery, such as dealing with communication failures between devices that are not represented in the Petri nets.

Emphasizing quality control in flexible manufacturing, Richard *et al.* (1992) propose a quality control loop architecture for a flexible manufacturing cell. They consider a cell to be composed of multiple stations that transform and verify part quality (these stations are the equivalent of cells as defined in this chapter). The architecture uses distributed control on several levels in order to effect solutions to quality problems as fast as possible. An information system is used to receive and report work-in-progress status to any control loop in the cell, and stores information on both products and equipment. Information gathered from machine processes is fed to the information system, which can analyze design problems and even highlight possible areas for modification in product design. In addition, the information system provides resolution strategies for defects detected in the system.

At NEC Corporation in Japan, functional modeling of cell controllers is being accomplished using a functional matrix approach (Odajima and Torii, 1992). The matrix is broken into basic functions and external objects. The four basic functions, used as the columns in the matrix, are defined as:

1. giving machines production commands;
2. transmitting technical/resource data;
3. collecting and processing production data; and,
4. collecting and processing resource status.

The external objects, used as the rows in the matrix, include an upper-level shop controller, a lower-level station controller and a supervisor of the cell level. Each functional element of the matrix is developed as a software module, using shared data areas for information needed for multiple modules. The cell controller system was applied to a printed wiring board assembly shop.

A generic cell controller has been developed for automated VLSI manufacturing facilities (Moyne and McAfee, 1992). The controller uses an extensive relational database with a generic schema which embodies many of the control aspects of the cell, thus reducing the burden of data maintenance by the control software. The generic SECS-II message format is used for message passing, and messages that are sent and received outside the scope of the cell controller use an interpreter that attaches or removes communications protocol-specific information. Thus, cell information can be communicated in a standardized format both inside and outside the cell. The cell controller receives incoming requests to the cell from parent levels in the facility through I/O interpreter modules, then deciphers the messages with a message parser module. A main program module interfaces with the database to find the response or action which matches the request received. Actions may include the invocation of specified routine modules, which are called to perform the required processing. These routines may also update the database.

A joint effort between researchers at the University of Illinois and Motorola Inc. has produced a fractal architecture for modeling and controlling flexible manufacturing systems (Tirpak *et al.*, 1992). The fractal architecture is used to hierarchically decompose the FMS into units with similar structure and control. The basis of the architecture is the basic fractal unit (BFU, or unit), which is applicable to the structure of all levels of the FMS hierarchy. Each fractal unit accepts incoming jobs from the next higher level in the hierarchy through an input queue that is enabled or disabled by an inhibit flag used to regulate incoming job flow. Similarly, outgoing jobs are delivered to higher levels via an output port. The FMS is progressively decomposed from one level to the next by the inclusion of subunits in each unit. Each unit also contains four analytical and control software modules including an observer, an analyzer, a resolver and a controller. The system is highly object-oriented, as each unit's internal operations are transparent to the subunits and superunits with which it interacts. The system is currently being implemented in the C++ language.

In summary, many approaches to cell control and modeling have been presented in the literature that collectively verify the complexity of implementing flexible cells. Although solutions have been applied to various domains, the need exists for a generic architecture which can be used as a template for the organization, design and eventual implementation of cell control systems across a range of applications. The emerging dominance of the object-oriented design/object-oriented programming (OOD/OOP) approach to software design and development is a compelling reason to explore its potential use in real-time cell control systems.

## 16.3 GENERIC CELL CONTROL ARCHITECTURE

In this section a generic, object-oriented architecture for control of a flexibly automated cell is presented. While the architecture can be applied in a variety of scenarios, the design and development is driven by the requirements of a class of flexible cells which involve the use of robotic devices for material handling. A particular implementation of this type engaged in the production of furniture parts is used to verify the design. This approach provides meaningful context, and avoids considering a highly abstracted 'model' cell. This flexible furniture manufacturing cell (FFMC) prototypes a feasible application of flexible automation in the furniture industry.

### 16.3.1   Use of object-oriented design and programming (OOD/OOP)

This research specifies the use of C++, an object-oriented programming (OOP) language, to code the cell control system. Thus, the first issue to be addressed is the choice of an OOP language over a conventional, procedural language such as C. The basic philosophy of OOD/OOP is to devise a problem solution in terms of a set of largely autonomous agents, called **objects**. Each object is responsible for a specific set of tasks associated with the problem (Budd, 1991). When an object requires the execution of some task which is not within the scope of its responsibilities, a request is made to another object designed to carry out that task. These requests are made in the form of a **message** specifying the task to be performed, including any pertinent, specific information that is required for task completion.

Every object is an instance of an object **class**, such that all objects of a given class will respond similarly to a request for services. Classes can be formed into a class hierarchy from the most general **superclass** to more specific **subclasses**, or **derived** classes. Layers of more specific classes can be built without the need for duplication of common traits; both data and **methods** (the means by which tasks are carried out) in the higher-level, or parent classes are accessible to (inherited by) the derived classes.

This decomposition of the problem into individual objects responsible for well-defined tasks differs from conventional procedural programming in that objects are responsible not only for performing certain tasks but also for independently maintaining the data associated with those tasks. Thus, the data structures for each class used in an OOP solution are determined after a problem has been partitioned and related to classes which are designed to perform specific functions. Moreover, objects requesting services from objects of another class do not require knowledge of the underlying data structure of the object to which the request was sent; only knowledge of the message format is required to successfully interface. This attribute, commonly referred to as **information hiding** and **data encapsulation**, allows for application-specific changes to a generic class structure to be transparent to collaborating objects requesting services (Budd, 1991, Stroustrup, 1991).

Since related knowledge is grouped together, classes have the effect of structuring knowledge in an organized and manageable way. Object-oriented systems also support hierarchical knowledge representations and inheritance of values. Consequently, classes allow for simple addition of new components through the creation of derived classes, and offer enhanced readability and ease of comprehension. These OOD/OOP characteristics support the creation of a generic architecture that can be refined to provide control for a range of specific cell configurations.

### 16.3.2 Generic cell components

For the purposes of this discussion, a generic flexible cell consists of the following components:

1. part(s) to be processed;
2. machine(s) to perform processing;
3. cell monitoring and control devices such as PLCs;
4. transport devices to deliver parts into/out of the cell;
5. material handling devices such as robots to effect part placement and transfer between machines and entrance/exit transport devices.

As stated, the class of applications considered is that of flexible manufacturing cells employing robotics for material handling. More general applications would require that the term robot be translated to mean 'material handling device'.

### 16.3.3 Generic cell control class structure

The system design process begins with the development of classes. Each class is assigned responsibilities for certain tasks required for the problem solution. As the classes are designed, relationships between classes are defined. The approach used in this system design involved the creation of CRC cards. 'CRC' stands for the system of developing classes as defined by Budd (1991) where:

C = Class (may represent a class or an instance);

R = Responsibilities of the class; and,

C = Collaborators (other classes with which this class interacts).

This method of organization promotes the object-oriented philosophy of data hiding and abstraction. The design is developed based upon objects and their responsibilities in the system rather than being built around preconceived data structures. The class, responsibilities and collaborators are often written on index cards. When the responsibilities have been defined and the collaborator classes have been identified, then the appropriate data structures are defined for each class and written on the back of the card.

For the proposed cell control architecture, the following superclasses were defined.

1. **Part.** Parts are grouped into major part types based upon common machine/robot visitation patterns. Every part type requiring processing in the cell is derived from the base class part. The responsibilities for each part are to be moved to and from machines by material handling devices and to be processed as required on the machines by requesting services from the appropriate resource managers.

2. **Manager.** A resource manager is required for transport devices (e.g. conveyors, AGVs), material handling devices (e.g. robots), cell monitoring and control devices (e.g. PLCs), and machines (e.g. routers, milling machines). Managers coordinate requests from parts for movement and processing, handle scheduling of the device that they manage and send processing requests to the facilitators.

3. **Facilitator.** Facilitators perform the actual communication with the devices in the cell based upon requests from the managers. They coordinate with error handlers to attempt resolution of problem states.

4. **Error_handler.** Error handlers attempt to resolve problem states in the cell and coordinate with the facilitators and operator console as necessary.

5. **Op_console.** The operator console is the interface to the human operator. Status and error messages are processed here, as well as requests for operator intervention.

6. **Communicator.** Communicators allow for operating system-level communication between independently running tasks. They are required to provide the actual interprocess communication between independently operating programs in the control system.

7. **Supervisor.** The supervisor keeps track of the number of parts in the cell; it creates new child processes and allocates communication channels for incoming parts and reclaims channels from exiting parts when their corresponding child processes terminate.

The hierarchy of these superclasses and their derived classes is shown in Fig. 16.3. The individual class definitions in the CRC format previously discussed appear in the following section with recommendations concerning appropriate modifications for different cell configurations.

Although each implementation will differ somewhat in structure, all systems will require a multi-tasking operating system capable of inter-

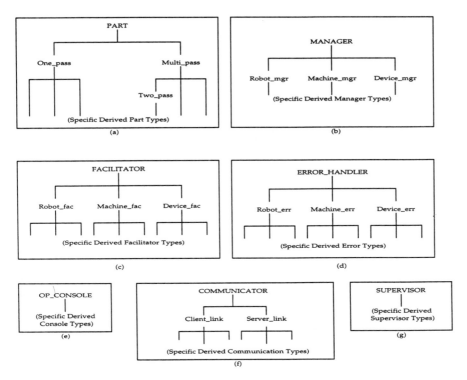

**Fig. 16.3** Generic cell control class hierarchies.

process communication. Conceptually, the system is built upon the interaction of different objects. Practically, these objects are separated into individually compiled and executed modules to allow for simultaneous execution in a multi-tasking environment.

The suggested implementation involves a single supervisor process that creates 'part' objects and child processes for each part that enters the cell. These child processes exist only while the part resides in the cell, therefore the number of child processes existing at any moment is limited by the number of parts that the cell can physically accommodate at one time. This is not restrictive since most implementation platforms, e.g. UNIX, allow the number of processes to be quite large.

A manager module, a facilitator module and an error-handling module are recommended for each of the robot, machine and device subclasses. A single operator console module serves as the human interface once the system is complete, although separate interfaces are useful for the monitoring and control of individual processes during development, integration and debugging of the system.

*Client_link* and/or *server_link* objects of base (parent) class communicator should be created within each module to provide the required interprocess communication (IPC). In cases where the client/server model is used, as is typically the case with multi-tasking systems, a server process waits for a connection from a client process that is requesting service. Individual client processes can be a client to more than one server, and server processes can have multiple clients.

Servers are of two basic types: iterative and concurrent. Iterative servers wait for a client request, process the request when received, and resume waiting for the next client request. Concurrent servers invoke another process to handle each client request, so that multiple requests are handled in a concurrent fashion (Stevens, 1990). This format is followed for the FFMC implementation and is described in more detail in the following section.

### 16.3.4  Description of class structure

This section presents the characteristics and associated hierarchy of each of the seven classes previously defined. This collection of base classes has sufficient utility to adequately construct a control architecture for cells with robotic material handling. Any application-specific classes can be derived from this set.

*Specific part classes derived from base class part*

Parts are the most obvious candidates for cell objects. After identification as a valid part upon entry to the cell, a part requests the required series of movements and machining operations from the robot, machine and device managers. All part types are derived from the base class *part,* which is responsible for maintaining the data associated with the minimal common part movement and machining operations.

All part types are loaded onto and processed by at least one machine and are offloaded from the final machine onto the exit transport device; therefore this subset of operations is put in the superclass. Additional data and methods required for the more specific part types are provided in the derived classes. Relevant data for this class includes robot and machine program parameters.

At least one 'pass' on one or more machines is required to complete processing of the part. A pass in this context is defined as a set of operations contained in a single NC program file. Some parts require multiple passes (NC programs) with intervening material handling operations (such as scrap removal). Classes are derived from the base class *part* based upon the number of machining passes required.

Table 16.1 shows the classes derived from the base class *part.* *One_pass* parts do not require any additional data beyond what is maintained by the superclass. The class *one_pass* serves as the logical

**Table 16.1**  CRCs for part class

| class | superclass | sub classes | responsibilities | collaborators |
|---|---|---|---|---|
| Part | none | One_pass<br>Multi_pass | • Request pickup and loading onto first machine<br>• Request first machining operation<br>• Request offload by robot<br>• Request system exit on transport device | Robot_mgr,<br>Machine_mgr,<br>Device_mgr |
| One_pass | Part | Application-specific derived classes of single pass, single machine parts | • Pass all information to Part class | (see base class) |
| Multi_pass | Part | Derived classes based on the number of machining passes (see Two_pass) | • Pass all load, offload and first pass information to Part base class<br>• Request additional machining operations<br>• Request any intermediate robot operations | Robot_mgr,<br>Machine_mgr |
| Two_pass | Multi_pass | Application-specific derived classes of dual pass, single/multiple machine parts | • Pass all information to Multi_pass class | (see Multi_pass class) |

base class for application-specific derived classes of parts which require a single load/machine/unload operation sequence. Upon creation of an object of a specific part type derived from this class, actual data values are simply passed to the base class variables. Similarly, objects of class *multi_pass* transfer initial load, first pass machining and cell exit data up to the *part* class. However, this class also has member functions and data for additional machining operations and robot movements. *Two_pass* parts are derived from the *multi_pass* class. Objects created from this class are similar to *one_pass* parts in that they simply pass parameters up to the parent (*multi_pass*) class which contains secondary machining and intermediate scrap removal member functions and data. The complexity of data maintenance increases with each additional machining operation required; thus, additional classes are derived depending upon the requirements for the parts being processed. This allows the more simplistic case of dual pass parts to be isolated from the more complex cases.

### The manager base class and derived classes

The next group of classes is derived from the base class *manager*. For each application, there is a robot manager, a machine manager and a device manager specific to each major component of the cell. Scheduling within each manager varies in complexity dependent upon the number of devices to be managed in the cell. However, emphasis is placed on restricting cell size and complexity to make automation more practical and to realize the goal of real-time processing.

Table 16.2 also shows the specific responsibilities of each main subclass of manager, namely *robot_mgr*, *machine_mgr* and *device_mgr*. Device managers will either schedule and communicate part transfer requests to the facilitator for one low-level I/O device (such as a PLC) or to multiple facilitators for devices with more intelligent interfaces.

### The facilitator base class and derived classes

The *facilitator* class and its derived classes (Table 16.3) control the passing of manager requests to the actual physical devices in the cell. Facilitator is a logical class from which the robot/machine/device-specific classes are derived, and consequently maintains only inter-process communication data. The derived facilitators translate requests from their respective (robot/machine/device) managers into the required format for the specific controller being used.

Completion status messages for error/no error states from the robot/machine/device are sent to the error handler. No error messages result in the error handler returning acknowledgment of successful completion of the task. Upon receipt of this acknowledgment the facilitator

**Table 16.2** CRCs for manager class

| class | superclass | sub classes | responsibilities | collaborators |
|---|---|---|---|---|
| Manager | none | Robot_mgr Machine_mgr Device_mgr | • Logical base class for all derived manager classes for devices performing scheduling of part requests and communication with other managers | (see derived classes) |
| Robot_mgr | Manager | Application-specific derived classes | • Perform robot scheduling to service requests for part movement received from Part processes<br>• Request services required for part placement from other Managers<br>• Request services from Facilitators once scheduled and process Facilitator responses to requested services<br>• Send part movement status back to Park processes | Robot_facs, Machine_mgr, Device_mgr, Parts |
| Machine_mgr | Manager | Application-specific derived classes | • Perform machine scheduling to service requests for part processing received from Part processed<br>• Report part movement status to Robot_mgr<br>• Request services from Facilitators once scheduled and process Facilitator responses to requested services<br>• Send part processing status back to Part processes | Machine_facs, Robot_mgr, Parts |
| Device_mgr | Manager | Application-specific derived classes | • Perform device scheduling to service part transport requests received from Part processes<br>• Request services from Facilitators once scheduled and process Facilitator responses to requested services<br>• Send part transfer status back to Park processes | Device_facs, Robot_mgr, Parts |

reports successful completion to the respective manager. Error messages result in the error handler attempting to resolve the problem by, for example, sending either a retry request or error confirmation back to the facilitator. Retry requests begin the process again and error states are reported to the managers.

### *The error-handler class and derived classes*

For error handling related to physical devices in the cell, a hierarchy of classes was developed starting with the logical base class *error_handler*. Because all error handlers serve the same purpose but deal with different error states, device-specific error handling is delegated to the derived classes *robot_err*, *machine_err* and *device_err*. The responsibilities for each class are shown in Table 16.4.

Error handling procedures vary with the complexity and severity of the error. In cases where the solution is to resend a message to a physical device, the error handler sends a request to the facilitator to retry the failed operation. Errors requiring operator intervention are referred to the *op_console*. No message is sent to the facilitator until a reply is received from the *op_console*. This suspends any new requests for action by managers from reaching the facilitator until the problem is resolved.

### *The communicator base class and derived classes*

For the generic architecture, the part supervisor process (and associated child part processes), operator console process and the facilitator processes are appropriate servers, with managers, error handlers and physical devices serving as the clients. The supervisor process acts as a concurrent server, creating a new child process to handle each new part. The operator console and facilitator processes act as iterative servers. In addition, if communication is required between the managers, one is chosen as a server to which the others connect as clients.

Table 16.5 shows the *communicator* base class and the derived classes *client_link* and *server_link*. Objects of this class family are embedded in each of the other classes to allow communication between all processes. Classes that act as client processes have member *client_link* objects and those that act as servers have member *server_link* objects. The data and methods used vary with the information requirements of the operating system under which the control system is running. In general, a file descriptor, address and name are required for each server to which a process is a client and additional data is required by server processes to keep track of multiple clients.

**Table 16.3** CRCs for facilitator class

| class | superclass | sub classes | responsibilities | collaborators |
|---|---|---|---|---|
| Facilitator | none | Robot_fac<br>Machine_fac<br>Device_fac | • Logical base class for all derived classes that facilitate communication to actual physical devices in the cell | (see derived classes) |
| Robot_fac | Facilitator | Application-specific derived classes | • Format requests for robot services received from the robot manager and send to the robot controller<br>• Report status of request error/no error states to the robot error handler<br>• Return completion/error status of requested processes to the robot manager | Robot_mgr,<br>Robot_err |
| Machine_fac | Facilitator | Application-specific derived classes | • Format requests for machine services received from the machine manager and send to the machine controller<br>• Report status of request error/no error states to the machine error handler<br>• Return completion/error status of requested processes to the machine manager | Machine_mgr,<br>Machine_err |
| Device_fac | Facilitator | Application-specific derived classes | • Format requests for device services received from the device manager and send to the device controller<br>• Report status of request error/no error states to the device error handler<br>• Return completion/error status of requested processes to the device manager | Device_mgr,<br>Device_err |

**Table 16.4** CRCs for error_handler class

| class | superclass | sub classes | responsibilities | collaborators |
|---|---|---|---|---|
| Error_handler | none | Robot_err<br>Machine_err<br>Device_err | • Logical base class for error handlers for each of the devices in the cell (robots, machines, transfer devices) | (see derived classes) |
| Robot_err | Error_handler | Application-specific derived classes | • Process robot error messages received from the robot facilitator and attempt to resolve<br>• Request assistance in resolution from operator console if needed | Robot_fac,<br>Op-console |
| Machine_err | Error_handler | Application-specific derived classes | • Process machine error messages received from the machine facilitator and attempt to resolve<br>• Request assistance in resolution from operator console if needed | Machine_fac,<br>Op-console |
| Device_err | Error_handler | Application-specific derived classes | • Process device error messages received from the device facilitator and attempt to resolve<br>• Request assistance in resolution from operator console if needed | Device_fac,<br>Op-console |

*The op_console class*

In cell operations, communication is necessary between the cell controller and the cell operator. This communication may be initiated at either end and is accomplished by an object of the *op_console* class (Table 16.6). Implementation of data and methods is highly application dependent, but a base class of operator console could be standardized for different cells using the same operating system platform. Dependent upon the desired sophistication of the system, this class, at a minimum, displays status and error messages and receives input from the operator through simple I/O routines.

*The supervisor class*

Finally, a class for the creation and supervision of child processes for individual *part* objects is required (Table 16.7). When a new part enters the cell, the *supervisor* is notified by the *device_mgr*. The *supervisor* creates a child process, passing to the process the type of the new part and allocating to the process a communications link to each of the managers. In the child process, an object of the required part-derived class is created which then requests services from the managers. When the last request has been satisfied, the child process terminates and the *supervisor* reclaims its communications links so that they can be used again for another part. In addition to resource allocation for the child processes, if the operator wishes to terminate all part processing, a terminate message can be relayed from the operator console through the *supervisor* to all child processes.

## 16.4 FFMC ARCHITECTURE SPECIFICS

The object-oriented control system for the FFMC was developed as an alternative to the original state table-based system which had been implemented and tested (Culbreth, King, and Sanii, 1989). This system contained a number of operational limitations which will be described. A description of the new system follows, with emphasis on the improvements obtained through the OOD/OOP approach.

### 16.4.1 FFMC cell operation

The general operational flow of a single part in the cell (Fig 16.1) is as follows.

1. A part arrives in the cell on an input conveyor and is moved into the field of view of a vision system.
2. A vision system establishes the identity and location of the part.

**Table 16.5** CRCs for communicator class

| class | superclass | sub classes | responsibilities | collaborators |
|---|---|---|---|---|
| Communicator | none | Client_link<br>Server_link | • Logical base class only; derived classes may differ if Client/Server model of IPC is not used | none |
| Client_link | Communicator | Application-specific derived classes | • Initiate interprocess communication<br>• Enable and handle message passing and buffering for messages to and from processes to which the current process is a client | none |
| Server_link | Communicator | Application-specific derived classes | • Initiate interprocess communication with client processes<br>• Enable and handle message passing and buffering for messages to and from client processes<br>• Handle client requests, initiating child processes when appropriate | none |

**Table 16.6** CRCs for op_console class

| class | superclass | sub classes | responsibilities | collaborators |
|---|---|---|---|---|
| Op_console | none | Application-specific derived classes | • Logical base class for the operator to cell controller interface | Supervisor, Error Handlers |

**Table 16.7** CRCs for supervisor class

| class | superclass | sub classes | responsibilities | collaborators |
|---|---|---|---|---|
| Supervisor | none | Application-specific derived classes | • Receives new part messages from Device_mgr and creates child processes<br>• Manages communication link resource allocation<br>• Maintains status of child processes<br>• Accepts cell init/end messages from Op_console | Device_mgr, Op_console |

3. The robot changes end-effectors if necessary, then moves the part from the conveyor to the router bed.
4. The part is machined on the router.
5. If scrap removal is required between machining operations, the robot intervenes, removes solid scrap, and returns to the perch position. The router then proceeds with remaining operations.
6. When all router operations are complete, the robot transfers the finished part from the router bed to an output conveyor and the process repeats.

A new part can enter the cell once the preceding part has been removed from the input conveyor and a message has been sent to turn the input conveyor back on. Since the FFMC is configured as a tightly coupled system without buffers, the maximum number of parts that can reside in the cell at the same time is three: the oldest part is either on the router bed or leaving the cell, the intermediate part is either being held by the robot or is on the router and a new part is on the input conveyor.

The router is equipped with a flexible workholding fixture that applies vacuum independently to individual part and scrap zones under computer control. Prior to the robot loading a part onto the router, the appropriate part and scrap zones are activated. If solid scrap removal is required between machining passes, the scrap vacuum zone for the part is deactivated and the robot removes the scrap. When all machining is complete, the part vacuum zone is deactivated and the robot transfers the part from the router bed to the output conveyor (Culbreth and King, 1992).

### 16.4.2  Limitations of the original system and their solutions in the proposed system

The original cell control software was written in the C language and was based on a state transition table. Figure 16.4 provides an overview of the possible system states. There are three types of software modules: a controlling manager, action tasks which perform specific actions in the system, and support functions which interface to the cell devices. Messages are passed via a set of mailboxes from the control manager to the action tasks to perform the appropriate tasks as dictated by the state table. Each action task, in turn, sends back a message when its task is completed. The responsibilities of each of these tasks are given in Table 16.8.

A number of limitations exist in the original cell control software. Although functional, the original code is somewhat difficult to maintain because it is implemented in a procedural language. The interactions of concurrent processes are extremely hard to grasp based

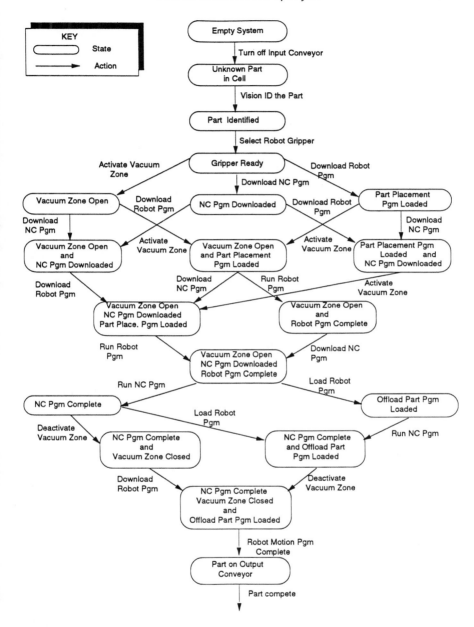

**Fig. 16.4** Original system state transition diagram.

**Table 16.8**   Original cell action task responsibilities

| action task | responsibilities |
| --- | --- |
| 0 | Turn on input conveyor |
| 1 | Vision identify a part |
| 2 | Activate part vacuum zone on fixture |
| 3 | Load robot part placement program into robot controller memory |
| 4 | Select robot end-effector |
| 5 | Activate scrap vacuum zone on fixture |
| 6 | Run robot part placement program |
| 7 | Download NC (router) program 1 |
| 8 | Run NC program 1 |
| 9 | Deactivate scrap vacuum zone on fixture |
| 10 | Load robot scrap removal program into robot controller memory |
| 11 | Run robot scrap removal program |
| 12 | Download NC program 2 |
| 13 | Run NC program 2 |
| 14 | Load robot off load part program into robot controller memory |
| 15 | Run robot off load part program |
| 16 | Turn on output conveyor |

on a reading of the procedural code. Another issue is the level of redundant programming in the software because similar sections of code are needed in several action tasks. During software development a continual tradeoff existed between the level of hard-coded, redundant modules and the degree of parameterization employed for each module. Using a number of similar, but specifically-written modules simplifies the logic and readability of the overall system, whereas use of a single, heavily parameterized module reduces code but increases execution time.

A further and related limitation is the difficulty of adding new devices to the cell which requires revised software modules or that new modules be coded and linked to the system. Perhaps the most serious deficiency of the system is that it limits the cell to having a single part on the bed of the router at any time. Machine tool technology is such that multiple part placements are possible to achieve enhanced productivity. Multiple part processing is possible using a state table approach but explosive growth in the size of the table as parts and/or machines are added is a problem.

Finally, more robust error diagnosis and resolution functionality is needed. In the original implementation, there are several error conditions from which manual recovery is required. However, with an improved architecture many of these manual recoveries could be automated. Thus, as a continuation of the original research, the control system has been rewritten on a multi-tasking Unix platform, incorporating both C language low-level I/O routines and C++ language object-oriented design and programming.

### 16.4.3   The FFMC object-oriented system

The FFMC control program is composed of fourteen separately coded and simultaneously executing program modules, eleven written in the C++ programming language and three in the C language (the physical device modules) running in a Unix-derivative environment. Although the original cell implementation ran on a VMS platform, the Unix platform was selected for this version due to its availability, open architecture, real-time extensions, and the portability of C-based code between Unix-based machines.

It is assumed that scheduling of part arrivals to the cell occurs at the shop level, and, to demonstrate its design goal of maximum flexibility, the cell must be capable of handling the random arrival of parts for processing. Parts arriving on the input conveyor are recognized by a vision system and subsequent internal cell scheduling is accomplished through resource allocation rules at the level of the resource managers. If a production schedule (sequence of parts) replaced part random arrivals, the supervisor module would perform detailed scheduling based upon the pre-determined sequencing of parts.

*Class development*

The system design process begins with the identification and design of classes. A general structure must be built for each of the classes needed to represent the cell's parts and physical components. For the FFMC, two types of parts are processed: those requiring a single NC program on the router and those requiring two. Therefore, the subclasses of parts *one_pass* and *two_pass* are needed. Two-pass parts require that solid scrap produced during initial machining be removed by the robot before subsequent machining can take place.

Data required for the *part* class includes: the part types; the robot end-effectors; the robot motion programs, and the machine tool NC programs. Encoding of this data is required for implementation. In the FFMC the program and part type coding is a very simple alpha-numeric code; for a greater variety of parts, a multi-character code may be required. In the FFMC, three one-pass parts and one two-pass part were used for demonstrations. The sample parts included an end panel (one-pass), a long rail (one-pass), a drawer front (one-pass), and a table base (two-pass). Each part is handled by one of three robot end-effectors: single jaw, dual jaw, or configurable vacuum gripper.

Following the development of the generic architecture, four new classes were created to represent the sample parts. For the three one-pass parts, the classes *end_panel, long_rail* and *drawer_front* are derived from class *one_pass*. Likewise, the class *table_base* is derived from class *two_pass*, as the table base is a two-pass part.

Each derived class (*end_panel, long_rail, drawer_front, table_base*) has a single member function, or method, that has the same name as the class and is called a 'constructor'. When an instance of that class is declared, the constructor creates an object of that class for the name given, passing any required values up through the constructor to the parent (base) class to initialize the data. Once the object has been constructed, the member functions can be accessed using the member access operators.

For example, consider an entering part of type table base. Upon identification, an object (instance) of class *table_base* is created. Once created, the object can call any of the member functions (methods) of the superclass *part* (such as initial pick and load request functions) or the grandparent class *multi_pass* (such as the scrap removal member function, which is not required for the general class *part*).

Application-specific subclasses of the *device_mgr, device_fac,* and *device_err* classes (*PLC_mgr, PLC_fac,* and *PLC_err*) are also required for the PLC which is used to monitor and control such things as the input/output conveyors. Similarly, subclasses of *machine_mgr, machine_fac,* and *machine_err* classes (*router_mgr, router_fac,* and *router_err*) are required for the router which is the specific cell machining device.

Parts request services from the *robot_mgr, router_mgr,* and *PLC_mgr* using member functions. The messages are passed using Berkeley sockets (Stevens, 1990), and communication between tasks is handled by objects of *client_link* and *server_link* classes as described earlier. These managers process and schedule the requests based upon decision rules for one, two or three parts in the cell at one time. The number of parts that can physically reside in the cell at one time for the FFMC is three, allowing simple state transition tables to be used for control. For a more complex cell, this number could be higher. As the number of parts and/or machines in the cell increases, the feasibility of using state-table control diminishes. However, by its nature, OOD/OOP allows the incorporation of other scheduling/control procedures of varying format and complexity as appropriate for a given application.

Figures 16.5, 16.6 and 16.7 show the state transitions for a single part in the cell for the robot, router and PLC managers, respectively. The robot manager drives the scheduling for the rest of the cell. As mentioned in the previous section, a new part can be input as soon as the part preceding it has been removed from the input conveyor and the conveyor has been turned back on. When a new part arrives in the cell an instance of the appropriate part type is created which then sends a request to the robot for a combination pick and load operation to effect

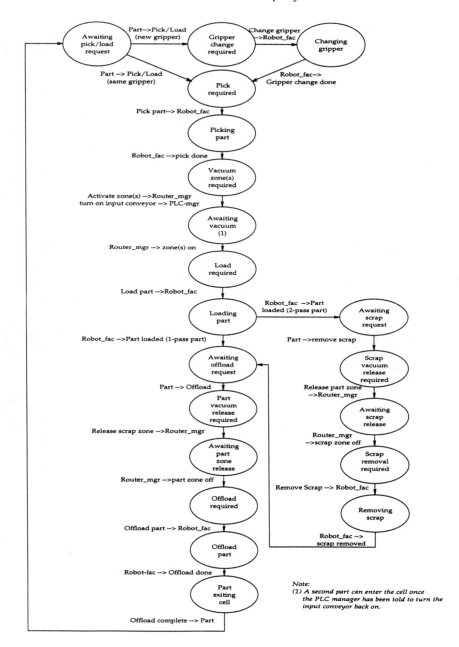

**Fig. 16.5** Robot_mgr state transitions.

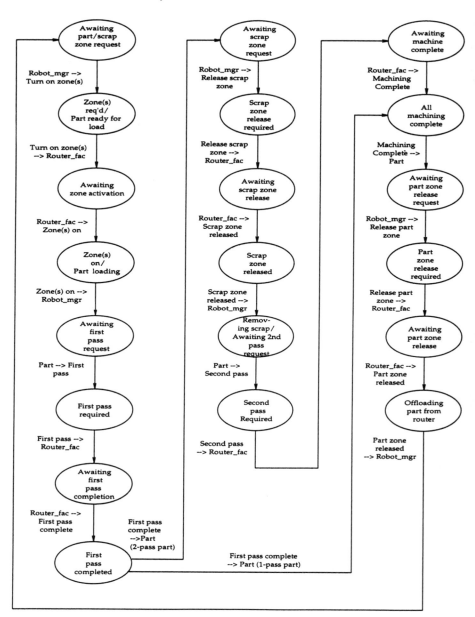

**Fig. 16.6**  Router_mgr state transitions.

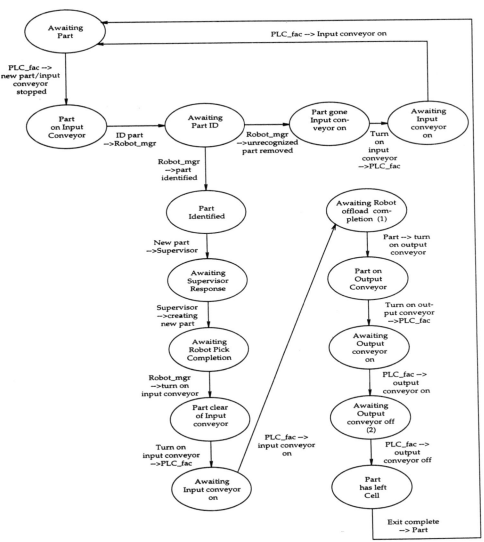

NOTES:
(1) As soon as robot has finished pick motion, it is safe to turn conveyor back on. Additional parts may arrive at this state.
(2) The output conveyor is turned off automatically when part reaches the end of conveyor. An interrupt is sent from PLC and passed through facilitator to the manager.

**Fig. 16.7** PLC_mgr state transitions.

a transfer from the input conveyor to the router. It is the robot manager's responsibility to schedule both the pick operation to remove the part from the conveyor (taking it to an intermediate holding position) and the load operation to move the part from the holding position to the router bed.

In addition to pick and load operations, the robot must schedule end-effector changes, scrap removal operations and offload operations. Part placement, scrap handling and offloading operations also require communication with the router manager to request activation or deactivation of part and scrap vacuum zones on the router bed. Finally, once a part has been picked, the robot manager must inform the *PLC manager* that the part has been removed from the input conveyor and that the conveyor can be turned back on.

Two conditions cause a part to wait until other parts have left the cell before it can be processed. First, since the robot uses multiple end-effectors, the current end-effector is compared to the end-effector for each arriving part. If a different end-effector is required, the new part must wait on the conveyor until the current part(s) has completed processing and an end-effector change is completed. The second condition occurs because there is a single dedicated position for each part type on the router bed. If the new part and its most recent predecessor are of the same type, the new part must wait for the previous part to be placed on the output conveyor before it can begin processing.

If neither of these two blocking conditions applies, there are two general cases to consider when a new part arrives prior to the completion of machining of the preceding part. (Note: assume both require the same robot end-effector.) First, if the preceding part is a one-pass part, then the new part will not be picked up off the input conveyor until after the preceding part (part 1) is loaded onto the router. Otherwise, part 1 is a two-pass part and the new part (part 2) will not be picked off the input conveyor until after the scrap removal operation for part 1. When machining is complete for part 1, workholding vacuum is requested for part 2 and it is placed on the router. Vacuum release is then requested for part 1 and it is offloaded. After part 1 exits the cell, the cycle repeats for each new arriving part.

The *manager* classes communicate with *facilitators*, which in turn instruct the actual physical devices on the operations to perform. There are three facilitator classes. *Robot_fac* handles robot operations, *router_fac* handles router operations and *PLC_fac* handles PLC operations that interface with the input and output conveyors and sensors.

In the specific FFMC implementation the vision system is integrated into the robot controller. Therefore, the robot manager receives requests from the PLC manager to identify newly arriving parts. These requests are scheduled whenever there is no robot facilitator response pending for a previously requested robot service. If a new part is successfully recognized or determined to be unrecognizable, the robot facilitator informs the robot manager, which reports this information back to the PLC manager. For successful recognition, the PLC sends a message on a dedicated communications link to the supervisor process to create a new child process for the part. Contingency in the event of a recognition failure is discussed in the next section.

*Error handling*

Error handling in the new system combines two approaches. When a member function call fails, an attempt is made to rectify the situation locally using the exception handling methods which are defined for the class in use (Stroustrup, 1991). However, when an error message is received or an error state occurs at the facilitator level that involves physical devices, control is passed to objects (instances) of the class *error_handler: robot_err* for robot errors, *router_err* for router errors and *PLC_err* for PLC errors. These error handlers attempt to resolve problems themselves. If they cannot, then they request help from the operator by communication through the *op_console* class. The *error_handler* objects disable further requests from the *manager* objects such that no new requests can be passed through the facilitator to the physical device when an error state is present.

Error handling is accomplished in much the same way as communication with external devices. Rules govern the execution of error handling on a priority basis and by having a higher priority, potentially severe problems are dealt with first. The priority scheme, coupled with the implementation of error handlers as independent, concurrent processes ensures timely response to errors.

Errors encountered in the FFMC system and their resolution procedures are listed below. These errors are only a subset of the many possible errors that can occur in a manufacturing cell, but are a representative set of common types of errors and their recovery strategies.

1. **Router errors.** A message is sent from the physical router controller to the *router_fac* process, which directs the error to the *router_err* process for handling. Router errors are generally not software recoverable, therefore the standard response is to inform the operator that the router error has occurred. Potential errors include an automatic tool changer alarm that occurs when an improper tool number has been requested, a problem with the dust collection system (a requirement in woodworking), a broken tool, or loss of workholding vacuum during cutting. The router has an intelligent controller that sends a message that includes information about the type of error. The *router_err* process deciphers the message, suspends operation of the cell and informs the operator of the error and the steps

needed to correct it. Cell operations are resumed upon receipt of a command given by the operator.

2. **Robot errors.** There are several types of error messages that the robot can send to the control software. The first is a program download error message; i.e. the control software attempted to run a robot program that was not successfully loaded into the robot controller, or a download error occurred. In these cases, the control software retransmits the download command and then attempts to run the robot program again. If the error message is repeated the control software sends a robot controller error message to the operator.

   A second type of error is due to an unscheduled impact, which is signalled by the force-torque sensor on the robot's end-effector, or by a overload error on the robot. This could occur when attempting to pick up a part in the wrong position or when a collision occurs with an object improperly located in the path of the robot. The robot controller automatically halts the robot and sends a message to the control software about the error. The control software informs the operator of the error and the operator corrects it. In the case of a positioning error, the robot must be recalibrated with the vision system, and in the case of a collision the operator must remove the foreign object from the cell. Either of these cases results in a shutdown of cell operations, and requires the robot to be reinitialized.

   A third type of error is due to a vacuum loss on the robot gripper. When vacuum is determined to be insufficient, the robot stops movement and waits for a resume command. The *PLC-err* module refers the error to the *op-console*, where the operator is given manual recovery instructions.

3. **Time-out error.** Facilitators take into account a 'time-out' factor, such that if a process-completed message is not received from the robot/router/PLC within a reasonable time frame, a message is sent to the operator to investigate. The operator then enters a resume command if the problem has been resolved, or a terminate command which shuts down the cell for more serious problems.

4. **Unrecognized part on input conveyor.** Changes in ambient lighting conditions can cause a false determination of 'unknown part' by the vision system. In the original system this error was handled manually by resetting the control software. The new system initiates a second search, and if the part is still unrecognizable sends an error message to the operator and waits for a resume command once the operator has removed the part. Having this functionality of new part recognition separated from the *supervisor* process allows part recognition errors to be handled before a new *part* process is created. In this way, the other operations of the cell are not interrupted.

5. **Other errors.** Other errors are handled by the controllers on the

individual devices. These controllers are intelligent and have the capability of recovering from minor problems on their own.

### System structure and interprocess communication

All modules in the system are separately compiled and concurrently executing processes. All interprocess communication is performed using AF-UNIX sockets. The interconnection of the modules is illustrated in Fig. 16.8.

The system is initiated by starting each of the processes in an order such that all of the socket connections are initialized before any work is actually done in the cell. Communication follows the client/server model, in which some of the processes act as servers to which the other client processes connect. For the purposes of initialization, the *supervisor* and *facilitators* (*robot_fac*, *router_fac*, and *PLC_fac*) are the servers, and two of the *managers* (*router_mgr* and *PLC_mgr*), *error handlers* (*robot_err*, *router_err*, and *PLC_err*) and *physical devices* (*robot*,

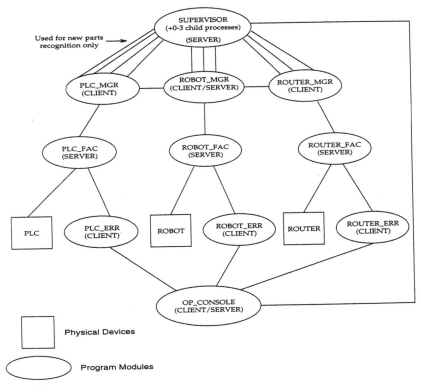

**Fig. 16.8** C++ system intertask communication.

router and PLC) are the clients. The *robot_mgr* and *op_console* are unique in that they act as both clients and servers. The *robot_mgr* connects to the *supervisor* and its facilitator as a client and then allows connection of the other managers as a server. Similarly, the *op_console* process connects as a client to the *supervisor* and accepts connections from the *error_handlers* as a server. The order of process initialization makes this dual role possible. Once the connections have been established, the notion of client versus server is no longer important; communication proceeds bidirectionally through the sockets.

The *supervisor* program establishes three links, or channels, to each of the *robot_mgr*, *PLC_mgr* and *router_mgr* processes (a total of nine channels). When a new part enters the cell, it breaks the beam of a photodetector monitored by the PLC. An interrupt is generated in the PLC that in turn sends a message to the *PLC_fac* process indicating that a new part has arrived. The *PLC_fac* informs the *PLC_mgr*, which requests part identification from the *robot_mgr* process. When the robot is free, the *robot_mgr* requests part identification from the integrated vision system on the robot. Recognition results in a message back to the *PLC_mgr*, which in turn sends a message to the *supervisor* to create a new *part* process for the entering part.

The *supervisor* assigns one of each type of channel to the new child *part* process (*robot_mgr, router_mgr,* and *PLC_mgr*). Because there can physically be only three parts in the cell at one time, the maximum number of channels needed will not exceed the nine channels initialized. The *part* process then requests all required services from the *managers* through these assigned channels. When the last operation has been completed, the child process terminates and the channels are returned to the parent *supervisor* process for reallocation to new parts entering the cell.

## 16.5 DISCUSSION OF RESULTS

A generic object-oriented cell control architecture has been developed. A specific manufacturing cell consisting of a single machine for part processing, a single robot for part handling and input and output conveyors for part transfer was then used as a test application. The method of CRC cards to decompose cell characteristics and requirements into classes with well-defined, finite responsibilities was used and found to be logical and straightforward.

There are two major points of discussion in terms of the applicability of this architecture for different cell configurations. First, not all flexible manufacturing cells are as simplistic in configuration as the FFMC. The question arises as to the conceptual validity of applying the architecture to more complex configurations. Second, the use of state-table control for the object-oriented FFMC suggests the same problems of combinatorial explosion that existed in the original cell. The ability of the proposed architecture to support scheduling of multiple devices by managers also becomes an issue.

The topic of architecture extensibility is addressed in the following manner. For more complex cells, the *part* classes require only minimal modification in the structure of the 'service request' messages sent to the managers. For example, parts requiring processing on more than one machine would need to have a code to represent the machine used for each operation added to the message structure used in the member functions that send requests to the *managers*. The other necessary modifications are also straightforward. *Facilitator* classes would be derived for each physical device, with associated *error_handler* classes being derived to deal with error states generated by the respective devices. In turn, the operator console would be modified to process any additional error messages requiring operator intervention.

The most extensive changes required would be in the scheduling of the additional devices by their respective managers. Increasing complexity in scheduling requirements will eventually necessitate an approach to control that is different from the state table system applied in the FFMC. Although scheduling of multiple devices by a single manager would be more complicated than single machine scheduling, many appropriate solutions exist for multiple part, multiple machine scheduling such as traditional mathematical programming-based algorithms, knowledge-based techniques, or neural network approaches. Regardless of the approach used to achieve synchronization and co-ordination of cell activities, the overall structure of the system remains intact. The real power of the object-oriented approach is that the basic responsibilities of the manager classes do not change: they receive processing requests from parts, schedule these requests and communicate these requests to the appropriate facilitators. It is simply a matter of modification of the methods associated with the scheduling of requests – the architecture does not change.

The very nature of object-oriented design and programming is rooted in the separation of class responsibilities and the interaction between classes by means of messages requesting services. A knowledge of the requisite message structure is all that is needed to request a service from any class. The data used and the implementation details of functions in each class are encapsulated and hidden from collaborating classes. Only the interfaces to the classes are public.

## 16.6  CONCLUSIONS AND FUTURE RESEARCH DIRECTIONS

The goal of this research was to design and develop a generic cell control architecture and then to demonstrate its viability by improving the flexibility and performance of a class of flexible cells involving robotic material handling. The control software was made more flexible and conceptually coherent through use of a generic architecture from which new classes could be derived to represent various components of the cell. The intuitive aspects of classes provide for more rapid system development, comprehension, maintenance and modification.

The system was also made more robust by increasing the level of automated error recovery. An improved operator interface provides for more intelligent and interactive communication between the operator and the cell control software.

The architecture was implemented for the FFMC as a specific instance of a flexible robotic cell. Although the specific implementation of the OOD/OOP cell control system was designed for furniture manufacturing, the modular style of the programming and the generic applicability of many of the features to other manufacturing processes broadens the relevance of the research. Many types of manufacturers could (and do) benefit from the use of robotic work cells, thus a flexible and reasonably portable software architecture could be an advantage for applying automation techniques in a variety of domains.

## ACKNOWLEDGMENT

This research was funded in part by the International Woodworking Fair and the American Furniture Manufacturers' Association.

## REFERENCES

Bourne, D.A. and Fox, M.S. (1984) Autonomous manufacturing: Automating the job shop, *IEEE Computer*, September, pp. 76–86.

Buckley, J. *et al.* (1988) An integrated production planning and scheduling system for manufacturing plants, *Robot and Comp-Integ Manuf*, **4**(3), pp. 517–23.

Budd, T. (1991) *An Introduction to Object-Oriented Programming*, Addison-Wesley.

Camurri, A. and Franchi, P. (1990) *An approach to the design and implementation of the hierarchical control system of FMS, combining structured knowledge representation formalisms and High Level Petri Nets*, Proceedings of the 1990 IEEE International Conference on Robotics and Automation, **1**, pp. 520–25.

Culbreth, C.T. and King, R.E. (1992) Design and development of a flexible manufacturing cell control architecture, *NCSU-IE Technical Report 92-2, Industrial Engineering Department*, N.C. State University.

Culbreth, C.T., King, R.E. and Sanii, E.T. (1989) A flexible manufacturing system for furniture production, *Manuf Rev*, **2**(4), pp. 257–65.

Drolet, J.R. and Moodie, C.L. (1989) A state table innovation for cell controllers, *Comp and Ind Eng*, **16**(2), pp. 235–43.

Gliviak, F., Kubis, J., Mocovsky, A. (1984) A manufacturing cell management system (CEMAS), *Artificial Intelligence and Information Control Systems of Robots*, North Holland, Elsevier Science Publishers B. V., pp. 153–56.

Hu, D. (1987) *Programmer's Reference Guide to Expert Systems*, Howard W. Sams and Company, Indianapolis, Ind.

Jafari, M.A. (1990) *Petri Net based shop floor controller and recovery analysis*, Proceedings of the 1990 IEEE International Conference on Robotics and Automation, 1, pp. 532–37.

Kim, T.G. and Ziegler, B.P. (1990) AIDECS: An AI-based, distributed environmental control system for self-sustaining habitats, *Artif-Intell in Eng*, 5(1), pp. 33–41.

Manivannan, S. and Banks, J. (1991) *Real-time control of a manufacturing cell using knowledge-based simulation*. 1991 Winter Simulation Conference Proceedings, Phoenix, Ariz., pp. 251–60.

Menon, S.R. and Ferriera, P.M. (1988) *A Colored Petri Net based architecture for coordination control of flexible manufacturing systems*, Proceedings of the ASME Annual Winter Meeting: Advances in Manufacturing Systems Engineering, 31, pp. 69–88.

Moyne, J. and McAfee, L. (1992) A generic cell controller for the automated VLSI manufacturing facility, *IEEE Trans on Semicond Manuf*, 5(2), pp. 77–87.

Odajima, T. and Torii, T. (1992) *Functional modeling of the cell controller in computer integrated manufacturing systems*, Eleventh IEEE/CHMT International Electronics Manufacturing Technology Symposium, San Francisco, CA, pp. 105–09.

O'Grady, P.J. (1989) Flexible manufacturing systems: present development and trends, *Comp in Ind*, 12, pp. 241–51.

O'Grady, P.J. and Lee, K.H. (1988) An intelligent cell control system for automated manufacturing, *Int Jnl. of Prod Res*, 26(5), pp. 845–61.

Olsen, H.B. (1990) *A flexible robotic work cell for the assembly of airframe components*, Proceedings of the 1990 IEEE International Conference on Robotics and Automation, 2, pp. 1278–83.

Rauch-Hindin, W.B. (1988) *A Guide to Commercial Artificial Intelligence*, Prentice Hall, Englewood Cliffs, N.J.

Richard, J., Idelmerfaa, Z., Lepage, F. *et al.* (1992) Quality control in a flexible manufacturing cell, *CIRP Annals*, 41(1), pp. 561–64.

Sepulveda, J.M. and Sullivan, W.G. (1988) Knowledge-based system for scheduling and control of an automated manufacturing cell, *Comp and Ind Eng*, 15, pp. 59–66.

Sheng-Hsien, T. and Black, J.T. (1989) An expert system for manufacturing cell control, *Comp and Ind Eng*, 17, pp. 18–23.

Smith, S.F., Fox, M.S. and Ow, P.S. (1986) Constructing and maintaining detailed production plans: investigations into the development of knowledge-based factory scheduling systems, *AI Magazine*, Fall, pp. 45–61.

Stevens, R.W. (1990) *UNIX Network Programming*, Prentice Hall, Englewood Cliffs, N.J., pp. 5–6, 258–341.

Stroustrup, B. (1991) *The C++ Programming Language*, Addison-Wesley, Reading, Mass.

Swyt, D.A. (1989) AI in manufacturing: The NBS AMRF as an intelligent machine, *Robotics and Autonomous Systems*, 4(4), pp. 327–32.

Tirpak, T., Daniel, S., LaLonde, J. *et al.* (1992) A note on a fractal architecture for modelling and controlling flexible manufacturing systems, *IEEE Trans on Sys, Man and Cybernetics*, 22(3), pp. 564–67.

Weber, D.M. and Moodie, C.L. (1989) An intelligent information system for an automated, integrated manufacturing cell, *Jnl. of Manuf Sys*, 8(2), pp. 99–113.

Wright, P.K. and Bourne, D.A. (1987) The manufacturing brain, *AI Expert*, November, pp. 32–47.
Wu, S.D. and Wysk, R.A. (1988) Multi-Pass Expert Control System – a control/ scheduling structure for flexible manufacturing cells, *Jnl. of Manuf Sys*, **7**(2), pp. 107–20.

# Index

Page numbers appearing in **bold** refer to figures and page numbers appearing in *italic* refer to tables.